MW01228808

Genesis, Origins, Near East Myth, and Science

Biblical Christianity in the modern age

BRAD P. BALEK

For permission requests, please write to the author at:
bradpbalek@gmail.com

ISBN 979-8-9907153-2-5

Cover image: © 1xpert/stock.adobe.com

Table of contents

Introduction

Genesis 1:1

1 In the beginning God created the heavens and the earth.

John 1:1-5

1 In the beginning was the Word, and the Word was with God, and the Word was God. 2 He was in the beginning with God. 3 All things came into being through Him, and apart from Him nothing came into being that has come into being. 4 In Him was life, and the life was the Light of men. 5 The Light shines in the darkness, and the darkness did not comprehend it.

Hebrews 11:3

3 By faith we understand that the worlds were prepared by the word of God, so that what is seen was not made out of things which are visible.

Jeremiah 10:12

12 It is He who made the earth by His power, Who established the world by His wisdom; And by His understanding He has stretched out the heavens.

Jeremiah 32:17

'Ah Lord God! Behold, You have made the heavens and the earth by Your great power and by Your outstretched arm! Nothing is too difficult for You,

Romans 1:20

For since the creation of the world His invisible attributes, His eternal power and divine nature, have been clearly seen, being understood through what has been made, so that they are without excuse.

Psalm 19:1-2

1 The heavens are telling of the glory of God; And their expanse is declaring the work of His hands. 2 Day to day pours forth speech, And night to night reveals knowledge.

Psalm 33:6-9

6 By the word of the LORD the heavens were made, And by the breath of His mouth all their host. 7 He gathers the waters of the sea together as a heap; He lays up the deeps in storehouses. 8 Let all the earth fear the LORD; Let all the inhabitants of the world stand in awe of Him. 9 For He spoke, and it was done; He commanded, and it stood fast.

The Word of God is clear. The God of the Bible created all things through the Word according to the counsel of His own will. God will share His glory with no others. So why do so few people give Him the glory due His name when they look at this created world and universe? I believe God desires for us to have our joy made more complete in Him by marveling at the work of His hands in creation. He wants us to look up into the heavens and sing His praises.

While Bible-believing Christians recognize that spiritual things are only appraised by spiritual people, all people everywhere have a knowledge of God and His attributes by studying the creation. Or what today people would usually call nature. This knowledge may be fuzzy, veiled, and suppressed, but the capacity for this knowledge remains. I believe that God's truth in the Bible will be compatible with God's truth in nature. I think that the correspondence theory of truth is presupposed by the Bible. That is, the Bible takes it as a given that true knowledge is that which corresponds to how the world really is. Something is true knowledge if it describes what is happening or has happened. The Bible assumes logic, rationality, and the correspondence theory of knowledge without necessarily proving them. But then again, so does everyone else.

As a conservative Christian, I believe that the Bible is God's word. It is infallible, inspired, and inerrant in its original manuscripts. And that it is rooted in history, in actual events that take place in space and time. Its propositional statements correspond to reality. The Bible is eminently falsifiable due to its claims of reality and history. But the word of God does not blush at this assertion. That is, assuming we properly interpret the different genres of the books, passages, and find the meaning in the way that the Author intended them.

This is not to say that the Bible is a book of science. Many Bible commentaries over the centuries have warned about finding our science through a study of Scripture. John Calvin warned in his commentary on Genesis regarding the creation of the firmament or expanse in 1:6 that Moses is describing how things looked and we should not expect to learn astronomy from the text. I believe that he was commenting on the fact that Moses was not relaying the account in scientific or astronomical terms by which we could learn whole technical subjects from the text. We should be careful not to think that we will find Newton's classical physics, Einstein's theory of relativity, quantum physics, or cell biology within the pages of the Bible. The Bible rarely gives us mechanisms underlying what God does.

With that being said, this book is not going to deny the historicity of Adam and Eve, nor discount the first 11 chapters of Genesis as allegory, proto-history, or ancient near eastern plagiarism. Although many literary devices are contained in the early chapters of Genesis and throughout the Bible, this just goes to show that God the Holy Spirit is a master storyteller, communicating His truths in diverse ways using the styles of literature in which the inspired authors wrote. Many times, by reiterating theological truths in different contexts and forms.

This model of God's two books, the book of nature (natural revelation) and the book of Scripture (special revelation) is an often-used synopsis of how to relate what God has revealed to us through the Bible and how it relates to what we learn about nature through science, our senses, and reason. Some will claim that science is not an accurate source of knowledge about the past. They will claim that science can only address the things that we see in the present. Things that we can set up in a laboratory, by repeating experiments, or to validate or falsify different hypotheses. But many sciences deal with historical questions and come up with hypotheses and theories to explain the empirical data and then make predictions of what we should find from past ages. If something happened in the past it should leave evidence that can be discovered if the right methods and logic are used.

This is not to say that reason is more important to the Christian than faith. We believe that the Bible is self-authenticating and God the Holy Spirit opens our eyes to see its truth. Put differently, we don't have to put the scriptures through the lens of reason or gather evidence to prove its true. Most of us Christians know that God exists, that He loves us, that this world is His, and that Jesus died and rose from the dead for us. And it is by faith that we appropriate this knowledge and experience. But I also personally don't discount reason and evidence. We believe the living God has acted creatively and redemptively in history and through this material creation. Some Christians are suspicious of over thinking and intellectualizing the Christian faith, but I don't see it that way. I think God is honored when we use our minds for His glory and knowing Him better. But I will confess that there is a danger in turning our relationship with the living God into a cold intellectual study. But I don't think we are to remain infants and babes in Christ, being tossed back and forth by every wind of doctrine. And one of the purposes of this book is to address the rampant skepticism and criticism of the early chapters of Genesis, which often causes believers to doubt their faith.

I suggest the perceived conflict between the early chapters of Genesis and science is due to many of us misinterpreting the first chapter of Genesis and forcing it into a genre of literature that would have been foreign to the early Israelites to whom it was originally written. And then thinking that this interpretation is irreconcilable to the discoveries of science.

We may have grown up under teaching that tells us that there is only one way to interpret Genesis 1 and that if we abandon that interpretation then we compromise the entire Bible and won't be able to believe any of it. These tactics often make Christians fearful that if the interpretation that they have been taught is jettisoned, then their faith is in trouble. These same people often go off to college and are bombarded by courses in science and religion, often taught by professors that are hostile to Christianity. They are taught things that directly contradict what they believe to be the one and only interpretation of the creation account from the Bible. They are then left with a choice; follow the Bible and deny the science or accept the science and deny the Bible. One of my aims is to show that this is a false dichotomy.

The subject matter for this book is not a new endeavor, but I have tried to address many of the most pointed attacks on Christianity and questions that many believers have about the relationship of God's word to science, ancient near eastern myths, and biblical interpretation. For instance, many theologians have recognized the enigmatic nature of the creation account from the time of the early church. And they have sought to properly interpret it and understand its meaning. As conservative, orthodox Christians we believe that the word of God is inerrant, but our interpretations are not. Throughout church history, but much more often in the last 200 years with the advancement of science, theologians who held the commonsense conviction that special revelation and natural revelation will find and affirm the same truths have sought to reconcile the Bible with science (tentatively or not so tentatively) or to reexamine their interpretations. I believe this can be done without compromising the word of God or discounting the empirical data and conclusions obtained through science. But I think it requires that we determine the type of literature of Genesis 1, how to best interpret that type of literature, and try to ascertain how it was viewed by the intended original audience, who were the Israelites during the time of Moses. I believe, with good reasons, that Moses wrote the first five books of the Bible. I will not claim that Genesis 1 is Hebrew poetry nor straight historical narrative but instead a type of genre that is something unique. C. John Collins has done excellent work on this topic. (Collins C. J., 2006) We will look at this in greater depth in later chapters.

Take the book of Revelation for example. Most theologians recognize that it is a genre of literature in the Bible that requires a symbolic interpretation that corresponds to real events in space and time. These apocalyptic books like Revelation also have examples from sources outside the Bible, especially during the intertestamental period (approx. 400 BC – 30 AD). If we were to read Revelation assuming that it was historical narrative and applied that method of interpretation to it, we would have red dragons with multiple heads and horns, stars falling to the Earth being dislodged by a dragon's tail, beasts coming up out of the sea with multiple heads and horns, having the body like a leopard, feet like a bear, and the mouth like a lion. Most theologians recognize that these are symbols for something else, with much of the imagery coming from apocalyptic books or passages in the Old Testament. But conservative scholars don't claim that Revelation contains errors because they can't imagine such fantastic beasts. They recognize the literary structure of the book and try to interpret what the symbols represent, what God was saying to the original audience, and what He is saying to us today. Thus, how we interpret a book in the Bible largely depends on the type of literary genre that it is. We aim to read the text in a way that it was meant to be read and understood.

Essentially my goal is to use the grammatical-historical method and let the Bible interpret itself to properly understand what the early chapters of Genesis say. If the New Testament gives us an interpretation of something in the Old Testament, that is our authority for interpretation. We will then look at whether a literalistic reading of Genesis 1 is warranted for its genre and if we may be reading something into the text instead of out of it.

We will be asking questions such as: Should these first two chapters be viewed as one account? Does propositional truth have the same meaning across genres? Are we being consistent interpreting different genres across the Bible? Was God teaching transcendental truths about astronomy, geology, and biology that were far removed from what the Israelites could understand? Are we reading the text in the proper grammatical-historical setting of the ancient near east or are we reading it with 21st century eyes? Is the type of literature different between Genesis 1 and 2?

I hope my readers will see that once we deal with questions on Genesis 1, that there is good evidence to believe Genesis 2-11 is written as regular historical narrative and not proto-history, myth, or allegory, where many modern scholars want to categorize it. I think the Bible demands that Adam and Eve were real people created miraculously by God. I will argue that they were His first covenant representatives to bring His glory and kingdom to the ends of the earth, so that people could find God and worship the King. But that no person is excluded from having the image of God and the dignity and worth that it gives us. We will see that God has always chosen people because of His grace and mercy and not based on anything worthy that He foresaw in them. I think that the Fall described in Genesis 3 was an actual event in history, that sin entered the world through Adam's disobedience, that it had repercussions for the whole human race, and that because of these realities in space and time, we need an atonement and a righteousness outside of ourselves to cancel our debt and justify us.

We will see that some of the assumptions of science are unfounded and have a dramatic effect on the philosophical views that they espouse. We will also see that scientists are just like the rest of us, having biases, being confident when they should be humble, and holding on to ideas and not wanting to be wrong. But I will readily admit that most of them are honest, well meaning, and rational. And the methods that they employ for discovering truth are fruitful and powerful for explaining many things. I will aim to not set up strawman arguments or demonize people. I simply want to find the truth and be able to give God His glory.

We'll explore how the Bible speaks of God working through His creation in providence and how materialism, physicalism, logical positivism, and reductionism cannot explain nor disprove that God is behind the mechanisms in His created world. On the contrary, the natural laws used by God to operate for His purposes are a pointer to Him and His attributes that all can see (though dimly until redeemed) through nature. The fact that the universe is governed by means of finely tuned laws, constants, mathematics, and logic allow us to appreciate the attributes of God more fully. And it also helps us to understand ourselves.

This world is a wonderful and amazing place. We go for a walk in the crisp autumn air, looking at the leaves changing colors, the birds singing and soaring overhead, and the moon still visible in the morning sky. We come home and watch the latest astronomical documentary describing the incomprehensible size of our universe, the formation and destruction of stars, the birth of heavy elements, the ideas of a spacetime fabric, and the physical forces working within all these celestial bodies.

We go about our daily routine and see the mass of humanity, each in their own subjective experience, each having a personal identity and a story as unique as our own. The memories, beliefs, temperament, and ideas guiding their way through this world we call Earth. We come home to those we love, and they share their days with us. What they learned, trials, successes, and the vagaries of life.

But all of this can become second nature to us, and we find ourselves in a kind of auto-drive. We can find ourselves unwittingly being conditioned by the scientific materialism of the culture and forget that God is involved in our physical world at each moment in time. That there is a whole spiritual world behind the scenes. We may tend to attribute what is going on around the world to politics, sociology, psychology, or to material cause and effect without trying to discern what God is doing in the physical world or how the real battle is taking place in the spiritual realm. Occasionally we break out of this unconscious mode, and we look around the natural world to ponder why things are the way they are. We marvel at why there are so many kinds of plants and animals. Why so diverse and spread throughout the globe? How things came to be the way they are presently and how they all tie to the past? Why do we have a sense of self and are amazed at this creation, considering we are part of it? Why people are moral and spiritual beings and long for the transcendent? Why we experience guilt and look for ways to atone for it and ease our minds?

We assume we know what God has said about the universe, Earth, man, and the other creatures, but we realize that He rarely gives us information about the mechanisms He used to accomplish these things. As Christians we know that the secret things belong to God, but He has not left us without a witness and knowledge. Not only do we have nature, but we also have the early chapters of Genesis.

The past 200 years have made huge advances in science and its ability to describe how natural processes work, to construct a view of the past, and how things came to be. It is a very different world than 200 years ago. Christians in the past never had to deal with so many questions about creation, nature, and mankind. This science often seems to be in direct conflict with what we believe the Bible says about these things. These scientific theories are often followed by a metanarrative (a grand overarching storyline) about why things are the way they are. Often with a philosophy that accompanies such discussions that asserts purposelessness and random coincidences that have happened in the remote past to lead to our present state. It often speaks of the depressing absurdity of the cosmos and the fact that we need to find our own meanings and values in this random world. This is in direct opposition to what God teaches us in His word. And I would argue what we feel in our souls. That He is Creator, that history has a definite goal, and that history is going somewhere specific. That people are made in the image of God and have dignity and worth. That morals and values are objective and unchanging. That God had to become a man to save us. That man is longing for something he has lost.

If you are a Bible-believing Christian like me, your worldview is shaped dramatically by what you read in the pages of the Bible because you believe they are the very words of God. Written down by around 40 different inspired authors over the course of around 1350 years to be a revelation about who God is and what He requires of us. We believe the Bible is a source of knowing (epistemology). That is a critical question in life, *how do we know things?* Philosophers debate these issues all the time about which forms of epistemology are valid, warranted ways of knowing. Except for some skeptics and those with an idealist philosophy (nothing real and objective exists outside our minds, or we must remain agnostic on the issue), most people accept that we know things through our senses, our power of reasoning, credible history and witnesses, written records, what we learn from others, reliance upon authority, and our memories. But many people, including many scientists, take for granted that sense perception or raw data are meaningless unless they are systematized by the mind. We can talk about data and facts all day long, but they are useless without the intellect itself.

But we know that our senses can deceive us and lead us to wrong conclusions. Take for instance that the Earth is rotating about its axis at approximately 1000 miles per hour at the equator. Our senses give us no knowledge that we are moving. This argument from the senses was used for a very long time to "disprove" that the Earth was moving. Many would say, "If the Earth was really rotating, everything would be thrown off like objects placed on the edge of a wheel." But once Galileo developed the first ideas of relativity, it could be shown that we don't perceive motion when we are moving at a constant speed and everything else around us is also moving at that same speed. And the force of gravity overwhelms any miniscule centripetal force.

And our logical reasoning is only as good as the premises we develop for our argument. If we form an argument based on what we see and that knowledge from our senses is incorrect, then our conclusions will also be incorrect. This is one of the reasons that so many people for millennia thought that the sun, moon, and stars rotated around the Earth. It seemed obvious to the senses. Science can give us information that corrects wrong conclusions based on our senses and show us where our intuitions can go wrong.

There are other forms of knowledge people may debate about and certainty the spiritual illumination of the Bible as a source of knowledge is not accepted as a means of true knowledge outside of Christianity. But we must not lose the conviction that the Word of God is surer than reason itself. It comes from the One who is infallible and perfect. The One who is all knowing. You may have heard various scientists make the claim, "I don't believe what I can't measure". This is a common mantra for materialism based solely on empirical knowledge. It often escapes their notice though that they are using philosophical statements and not science to assert their views. The statement that only science can give us true beliefs is itself a philosophical statement. You can't measure that statement to prove it's true. It also ignores the fact that there is a great deal going on in the mind to measure something and then put it in its context to produce a coherent picture. What it also does is discount much of the knowledge we rely on every day. Such as what we read, learn from others, accept on authority, and from our own experiences and memories. All people, including scientists, believe tons of things they have never empirically verified.

So how do we know what we are supposed to believe about things like: where do we come from, what happened in the past, is there something besides the material world that I see and touch every day? This is especially true when it comes to the question of human origins and what science and the Bible say to the matter. At this point someone could respond, "The Bible is very clear about the creation of the universe, the origins of humans, plants, animals, and the age of the Earth." And I would have given that same response not too many years ago. I would just ask for your patience as we work through these things using the Bible itself to see what it claims, and to show that the Bible is indeed the infallible Word of God and contains no errors.

The world of literature is abounding with books, articles, and essays on the topics of creation, evolution, and the Christian faith. And there are many great ones out there. Most of us are dumbfounded by the sheer volume of perspectives, scientific data, and the rhetoric involved. And we throw up our hands in exhaustion, choosing either to ignore the scientific data or not think about it. Or else we get the sense that the tenets of the Christian faith and the data from biology, chemistry, physics, and other sciences cannot possibly be reconciled to one another. Many see only a couple of ways out of this tension. We either need to reinterpret the Bible so it fits with science or attack the science to show that it is coming to wrong conclusions. Or to just flatly deny that science has anything worthy to say about these matters.

This was my own personal experience. And it has led to many moments of doubt, frustration, reworking my theology, doubting the interpretation of the scientific data, doubting the Bible, and a general consternation about where to go from there. So, like most people who believe that objective truth is real and is something that God has enabled our minds to ascertain, I would buy or check out the latest books on astronomy or evolution written by Christians to help me reconcile how to make my faith and science work harmoniously. After this, I would spend hours online reading the latest scientific discoveries. I would watch documentaries and courses explaining nature, cosmology, astronomy, biology, or related fields. Yet, always at the back of my mind were the nagging feelings, doubts, and questions that all of this did not add up to what the Bible said about creation and human history.

Most Christians that I knew would talk about the Earth being 6000-12000 years old and that Adam and Eve were the progenitors of the entire human race. Others not in my immediate circle would say that we must take the story of Adam and Eve as an allegory that was meant to teach us some moral lessons about ourselves. Yet others would claim that God picked out a couple of the first anatomically modern humans between 50,000-300,000 years ago. Or maybe chose a couple of ancient human ancestors (hominins) prior to that. And that one of these couples is what the Bible is referring to as Adam and Eve. And still others would say evolution explains it all and Adam and Eve never existed. That Adam and Eve were only a prescientific fairy tale to make sense of where humans came from. After researching this topic for a long time and not finding a book that answered all my questions in an intellectually satisfying way, I decided to write this book.

At this point someone may be saying, "You are just going to use science as your hermeneutic (a method of interpretation) for the Bible and not let scripture interpret scripture." I am first and foremost concerned about what the Bible itself teaches. I believe in sola scriptura, a Latin phrase that means scripture alone. Or that the Bible is the Christian's highest authority for faith and practice. We readily admit that our interpretations are fallible, but God's Word is not. Let us then let scripture interpret scripture. But I think even Christians who use the most literalistic interpretative method for the Bible will find that scientific knowledge has made us rethink our interpretation of scripture is various places throughout the history of the church. Very few people today believe that the Earth is at the center of the solar system, with the sun and planets revolving around it. Very few people believe that the Earth does not rotate or move. Very few believe that the Earth is a flat circle or has four corners. Very few people think that the seat of the will and emotions is in the bowels or intestines. This knowledge was obtained through various means, including science. But we need to be clear here that a proper interpretation of the biblical texts and the type of literature they are could have avoided many of these issues in the first place. For the Bible never asserts such silly things.

Just left alone with the Bible and no outside knowledge of astronomy, I don't see how someone would have come up with the correct view of the solar system, apart from the Holy Spirit revealing that knowledge to them. In fact, this is sometimes what we find when we study church history. Very few people today think that the stars are fixed in the firmament. Many of us seem to be fine with acknowledging these findings and letting them guide our interpretation of the Bible, but we seem to hesitate in considering Genesis 1 as something other than historical narrative teaching us science and material processes. We may be uncomfortable with the apparent discrepancies between Genesis 1 and 2, but we need to look more closely at what the Israelites of Moses's day would have thought about any supposed incongruities.

What I am proposing is that we do not have to reject the Bible's account of creation or Adam and Eve, nor do we have to reject most of the conclusions from modern science. Although, I would caution that we be careful not to fall behind some scientific paradigm too strongly, as science is always updating its knowledge and tweaking its theories. Sometimes involving an entire paradigm shift. As of the time of this writing, scientists have claimed to have found anatomically modern human (AMH) fossils dated to 300,000 years ago in western Africa. Prior to this, the oldest known AMH fossil was dated to 195,000 years ago in eastern Africa. The point is that data like these can shift the theories significantly. As of the time of this writing, the James Webb Space Telescope is in the heavens gathering data. Early reports seem to point to galaxies that formed much earlier than current cosmological models allow. Whether the data is being misinterpreted is yet to be seen, but the cosmological models will have to be reimagined if the data are true. All this is to say that science by its very nature is provisional. It doesn't offer proof in the logical sense, so it is wise not to cling too tightly to the discoveries of the day.

People of every age have assumed that they "knew" what was right and often claimed that there was no more that science could possibly discover. I also propose that we try not to make the text of the Bible say more than it wants to say and try to force scientific theory where it does not belong. I ask that you follow along with me as we look through the Bible and the scientific data to formulate what I believe to be the truth. For those uncomfortable with evolution and the idea of an old Earth, just remember that I was once in your shoes too. I would ask for your patience as I lay out the scientific theories and the data allegedly supporting them so we can address the big picture. I will try to present the best up-to-date data and theories from science and then critique them where they are weak, where alternative hypotheses may prove better, or when scientists' confidence is unwarranted.

I also felt a strong need to include ancient near eastern creation and floods myths in this book and how they may relate to the Bible. There are various assaults on the word of God coming from the secular world, but in relation to the early chapters of Genesis none are more often leveled against the Bible than the age of the universe and Earth, biological evolution, and comparisons with other ancient near eastern literature.

But this all hangs on believing that the Bible is the infallible Word of God and showing that the facts from the natural world are consistent with His revelation. And in the end, my hope is that this gives us more confidence in the Bible and leads us to give God the glory through Jesus Christ for who He is and what He has done.

Part 1

The problem, deep beliefs, and Genesis

Chapter 1

Conflict between science and Christianity?

Is there really a conflict between scientific discovery and the Christian faith? In one sense, the answer is no. Most scientific discoveries and experimentation have a neutral impact on theology. The very basis of scientific inquiry depends on the fact that we believe the universe is discoverable, ordered, and knowable. There is the assumption that our minds can systematize the data and reach true beliefs. That through the scientific method and collaboration with others we can form hypotheses, test them, and either give more credence to or falsify our hypotheses. Science claims to speak about what can be discovered through the natural world and material processes. And it is hard to deny that it has been successful.

Immanuel Kant was a famous German philosopher in the 18th century who had a great deal of influence on later philosophy and eventually the common mind of society. One of his ideas was that we can only know things of phenomena, or what our senses perceive and the mind filters. He says that the thing in itself may or may not exist, but everything is filtered through our minds so that is all we have access to. I have always found this type of reasoning bizarre, but maybe I have misunderstood him. It seems to me that if we all agree that the apple is red, and we agree on what apple and red mean in our language, then the apple exists, as do the wavelengths of light that it reflects. Science is not possible without assuming the metaphysical reality of apple and red.

Theoretical physics is largely based on developing theories using thought experiments and mathematics. Einstein used thought experiments almost exclusively to initially come up with his ideas of relativity. He needed help to formalize them and put them in the language of mathematics, but he left the predictions to be confirmed by the empirical data of the applied scientists. I believe this was a triumph for the rationalists who claimed that we can understand things about reality from mental reasoning alone.

Therefore, most of us believe the commonsense idea that our experiences are usually trustworthy, and we can reason abstractly about things we are not observing. And this reasoning can often produce theories and predictions that are later confirmed through observations. Although psychologists have shown that this is far more prevalent in educated technological societies than in traditional ones. We have been conditioned in the modern educated world to think more in logical and abstract categories to generalize and infer things than we do by nature. But because we are talking about modern science and how it relates to the Bible, we are concerned about scientific cultures and what they claim to know. Although most of us Christians are not threatened by the findings of physics and chemistry, we often are by findings from cosmology and biology, because they address issues that seem to directly contradict scripture. But there has not always been a distrust of science among Christians. Eminent Christian scientists were at the forefront of scientific discoveries in the western world. And many people have written convincingly that the Christian mindset of knowing God's creation and exercising dominion were vital for the western scientific revolution. Let's look at a few of these pioneering scientists.

Nicolaus Copernicus (1473-1543) developed the model of the sun at the center of the solar system, reinventing the forgotten work of Aristarchus (3rd century BC). Copernicus posited that all the planets revolved around the sun, the retrograde motion of some of the planets, that the Earth rotated on its axis to explain the apparent motions of the stars, and that the stars were much more distant than the sun.

"To know the mighty works of God, to comprehend His wisdom and majesty and power, to appreciate, in degree, the wonderful working of His laws, surely all this must be a pleasing and acceptable mode of worship to the Most High, to whom ignorance cannot be more grateful than knowledge." - Copernicus

Johannes Kepler (1571-1630): suggested that the planets had elliptical orbits, the sun was at one focus in the ellipse, and that the planets did not experience constant speed through their orbits.

"The chief aim of all investigations of the external world should be to discover the rational order and harmony which has been imposed on it by God, and which He revealed to us in the language of mathematics." - Kepler

Isaac Newton (1643-1727): developed the laws of motion, universal gravitation, co-developed calculus, and was arguably one of the most influential scientists of all time.

"It is the perfection of God's works that they are all done with the greatest simplicity. He is the God of order and not of confusion." - Newton

Gregor Mendel (1822-1884): father of modern genetics.

Mendel was an Augustinian monk and abbot of St. Thomas' Abbey in Brno. He developed the understanding of inheritable characteristics through dominant and recessive genes that we now call alleles. He preached numerous sermons as a monk and abbot.

James Clerk Maxwell (1831-1879): formulated the classical theory of electromagnetism, unifying physics to a much greater degree. His contributions in optics, electromagnetic radiation, electromagnetism, and field theory were among the most influential in physics for his era and he was a devout Christian.

"Almighty God, Who hast created man in Thine own image, and made him a living soul that he might seek after Thee, and have dominion over Thy creatures, teach us to study the works of Thy hands, that we may subdue the earth to our use, and strengthen the reason for Thy service;" - Maxwell

William Thomson Kelvin (1824-1907): physicist and engineer, developed physics and math on topics of electricity and the laws of thermodynamics.

"The more thoroughly I conduct scientific research, the more I believe science excludes atheism."

"If you think strongly enough, you will be forced by science to the belief in God, which is the foundation of all religion." – Kelvin

These are but a few of the most prominent scientists who were Christians. Not to mention Francis Bacon, Galileo, Blaise Pascal, Robert Boyle, Gottfried Leibniz, Michael Faraday, Asa Gray, George Washington Carver, Arthur Eddington, Max Born and others.

So, we can see that scientists of tremendous intellect have forged the way in the discoveries and furthering of our scientific knowledge while giving God His due glory in the process. These men and women knew that it was of no use to dispose of the mind with which we are blessed above all the other creatures on this planet. We should be unashamed of following in their footsteps to better understand the world and universe in which God displays His glory. What does the Bible say about God's natural world and our response to it?

Isaiah 40:26

26 Lift up your eyes on high And see who has created these stars, The One who leads forth their host by number, He calls them all by name; Because of the greatness of His might and the strength of His power, Not one of them is missing.

Psalm 8:3-4

3 When I consider Your heavens, the work of Your fingers, The moon and the stars, which You have ordained; 4 What is man that You take thought of him, And the son of man that You care for him?

We should be humbled that God has condescended to us to provide life everlasting. And to give us the study and dominion over His created order on this planet. The natural world is a reflection of the power of God, and we can ascertain certain things about Him from the study of it. As part of the creation mandate given to Adam and Eve, humans are called to exercise dominion over the Earth and act as God's vicegerents (one who is appointed to exercise authority on behalf of a sovereign ruler or king), exercising wise and righteous rule.

These should be reasons enough for Christians to want to get involved in science and to further the fields of astronomy, physics, biology, and others. Another very compelling reason for advancing science is to alleviate pain and suffering in this world. Although we groan within ourselves waiting for the redemption of our bodies, we are called to fight the effects of living in a hostile world, sin, and the devil. To walk in the good works that God prepared beforehand for us. So much suffering is the result of people not having adequate technology to live better, safer lives. We as Christians should be very interested in advancing technology for this end. We simply cannot be content with living our lives without trying to make this world a better place for all human beings and by being a blessing to the nations. In God's common grace and forbearance, He provides the things that make the world a better place to live, even for those who do not have eyes to see or give Him thanks. Christians should work within His sovereign framework to reflect His goodness and mercy, so that people will see these good works and give Him glory. As image bearers who are being renewed day by day, we should be striving to reflect His character to the world, and that should certainly include offering grace, mercy, and life to those around us.

This brings us back around to the question of whether science and Christianity are in conflict? We have already answered no, in the sense of studying the natural world and letting it teach us about God, His creation, and ourselves. And by using scientific knowledge and its application for the betterment of humanity to fulfill the creation mandate, to love others, and reflect the image of God to an unbelieving world. But there is another sense of conflict where the answer is yes. And that is really where the heart of this book is aimed.

When we talk about the beginning of the universe, that has direct theological implications regarding creation. There is direct conflict between an interpretation of the Bible that says the universe must be between 6000-12000 years old and the consensus of scientists who are telling us it is somewhere around 13.8 billion years. Both propositions cannot be true. They could both be false, and another answer the correct one, but one rules out the other. The other direct conflict is the theory of biological evolution. This theory states that life has been on the Earth for about 3.5 billion years starting with simple one-celled organisms and through genetic mutation, natural selection, genetic drift, and other mechanisms of evolution working on those mutations through populations isolated from one another, species branch off and develop in their own divergent ways with further branching down the line. It claims that all organisms from bacteria, to plants, to fish, to reptiles, to land mammals, to monkeys, to apes, to hominins, and even human beings all share a common ancestry. And that if we follow the evidence from genetics and the fossil record, we can determine where the most recent common ancestor (MRCA) occurred in that branching pattern to develop what is called a phylogenetic tree. There is direct conflict between that theory of evolution and Adam and Eve being the first human beings on Earth around 10,000 years ago and by them populating the entire planet. As well as the idea that God created all the plants and animals fully formed with no predecessors in the history of Earth. These are tough issues to deal with and to think through, but I hope to show that we do not have to compromise on the word of God or the substantiated findings and claims of science to get to the truth. But we should be humble with our brothers and sister in Christ who hold other views relating to the doctrines of creation and mankind and how they apply to the claims of science. We will look at some of those views next.

Chapter 2

Five perspectives on creation

Let me say right up front that I do not think someone's interpretation of the first few chapters of Genesis determines whether they are a born again Christian or not. I am humbled by the different interpretations out there from people who love Jesus Christ, His church, and trust Him alone for their salvation. People in whom the Spirit has regenerated them, and they are in union with Christ. People who follow Jesus as Lord and love others. However, I do acknowledge that many of these views have implications for orthodox Christian theology, especially doctrines about the goodness of creation, evil, Adam and Eve, the image of God, the Fall, original sin, guilt, corruption, death, and redemption through Christ. I plan on addressing all these issues in future chapters.

I believe the Bible is a self-authenticating book. Which simply means that it speaks for its own truth. We can read it and know it is true by the illumination of the Holy Spirit. The belief that the Bible is true can be a true, warranted belief without having to prove it. Just like we assume our senses, our rational minds, logic, and reason itself. Some things must be assumed and fundamental to build any knowledge.

Having a correct biblical model of interpretation for Genesis does not save anyone without God giving us a heart of flesh for our heart of stone. We can have more solid hermeneutics and exegesis (getting the correct meaning from the text) than someone else, but these things in and of themselves are impotent to save. We do have to believe certain propositions for saving faith, but intellectual assent alone is not enough to save. God the Holy Spirit works through the means of the Bible to produce saving faith.

I am going to briefly discuss five different perspectives on creation in this chapter, but I make no claims about representing all the views of Christians or non-Christians. Even within these five there are numerous variations, flavors, and ideas. And my thesis in this book will clearly show that I do not fit into any one of these categories. The five perspectives that we will summarize are young earth creationism, progressive creationism, evolutionary creationism (or theistic evolution), intelligent design, and materialist naturalism (either agnostic or atheist).

Young Earth Creationism (or Biblical Creationism)

This is the view that I held most of my younger life. It is what I heard preached from the pulpit and most of my Christian friends believed. It is still what many of them believe today. It is the view that seemed to be held by the most conservative Christian teachers I listened to, although I came to find out that this wasn't always the case. This has also been the dominant doctrine of creation in the history of the Christian church. This alone should give us pause about too quickly rejecting it. And from my own reading of Genesis this seemed like the most straightforward reading. But to determine an age of 6000-12000 years for the universe was something that required more digging from the text of the Bible. Nowhere does a plain reading of Genesis tell us that the universe is that old. It comes from the work of biblical scholars putting together the genealogies from the Old Testament and coming up with that number. It assumes that Adam was created six days after the beginning of the universe and that the genealogies are compiled with either no gaps or very few gaps. The man most famously associated with this chronology was Bishop James Ussher (1581-1656) who came up with the date for creation on October 23rd, 4004 BC. Many others have come up with very similar dates using comparable methodologies.

Young Earth creationists believe that Adam and Eve were literal, historical people whom God created supernaturally as fully formed, talking, thinking humans. They believe that God created the plants and animals as fully formed and having no antecedents. The formation of the universe, Earth, and all its inhabitants happened in six 24-hour days. The Garden of Eden was an actual geographical place in history and Adam was placed there as the very first human being on the planet.

Many, but not all, believe that Noah's flood occurred within the last few thousand years, and that it was a flood that was global in scale (covered the entire planet). That the only surviving land and flying creatures that survived on the Earth were the ones kept safe in the ark and that all of humanity, besides

Noah, his wife, his 3 sons and their 3 wives, were drowned in the deluge. From these 8 people and animals aboard, they repopulated the planet with humans and animals again. And that most of the sedimentation, geology, and fossil record is explained by this cataclysmic event.

Most in this camp are fine with the concept of "microevolution", or species or "kinds" changing within their groups. They agree that we can see things like the results of animal and plant breeding, bacteria developing resistance to antibiotics, and novel characteristics of change within a species. Although species is probably not where they would draw the line of biblical kinds. Species do not even have an agreed upon definition in biology, many of whom can mate with each other and produce fertile offspring. Many of the young Earth creationists would place biblical kinds somewhere closer to family in taxonomy. What they almost universally reject is "macroevolution", kinds changing over time by branching out into new kinds.

Progressive Creationism

They agree with the scientific consensus for the age of the universe and Earth of billions of years. They agree that the Big Bang (currently called the Lambda Cold Dark Matter model) is the best model for the empirical data to explain the age and expansion of the universe. They believe that the days in Genesis 1 are not 24-hour days but instead represent unspecified long periods of time (perhaps millions or billions of years). This view is often called Day-Age theory.

Most believe that God created each kind supernaturally and then let microevolution take place within each kind dictated by the constraints of each kind's genome for change. They are fine with the idea of microevolution and animals evolving within their own kind, but they deny macroevolution, or that kinds evolve into other kinds. They note that the fossil record seems to indicate long periods of stasis (without change) punctuated by periods of rapid change, with the result being the emergence of new organisms. They attribute these fully formed new body plans and creatures to instances of special creation by God for each new novel animal or plant kind interspersed with periods of extinction. They believe mankind was also miraculously created by God and did not have any biological predecessors, including apes. Some believe that hominins existed, but that they were themselves created miraculously by God and did not evolve into modern day human beings. Or that hominins and human beings are really the same kind, just with variation within the human kind. Some of them posit that Adam and Eve lived between 50,000-150,000 years ago and were the first human beings from which all other humans descended. They were a special, miraculous creation by God endowed with the image of God and all the attributes that separate us from the other animals. Many believe that Noah's flood was a local event and not global. And that the population of humanity at the time was localized to the Mesopotamian region, so that all humanity perished save the eight aboard the ark.

Evolutionary Creationism (or Theistic Evolution)

This perspective believes that God created the universe and all that is in it. Everything created has a teleological (a certain end, purpose, or goal) aspect to it. The universe is very old (around 13.8 billion years) and has come about by the mechanisms discovered by astronomy and astrophysics. The Big Bang is the best current model for the evolution of the universe given the data and current knowledge. God chose to use the process of evolution to develop the species that are on this planet, including human beings. The origins of self-replicating life are still unknown to science. There is no consensus of any dogmatic view on the origin of life except that science most likely will explain how it arose from non-life someday. Humans are descended from the great apes and share common ancestry with all other living things if you look far enough into the past. They are comfortable with saying that God directed the path from non-life to humans but at no stage in the steps of evolution is God's direct miraculous action required to explain how the organisms developed. They feel that the natural mechanisms of physics, chemistry, and biology can account for the universe and life evolving. There are several views in this camp on whether God is the source of causation or is hands off, letting the physics and biochemistry take over.

There seems to be no clear-cut agreement on whether the Garden of Eden was an actual place. They differ amongst themselves on Adam and Eve, but most seem to take an allegorical view, as opposed to them being historical people that God created from the dust of the earth. Some take the view that God chose two Homo Sapiens that were developed enough mentally to instill in them immortal souls, the knowledge of good and evil, and the longing for spirituality and worship. Some push this back to Neanderthals or earlier Homo species. Some seem to view Genesis 1-3 more like a story to describe the progression of evolutionary psychology in the human mind. There is no consensus on what the six-day week of creation in Genesis 1 refers to. Some take the view that God was presenting a narrative to the people of the ancient near east as a story that they could grasp because if He were to present genetics, cosmology, and geology it would have been little help to their minds without our current scientific knowledge.

Noah's flood is not a prominent part of their theory, but most would say that it simply did not happen or that the flood was a local event and that not all the humans and animals on Earth died.

Intelligent Design

Intelligent Design teaches that the abundance of flora and fauna on this Earth cannot be explained by random genetic variation and evolutionary mechanisms. In this respect, the overlapping creeds of intelligent design and evolutionary creationism get somewhat muddled. They both agree that evolution happened and that there is a telos (purpose, goal) behind the evolution of life. But intelligent design would say that natural processes, physical laws, random mutations, and the mechanisms of biological evolution are not enough to account for the complexity of life. For biological information like RNA and DNA, of irreducibly complex cells, tissues, and organs that make life possible, or as satisfying explanations for the history of life and the fossil record like in the Cambrian Explosion. They would invoke an intelligent designer to create new information and causality not explained by natural laws and processes into the system to get to life as we know it today. Evolutionary creationists would not embrace this idea, as they would claim that the various physical mechanisms are adequate to explain the outcomes. Intelligent design folks often like to point out that many of the explanations used by evolutionary scientists for complex biochemistry, novel biological features, and the speciation of organisms employ a lot of hand waving and black box faith.

Intelligent design does not seem to be tailored to address specific questions regarding the Bible, and its adherents will generally tell you as much. They are more concerned about the improbabilities of a materialist explanation for the bio-friendliness of the universe, the information in DNA, the complexity in life, and the rapid appearance of life in the fossil record at points in the geologic column.

Materialistic Naturalism (either atheistic or agnostic)

Materialistic naturalism is a worldview that no God or gods are necessary to explain anything about reality. Many others would claim that there is no way of knowing and therefore remain agnostic.

The atheist branch of this worldview generally holds that the universe is here either by necessity or from random processes of an infinitely vast multiverse. I would argue that neither of these are physical claims but instead philosophical ones. Most don't claim to have a deductive argument that God doesn't exist; they just believe that there is not sufficient evidence to believe in God and therefore a god or gods is an unnecessary hypothesis. Many of them will honestly acknowledge that the existence of the energy for the Big Bang, the initial conditions in the very early universe, the laws of physics still to be described, the fine tuning of the universe with its free parameters, the emergence of life from the non-life, and the detailed steps evolution needed to take to reach the current diversity of life are genuinely hard problems that have yet to be fully explained.

They generally believe that the physical laws of nature acting without any intelligent mind or goal-oriented process resulted in the expansion and evolution of our universe. Most cosmologists and astronomers believe that the current data and evidence can only tell us about what happened after time=0 in the early universe and that speculating about time=0 cannot be supported by science, probably even in principle. Although many models attempt to offer hypotheses about what may have preceded this early time. Because people are naturally curious about what happened at the beginning, many cosmologists have gained popularity by developing and promoting models of the universe that claim to address these issues. There is no consensus as far as I can tell, and I think most scientists dealing with the universe are fine saying they aren't confident what happened at or above Planck density. Many sources will claim that we can describe what happened all the way back to the Planck density, which is still after t=0, but these theories cannot be tested directly, even in principle. I'm not saying they are uneducated guesses. There may be indirect evidence that can give confidence or rule out certain models, I don't really know. But it seems to me that theories about the origin of the universe and grand unified theories will need a lot of support and data to be called science. And almost all the cosmological models seem to incorporate inflation in their theories, which essentially erases any information about what preceded it. So, I'm not sure how you decide between theories when they all make predictions based off inflationary models and the same evolution of the universe after it.

They believe that matter came from energy, the accumulation of matter came together to form stars, solar systems, and galaxies. One of these galaxies happened to be ours, in which our solar system coalesced from a disk of dust and gas to form our sun and the planets, including the Earth. The solar system and Earth were formed about 4.5 billion years ago and within one billion years of formation, the first life formed by some unknown chemical process into the first form of biological life.

They would deny that the information needed for the first cell came from intelligence or consciousness. Most would say that we just don't understand the chemistry yet. Through slow, gradual random mutations and acted upon by natural selection and other mechanisms of evolution over billions of years, this life (however it first came to be) branched off and diverged into the many forms of life we see around us today. Some in this worldview state that if we were to wind back the clock (say a billion years and start it again) we could end up with a whole host of different plants and animals, and that human beings may never have evolved at all. Others in this worldview think that convergent evolution would more or less produce the same kind of organisms we see today.

They think that human beings are essentially no different than animals (because they say we are one), except in the capacity of our brains for things such as abstract and symbolic reasoning, logic, and language. Humans have no immaterial parts like a soul. Morality has no objective reality. It can generally be explained by evolutionary psychology and cultural forces.

The atheists would generally claim God or gods are not needed for creation and Adam and Eve were not historical people. Some of the more aggressive and antagonist among them would side with Freud that religion is just wishful thinking for immature people who need comfort from the suffering of life and fear of death. And that religious faith is a form of neurosis that a cure should be sought for. Others would side with Nietzsche and claim religion is a way to keep the masses down under the power of others and not let them find their potential through self-realization. Among the more gentle and kind, some would say that if religion helps you to feel good and do good then it has utility. Maybe not only for the individual but maybe even for society as a whole. But all these views offering a psychological explanation for the origin of religious tendencies do not really address the question of whether it is rational to believe in the existence of God, nor does it follow that psychology should be detached from religion.

However we want to unpack it, atheistic naturalism says there is nothing besides the material universe operated on by physical laws. And the agnostic says that it is probably superfluous whether a god or gods exist. We will often hear this philosophy referred to as just materialism, naturalism, or physicalism.

And finally, there are other views which do not fit neatly into one of these five. Some in evolutionary creationism, or those who are practically, if not ideologically, materialists believe that if God or gods do exist then they may have got the ball rolling at the beginning of the universe (or multiverse) and then left it to the laws of nature. This type of belief in god is known as Deism (a supreme being/creator who does not intervene, does not uphold the natural creation by providence, nor seek a personal relationship with us). Another view of god sometimes held among people is one that makes god and nature the same substance or thing. This is generally called pantheism. God is everything and everything is god. These views of god would be the god of many contemporary and past philosophers and scientists.

I have left out some other views like the gap theory which posits a huge gap of time (millions or billions of years) in between Genesis 1:1 and 1:2. This theory states that Genesis 1:1 describes an initially perfect universe and an Earth without sin that contained animals (and sometimes people). And then the destruction of that world with a global flood as a judgement usually attributed to the fall of Satan and his reprobate angels, people sinning before Adam, or both. This then purports to explain the evidence for the age of the universe and Earth and the presence of the fossil record, but that the animals and people during this time have no biological connection to present day species. And then beginning in

Genesis 1:2 after the judgement of the global flood, the Earth is in a state of being covered by water, formless and void, and God begins again by re-creation. This new creation is once again made without sin and is perfect and corresponds to events that happened in the recent past. And it also provides an explanation for why Satan is already there in the garden.

Many tomes have been written on each of these views and could easily be expanded upon by those wishing to study them further. In the next chapter, we explore what I believe to be at the heart of modern scientific skepticism regarding the Christian faith.

Chapter 3

The foundations of skepticism

Most of us have sat down on the couch at night to watch some TV and came across a documentary on the Bible or on the life of Jesus. And felt like it might be worth tuning into. At first all seems well, and we are happy that amongst the cultural garbage, that a show about the word of God in still on the air. They begin talking about Jesus and how He was a good man and a great moral teacher. We agree and nod our head in approval waiting for them to give Him credit as being God incarnate, which predictably never happens. But we continue to watch as surely they will talk about the miracles He performed. And how the early Jewish sources like the Babylonian Talmud attributed His workings to sorcery, giving extra-biblical confirmation for his miracles. Surely this show will give an account as to why other extra-biblical sources attribute a solar eclipse and darkness during the hours of Jesus' death on the cross, even if they don't reference the Bible as a reliable source? But the program mentions none of this. Instead, they talk about the demon possessed men in the Gospels as having conditions such as epilepsy, the disciples hallucinating when they claim to see Jesus walk on the water, and how the people supposedly brought back to life by Jesus were not actually dead but appeared that way to the onlookers.

Then things really turn for the worse as we continue to watch, as the narrator claims that most of the accounts of miracles in the Gospels were just made up a long time after the death of Jesus as a way to claim that Jesus was God. And to convince the early followers of Him to have conviction in what they believed so that this religion would spread. They assert that the transfiguration never took place, nothing miraculous happened on the cross when He died, and that He did not actually rise from the dead. We know we should recoil in horror from the unbelief being presented in the show, but we have become so accustomed to it that it doesn't seem to surprise us.

We change channels to another show that is about the exodus from Egypt. We figure this one will hopefully be more charitable and accurate to the Bible, as it is on a channel dedicated to history. They begin by talking about the lack of archaeological evidence for the Israelites even being in Egypt. We think to ourselves, "Why aren't they mentioning the internal consistency of the account of Egypt in Genesis and Exodus? They must know that the biblical accounts give an accurate description of Egypt and Egyptian life? Don't they know that ancient empires, including Egypt, never mentioned any military defeats or anything that would look embarrassing for the kingdom on their tombs, temples, and monuments? Or that if such documentation did exist, that later kingdoms tried to remove any trace of them by erasing them from history? Why aren't they talking about the relief of Khnumhotep II in the 20th century BC showing Semitic traders wearing multi-colored tunics in the land of Egypt? Why are they not talking about the archaeological finds in Tell el Dab'a, which was the capital city of the Semitic Hyksos kings, named Avaris, in the land the Bible described as Goshen? They must give some credence to the idea that Hebrews living in the Nile delta land of Goshen would have multiplied and prospered under the Semitic Hyksos rulers? When the Hyksos were finally expelled from Egypt by Ahmose I, c. 1550 BC, would he and the Egyptians not have had much animosity towards the remaining Semitic peoples and put them to forced labor? Don't they acknowledge the evidence for slavery in the New Kingdom? Why don't they acknowledge the storage cities of Pithom and Raamses under Raamses II? How about the introduction of the horse and chariot into Egypt under the Hyksos rule? What evidence were they expecting to find from a nomadic people wandering in the deserts of Arabia for 40 years with the sands of time literally and figuratively covering their tracks? Most certainly, they would mention the Merneptah Stele, c. 1205 BC, in which most scholars agree is the one of the earliest mentions of an ethnic people group called Israel in the promised land? Perhaps, they would bring up the Egyptian statue pedestal relief in a Berlin museum that most likely mentions Israel on hieroglyphic name-rings dated to around 1400 BC?"

We sit there patiently listening and they continue to elaborate the details of the story, in which the narrator makes perfectly clear that it is all a myth. Such as the plagues of Egypt, where they go to painful lengths to explain them as a series of natural events. They talk about the Red Sea parting as a local lake with barely enough water to cover your ankles. And we think, "How could Pharaoh's army drown in a few inches of water?" And then they boldly claim that Moses did not receive the covenant from God in any shape or form, nor was he even able to write the first five books of the Bible. We finally give up in exhaustion and turn off the TV.

Why such unbelief? Even for the parts of the Bible that the secular historians agree to be true and accurate, why do they always try to come up with natural explanations for things that the Bible clearly attributes to God? I will argue later that one of the reasons is that many of us have a narrower understanding than the Bible does as to how God works in His creation. But unfortunately, we see this type of thinking all the time, especially in academia. If we were to ask 10 people walking along campus at colleges and universities if they believed in miracles, we would most likely get a majority percentage that don't. And a much larger percentage of professors.

We see this phenomenon with professional archaeologists trying to dismiss the claims of people like their colleague Dr. Steve Collins who many believe have found the cities of the plain (Kikkar, Gen. 19:25), including the city of Sodom and their destruction described in Genesis 19. He postulates that Sodom and the cities of the Jordan disk met their demise around 1650 BC with a possible airburst from an exploding asteroid or comet that had enough heat and pressure to fuse materials together into glass similar to the Trinity test site for nuclear weapons in 1945. (Collins & Scott, 2016) An article in the journal Nature has argued for something very similar. (Bunch, 2021)

Another example is the attempted repudiation of Dr. Bryant G. Wood's conclusions of the destruction of Jericho being destroyed around 1400 BC according to the biblical account. This in spite of the evidence from the type and dating of the pottery, the evidence for a fortified city, the short siege, the walls being leveled, the city not being plundered, and that the city was burned. Wood claimed that a possible cause of the walls falling down was an earthquake. (Wood, 1990) I think a much stronger case can be made for a date around 1250 BC for the Exodus, but it makes the point about trying to dismiss findings and evidence that confirms the historicity of the Bible. If you want to read a fantastic book about the historicity of the Old Testament, I highly recommend the book *On the Reliability of the Old Testament* by K.A. Kitchen.

So here are two Christian archaeologists (Collins and Wood) that claim a possible natural cause for the catastrophes. But if the events correspond to the biblical account, the timing of such events would have to be miraculous. But I believe that to publish in a scientific journal and be given credence if your field, it is not permitted to claim a miraculous cause for the event you are trying to describe. So here we have the example of the walls of Jericho laid flat (Joshua 6:20). But the Bible does not tell us about the mechanism by which the wall fell, only that it happened exactly when the Lord said that it would, after the Israelites circled the walls seven times and blew the trumpets. God could have used an earthquake as the means to topple the walls. In the case of Sodom, we are told that God rained down brimstone and fire out of heaven (Gen. 19:24). I also believe that this could be interpreted as a natural phenomenon like an asteroid or comet airburst. But make no mistake, according to God's word, it came at a precise time to judge the cities of the valley and its inhabitants for their wickedness, making it an undisputed miracle. But Lot's wife turning into a pillar of salt, that has no natural explanation (Gen. 19:26). These examples are a few among many, many miracles in the Bible in which nothing less than a supernatural explanation will do.

The point I am trying to make is that if God uses natural phenomena that He is in control of to cause events and their effects (think the strong east wind blowing all night prior to the Israelites passing through the Red Sea, or Sea of Reeds), this does not lessen the fact that it is called a miracle because it is all based on miraculously timing and the desired effects according to God's will. Later we will explore how the Bible describes God's providence and control over His creation for His ends.

Many liberal scholars, when applying some form of higher criticism to the Bible, will propose dates about when certain books of the Bible were written based on if they predicted future events that came to pass. They will claim that it is impossible to predict a future event accurately (which would be a miracle), so they assume the book had to be written after the time of the fulfillment. They will claim for example that the synoptic gospels (Matthew, Mark, and Luke) had to have been written after 70 AD, because Jesus predicts the destruction of the temple and Jerusalem by the Romans.

Most are aware that the discovery of the Dead Sea Scrolls in 1947 show that the books of the Old Testament were written before the time of Jesus. Even if they discount the idea that Isaiah (or two or three Isaiahs like some claim) wrote this book during his lifetime in the 8th and 7th centuries BC, they cannot give the book a date later than 125 BC. The only option left to them is to say that the life of Jesus was written to correspond to chapters such as Isaiah 52 and 53 (among numerous other Old Testament prophecies) which describes the suffering Servant in stunning detail. Scholars have taken liberty in trying to re-date the book of Daniel from the Dead Sea Scrolls on the grounds that the book could not foretell the rise and fall of empires in such detail, so it had to be written after the fact. They contend that Daniel, or at least the chapters dealing with the abomination of desolation and those dealing with the wars between the Seleucid and Ptolemaic Greek empires, could not have been written prior to 169-168 BC when Antiochus IV Epiphanes set up an image of Zeus in the Jerusalem temple and sacrificed a pig on the altar. Which led to the Maccabean revolt and the establishment of the Maccabean or Hasmonean kingdom. Which those practicing Judaism now celebrate in the holiday of Hanukkah.

What many of these critics fail to understand is how many, if not most, biblical prophecies have more than one fulfillment. Usually both a near fulfillment and a farther one in a future fulfillment. That is exactly the case with the abomination of desolation. It was fulfilled in the near term by Antiochus IV Epiphanes, but Jesus references this prophecy in Daniel (Matthew 24:15-22) and applies it to things still to come. And I believe our Lord was predicting the destruction and desecration of the Jerusalem temple by the Romans in 70 AD. And people did indeed flee to the mountains before the temple and city were destroyed.

But even in the book of Daniel itself, they do not recognize the precise nature of some of these prophecies nor their fulfillment. Take the 70 "weeks" of Daniel chapter 9, which most take as 70 "units of seven", or 70 times 7 years. From the decree of Artaxerxes to rebuild Jerusalem in 444 BC, one seven "week" period gets us to the rebuilding of Jerusalem. The next 62 "weeks" take us right to the death of the Messiah, who is to be cut off and who brings an atonement for sin (Daniel 9:24-27). The last seven "weeks" have various interpretations among Christians. From the destruction of the Romans in 70 AD, to a long period of time before the seventh "week". These critics know that Daniel was written in the centuries before Christ, because it found among the Dead Sea Scrolls.

It is really nothing more than unbelief. And this is one of the foundations of modern skepticism, that miracles cannot happen. But we know that unbelief is the default setting in natural man. The natural man does not have the spiritual faculties of perception to see God or appreciate Him. But I believe we need to go further and address the worldview of modern scientific man. Many would claim that miracles cannot occur, but does science really prove that miracles are impossible?

It is often assumed that this problem was put to rest by the 18th-century philosopher David Hume. Hume defined a miracle as a transgression of a law of nature. And that something so contrary to how the world normally works requires a great deal of evidence for a rational person to believe in it. And then he asserts, without justification, that no such evidence is available. Therefore, a simpler explanation not reverting to miracles is the more rational one. Others have built upon this definition and said that miracles are impossible because you cannot violate the laws of nature and extraordinary claims require extraordinary evidence. It is a kind of Bayesian reasoning argument.

First, I would agree that a miracle is generally something that does not occur in normal circumstances. But that is just defining what a miracle is. I would argue that a miracle can also be something natural that takes place but happens at a very specific and precise time that cannot be explained otherwise, such as our examples from Jericho and Sodom above. Or something that is predicted long before it occurs, even though its fulfillment may be explained naturally. But really, Hume is just defining miracles as impossible because he incorrectly assumes he understands the laws of nature, how they work, and that they can't be violated. And any contrary claim could be dismissed out of hand as not having huge amounts of evidence or credible witnesses. He would have been more modest in saying that miracles are improbable, which would be true based on Bayesian reasoning in most circumstances. Although if our prior probabilities included that God exists and the current context made our priors higher (Jesus's ministry and following Him around and observing Him), then the probability of a miracle would be quite high.

But most modern scientists are more persuaded by arguments from science and physics than from philosophy.

Classical physics is largely the discovery of Sir Isaac Newton. Newton's laws operate in closed or isolated systems. This means that we make the assumption of no outside influence of causation for the system we are studying. For example, if you drop a stone from the roof of a building on a calm day you can determine what the stone will do, provided that nothing acts on the stone until it hits the ground. This is your assumed closed system. If someone opens a window and deflects the stone on the way down, a new force and causation have been introduced into the system. Physics cannot give you a prediction of behavior based on factors from an influence from outside your perceived closed system. And the assertion that miracles cannot occur is sometimes based on the unfounded assumption that the universe is indeed a closed system. Newton himself thought God intervened on occasion to balance out the effects of gravity that planets had on one another in their orbits (which was later explained by his own laws). So obviously, Newton did not see the universe as a closed system as he developed his theories of physical laws.

For all practical purposes of doing science, we must make this assumption of closed systems to make sense of the data and make predictions. The claim that the universe is a closed system is not a physical claim based on evidence, but one based on philosophical assumptions.

Our logical argument to dispel this would go something like this.

1. The Earth and the universe are open systems.

2. The laws of nature are not violated in an open system.

3. Miracles could involve causation from outside the system being discussed.

4. Therefore, miracles would not violate the laws of nature.

Even using this logical argument, if God created the physical laws to operate an ordered, physical system, He is not bound by these laws. Science seems to intimate that the laws of physics could have been different. Also, science does not claim to have described all *possible* physical laws. So, if God wants to operate by introducing causation from outside the perceived closed system or with laws of physics besides the ones we have mathematically described in this universe, nothing in that statement is logically inconsistent or self-contradictory. Apart from this, we believe that Christ is holding all things together. If He were to remove His hand, the chain of causation would break down and the universe would be sent into chaos.

Many skeptics talk about miracles being impossible based on conservation laws, like conservation of energy, conservation of electrical charge, and the laws of thermodynamics. They claim that these laws show that the laws of physics cannot be changed or suspended to do a miracle. But that again just assumes that the system under question is closed and isolated, and that the system is also translationally symmetric. Such as under time-translation symmetry, space-translation symmetry, or rotation-translation symmetry. Once you break symmetry the law no longer applies. Even the universe does not obey conservation of energy because it is expanding and not symmetric under time-translation. To say that a miracle cannot occur, like the creation of new energy or electric charge from nothing, you would have to have perfect knowledge of the entire system to show that the laws of physics enforced this conservation throughout the entire perceived closed and isolated system.

As we scale down into the subatomic sizes, we get into the area of physics known as quantum theory. As far as quantum mechanics goes, if you subscribe to the Copenhagen interpretation, then no conflict should exist because everything is reduced to solutions based on probability. Where the wave function "collapses" to determine the location and momentum of electromagnetic energy and matter. Causation and outcomes at this subatomic level do not appear to be deterministic. It may be improbable that we observe a miracle, but that is part of the definition of what a miracle is.

Although quantum theory is much, much more complicated than this. Including wave functions and the Schrödinger equation that describes the evolution of wave functions in a quantum system. Not to mention quantum field theory. Most physicists seem to believe that macroscopic objects of everyday life do not show us weird quantum effects due to something they call decoherence. That is, interactions between quantum states and the surrounding environment decohere the superposition of particles into one state and particular observables. That is why we don't experience quantum effects in everyday life. But many of these concepts and equations seem to rely on the assumptions of closed systems and conservation laws. So, we are back to where we started. But there are many issues with the interpretations of quantum theory. Like what is the physical interpretation of the wave function and quantum states? What does it mean to make a measurement? From what I have read and tried to understand, it seems to me that we either must give up locality, which would undermine our experience of cause and effect and classical physics, or we must give up the idea of determinism. Unless someone adopts an interpretation like the many worlds interpretation, which basically says that a new branch of reality stems from each measurement and we happen to find ourselves in one of those branches getting definite values and results. I know, confusing right? I don't claim to understand quantum theory very well, but even physicists who are experts in the mathematics and theory don't agree on what it physically means. But in my view, quantum theory undermines determinism, meaning that we can't predict all future events due to the laws of physics.

So, if Jesus were to use quantum tunneling for all the subatomic particles in His body to walk through walls while at other times sitting down to eat fish with the disciples after His resurrection, is this too difficult for the glorified God-man? This is pure speculation, and we have no evidence from the Bible for such mechanisms, but it is just to illustrate a point.

Some physicists seem to be the closest of all the scientists to claiming that there is no room for God in their theories. This is because some of them think that the laws of physics make the entire universe and its evolution deterministic. Which is a pretty bold assertion, given what assumptions must be true and would have to be known comprehensively. Besides that, in practice there is only a small domain in which they can predict outcomes. And I believe that one decision of a small child frustrates their whole worldview. I think that a simple example of a child learning something new and then acting on that knowledge disproves physical determinism. I think if God inspired the book of Job today, He may have some harsh things to say to some overconfident physicists.

But the larger overarching point of science in general and physics in particular is that it doesn't prove anything in the logical sense. Physics takes many, many examples of how nature works and then generalizes them into laws. I personally think that law is an unfortunate term because even most of the laws have domains of applicability. They give us the right answer under certain conditions, but don't generalize to all situations. And even physicists cannot prove that these laws always hold. It is reasonable and rational to assume that gravity will function in the same way tomorrow because we have a lot of evidence for its regularity, but nothing in physics proves that it must. But more broadly, physics develops mathematical theories that mature over time, make predictions, and can be compared to observations. The best theories survive and are accepted as consensus and the others are set aside. But most physical phenomena can be described in several different ways with different mathematical equations. For instance, there are different mathematical ways to describe classic mechanics and the laws of motion.

Many (but not all) scientists are naturalists, and this has a substantial influence on our culture's worldview. Many of them would claim that miracles are impossible because we know the standard model of particle physics and all the fundamental forces and particles involved. If there were other forces and particles involved in miracles and God interacting with the world, we would surely have discovered them by now. But I don't find this line of argumentation persuasive. The biblical description of God is a Person who is transcendent above His creation and is *also* immanent within it. He is everywhere (omnipresent), all knowing (omniscient), and all powerful (omnipotent). He is spirit and does not occupy spacetime by extension. It would be odd to find God's mechanisms of interaction inside a particle accelerator studying high-energy physics. He is there, whether the scientists perceive Him or not, but He cannot be measured. And we will see that the events in spacetime evolving according to the laws of physics is clearly included in the description of God's providence.

The main takeaway from this admittedly complicated section is that we can see that miracles are possible, and science has not shown otherwise. Some people just do not want to believe they can happen because this implies an Agent distinct from creation. God is not one in substance with His creation like the pantheists would like us to believe. This brings us to an important point about how God manages His creation. The skeptic may say, "We know precisely how the natural processes work in a specific system from beginning to end, therefore God is not needed in our hypothesis." I think she has misunderstood how the Bible talks about the providence of God in the natural world. Let's see what the Bible says about this after looking at the Westminster Confession of Faith on Providence.

The Westminster Confession of Faith in Chapter 5 on Providence states:

I. God the great Creator of all things does uphold, direct, dispose, and govern all creatures, actions, and things, from the greatest even to the least, by His most wise and holy providence, according to His infallible foreknowledge, and the free and immutable counsel of His own will, to the praise of the glory of His wisdom, power, justice, goodness, and mercy.

II. Although, in relation to the foreknowledge and decree of God, the first Cause, all things come to pass immutably, and infallibly; yet, by the same providence, He orders them to fall out, according to the nature of second causes, either necessarily, freely, or contingently.

III. God, in His ordinary providence, makes use of means, yet is free to work without, above, and against them, at His pleasure. (Westminster, 1646)

So, although God is in control of all things in His universe as the first and ultimate Cause, He presumably governs and administers His creation by operating under normal secondary causes (laws of nature and causality). But He can and does freely work outside those ordinary means, either by miracles or causation that we are ignorant of.

Psalm 139:13-14

13 For You formed my inward parts; You wove me in my mother's womb. 14 I will give thanks to You, for I am fearfully and wonderfully made; Wonderful are Your works, And my soul knows it very well.

Here David is expressing his praise to God for forming him in his mother's womb. This is very important in our discussion about how it is that God works in His providence. We know quite well from biology how a human being is formed in the womb of a woman. We can trace the development of the gametes (egg and sperm), how the egg is fertilized, and how the combination of DNA of mother and father takes place in the cell. And with the rise of genetics, we are even getting a good idea of which sections of DNA are responsible for developing different parts of the body. We are gaining understanding of how the control genes like HOX and PAX tell other networks of genes when and how long to operate, how cell types differentiate, and what other epigenetic influences have on the expression of those genes. We know at what stage of the pregnancy the different parts of the body develop and essentially how the whole process unravels. All of this to say that it does not take away from the fact that God is behind it all. I don't think many of us would say that God supersedes His natural laws to develop a baby's arms or legs. He does create the soul and stamp it with His image, but modern science would just call this animating life biochemistry or molecular biology, while being ignorant of the immaterial nature of the soul and the divine image. But the point is that God is working through His natural laws and the language of genes and cellular biology to construct a unique human being. And we can proclaim with David that we are fearfully and wonderfully made. We are a unique person who is a unity of body and soul known by God before our conception.

Job 38:25-30

25 "Who has cleft a channel for the flood, Or a way for the thunderbolt, 26 To bring rain on a land without people, On a desert without a man in it, 27 To satisfy the waste and desolate land And to make the seeds of grass to sprout? 28 "Has the rain a father? Or who has begotten the drops of dew? 29 "From whose womb has come the ice? And the frost of heaven, who has given it birth? 30 "Water becomes hard like stone, And the surface of the deep is imprisoned.

Job 38:34-41

34 "Can you lift up your voice to the clouds, So that an abundance of water will cover you? 35 "Can you send forth lightnings that they may go And say to you, 'Here we are'? 36 "Who has put wisdom in the innermost being Or given understanding to the mind? 37 "Who can count the clouds by wisdom, Or tip the water jars of the heavens, 38 When the dust hardens into a mass And the clods stick together? 39 "Can you hunt the prey for the lion, Or satisfy the appetite of the young lions, 40 When they crouch in their dens And lie in wait in

their lair? 41 "Who prepares for the raven its nourishment When its young cry to God And wander about without food?

Psalms 65:9-10

9 You visit the earth and cause it to overflow; You greatly enrich it; The stream of God is full of water; You prepare their grain, for thus You prepare the earth. 10 You water its furrows abundantly, You settle its ridges, You soften it with showers, You bless its growth.

Psalms 104:10-11

10 He sends forth springs in the valleys; They flow between the mountains; 11 They give drink to every beast of the field; The wild donkeys quench their thirst.

Psalms 104:14-15

14 He causes the grass to grow for the cattle, And vegetation for the labor of man, So that he may bring forth food from the earth, 15 And wine which makes man's heart glad, So that he may make his face glisten with oil, And food which sustains man's heart.

Psalms 104:20-22

20 You appoint darkness and it becomes night, In which all the beasts of the forest prowl about. 21 The young lions roar after their prey And seek their food from God. 22 When the sun rises they withdraw And lie down in their dens.

Psalms 104:27-30

27 They all wait for You To give them their food in due season. 28 You give to them, they gather it up; You open Your hand, they are satisfied with good. 29 You hide Your face, they are dismayed; You take away their spirit, they expire And return to their dust. 30 You send forth Your Spirit, they are created; And You renew the face of the ground.

Isaiah 42:5

5 Thus says God the LORD, Who created the heavens and stretched them out, Who spread out the earth and its offspring, Who gives breath to the people on it And spirit to those who walk in it,

Matthew 6:26

26 "Look at the birds of the air, that they do not sow, nor reap nor gather into barns, and yet your heavenly Father feeds them. Are you not worth much more than they?

Jeremiah 10:13

13 When He utters His voice, there is a tumult of waters in the heavens, And He causes the clouds to ascend from the end of the earth; He makes lightning for the rain, And brings out the wind from His storehouses.

Acts 14:16-17

16 "In the generations gone by He permitted all the nations to go their own ways; 17 and yet He did not leave Himself without witness, in that He did good and gave you rains from heaven and fruitful seasons, satisfying your hearts with food and gladness."

Before we go on, it is important to note that Yahweh, the God of Abraham, Isaac, and Jacob, is incomprehensible. His judgements are unsearchable, and His ways are past finding out. A lot of trying to discern the mechanisms by which God ***may*** work through means are speculation if God has not clearly revealed them to us through His word. We risk being rebuked like Job if we claim we know how God works and if we claim that we have figured out the mind of God. Also, we need to be on guard against making God a force or cause while removing His Personhood. He is a God who speaks and acts according to the counsel of His own will. God is a Person and not a force. He is one in substance and three in Persons. Each member of the Trinity exercising a unique role in creation and providence, while the whole triune God works. This God cannot be domesticated and put in our box. He is not to be placed in the dock and questioned by His creatures. We are here for His glory and any questioning of His motives and means are the height of folly.

So, what can we infer about the way God operates in His providence over His creation? The word of God clearly portrays God in His providence working through His creation for His glory. And that He is ultimately the chain of causation in the natural world. Reformed theology has long acknowledged that the Bible teaches that God is fundamentally the first cause of all things by His eternal decrees. His ordinary means are thought to be through secondary causes within the natural world. Presumably, working through His ordained laws of physics and chemistry to govern the natural world. But whether we are talking about the development of a baby in the womb or the clouds and weather, God is governing this world by His providence.

When we look through the Bible for the workings of God's providence in the natural world, we see that He is in control. I think the more we can grasp this truth, the more we can rest. But we are given very little information about the means or mechanisms by which He is accomplishing His ends. We are told that He knows us before we are born, gives us life at birth, forms us in our mother's womb, appoints the times and places of our habitation, and takes away our spirit at death. We are told that He feeds the animals, makes the wind to blow, calls out the rain and hail, brings forth the lightning, and sets the boundaries for the sea. The wind and the sea obey His command, and He brings droughts, famine, and pestilence. We are told that He makes the grass, flowers, and crops to grow and produce their harvest. He gives food to the lion and feeds the beasts of the field and the birds of the air. He calls the stars by name and causes the rising and setting of the sun and moon by His fixed order.

It is quite an exhaustive list of God through His providence upholding, maintaining, and accomplishing His ends. And that is not even taking into account how He works in the spiritual realm or in the affairs on men. But as far as I can discern, the mechanisms between His decrees and causation are rarely given illumination by scripture. By the scientific method, especially over the last two hundred years, we have described the laws of physics and chemistry that contribute to our understanding of how the natural world works. So, should we do our best Pierre-Simon Laplace impersonation and in response to God ordering and causing events proclaim, "We have no need for that hypothesis"? As Christians, we cannot.

We can say that God is knitting us together in the womb, although we understand most of the genetics, the embryology, and how we go from a sperm and an egg to a human baby. God makes the grass and crops grow, although we understand photosynthesis and the biochemistry that goes into turning a seed into a plant. And that He commands the rain and lightning, although we understand meteorology. While we still have much to learn based on the methodology of science, I contend that it would not be possible without the providence of God. So just because science knows a lot of the details about why and how physical processes work, this in no way implies that God is not involved. In fact, God's word tells us just the opposite. Just that His operation of preserving and governing nature for His purposes and goals is invisible to our empirical methods. And for that He has condescended to give us special revelation. But the secrets things still belong to God.

Put differently, God's word rarely tells us about the mechanisms of how He is reaching His ends. If we have eyes to see, we can see attributes of God through His creation, including His power and wisdom. We can even be rebuked by the animals by watching their ways in displaying the wisdom of God. The ant can teach us not to be slothful and the birds of the heavens know their times and ways. The animals can teach us to depend on the God who controls all things and to look to Him in adoration.

Scientists (or the scientifically literate) are well acquainted with the idea of methodological naturalism. Which essentially says that we must always assume a naturalistic and empirically discernable cause for all things in which we are trying to describe and explain. And I agree that this is where they must start in science. If they were to come up with a hypothesis and instead of doing the rigorous work of coming up with an explanation of the why and how some phenomena operate, they said, "God must have done it.", that would stop the progress of scientific inquiry quite quickly.

But what we find is this methodology can quickly give way to the concept of philosophical naturalism, which makes dogmatic claims that there is nothing besides naturalistic causes and the material world. It is one thing to use naturalistic assumptions in our methodology for science but quite another to make broad sweeping claims that nothing exists apart from the physical world. Many scientists have no time or patience for such metanarratives or metaphysics. They want the most simple and parsimonious explanation in theories that make predictions and conform to observations. Which is all well and good, but I will argue that God and the Christian faith is the best explanation for the universe, the existence of objective morality, the longing for meaning and purpose, the dignity of man, ethics for human rights, and a need that we long for something we have lost. Christianity also has other lines of evidence to support its truthfulness including prophecy, events in history, eyewitnesses, and last, but definitely not least, personal experience and transformation.

I will also argue that science offers no good justification for our curiosity about nature and its workings, emotions that motivate learning and discovery, desires for a coherent and systematic understanding of things, what we ought to do, what is a good life, and meaning and purpose. I am sure most scientists believe these things themselves, but I don't think they have a scientific warrant for doing so. But we should be careful not to give into the temptation that God is superfluous in describing nature and its workings. God deserves to be glorified for His material creation and creating and sustaining life, not to mention His acts of redemption.

As Christians, we understand that fundamentally underlying all causes attributed to physics, chemistry, and history is God causing all things to come to pass. We may say that the reason for wars is human hatred, evil, and ambition. But is that fundamentally the reason behind them? Did God not use Assyria and Babylonia to bring chastisement to Israel and Judah? God used these evil pagan nations as His instrument of discipline and then He turns around and judges these nations for their evil motives, behavior, and pride. We may say that bitter circumstances were the cause of Job's questioning God's justice. Or we may say that Satan was the cause. But is this fundamentally the biblical view? Does God not allow Satan to afflict Job with bitter circumstances which lead to his temporary misery? But this same God is good and has the best intentions for his children, including weaning them off the world and its temporary comfort. After teaching Job (and mostly his theologically erring friends) a lesson about how He is not bound to tit-for-tat blessings, that His chosen people will continue to trust in Him when times get tough, and that questioning Him in His justice by the creatures He created presupposes that they know more than He does, He rewards and restores Job for holding onto his faith despite these bitter circumstances. And Job is better for it.

According to the Bible, the fundamental causes of reality are God's eternal decrees and His sovereignty. Now here I want to pause and say that I have found what I see as too much speculation when reading theologians discussing the eternal decrees of God and how God brings everything to pass according to His plan. I think they sometimes lack the biblical evidence to go as far as they do in their theology. So, I am more comfortable with saying God is sovereign, nothing can thwart His plans, and that He is good and wise in all He does. But as Christians, we need to be disciplined to have a worldview that looks through the lenses of faith to see God causing all things to pass according to His own counsel and for His ultimate glory. We should not be seduced by things like openness theology in which God does not know the future nor causes things to fall out in a certain predetermined way. In my view, this would be a god who could not fulfill prophecy and bring about redemptive history according to His timing. I think that a Christian's ability to live with certain doctrines in tension like sovereignty and free will, and eternal decrees and secondary causes, is the amount of mystery that we can live with. I think we must live with some mystery if we want to embrace the whole counsel of God in His word.

But unfortunately, this philosophy of materialism that denies supernatural causes is well rooted in the modern western consciousness. We are told through education, television, and other mediums that anything involving the supernatural can be called myth, fantasy, or magic. And this worldview is prevalent in some aggressive people seeking to actively undermine Christian faith. The ardent materialists assert confidence in their views that we can finally bury God and get on with rational life. They claim that no divinity exists, that religious belief is inherently irrational, and that it is even dangerous for culture and should be displaced. Their rhetoric is often meant to make the religious look foolish and irrational. Apart from arguments convincingly made to the contrary by people like historian Tom Holland, who argues that our moral and social norms and ideals in the west are largely derived from a legacy of Christianity (Holland, 2019), this philosophy of materialism is too bold and has to be taken by blind faith.

They claim that they have a lock on reality. And like the blind men feeling different parts of the elephant, they can stand aloof, like the man with sight, and claim that all the blind (religious) men grope in vain. This is the psychology of atheism for some. Thankfully, not all materialists have this combative attitude.

But again, we really should not be surprised when unbelievers do not accept spiritual things. God's word tells us that the natural man cannot discern spiritual things (1 Corinthians 2:10-16). So let us not be taken back by unbelief.

This presupposition that the miraculous is impossible then seems to me to be the very foundation of skepticism and secularism. Add to this, many people think that the only things we can truly know must be confirmed with science. The result of this foundation is that cultures who become more scientifically educated tend to gradually become less and less religious. Science itself is notoriously hard to define and I have tried to show that studying nature empirically does not preclude the supernatural or religious faith. Science cannot tell us the ultimate cause of things, that miracles are impossible, what is good, how we should live, and the purpose of life. And many important areas of knowledge do not come from science. Civilizations got along just fine before the rise of modern science.

If we recall, this skepticism is the reason that so-called higher criticism of the Bible denies that the gospels could have been written before 70 AD, because Jesus describes the Roman armies surrounding and destroying Jerusalem. This is why they feel many of the prophetic books were written much later than conservative scholars because they foretell events that were fulfilled. The skeptics don't seem to have a tenable explanation for the prophecies of Jesus dying the death of crucifixion amongst criminals with His hands and feet nailed to the cross. They can't see how a Psalm written 1000 years earlier could have foretold the death of Christ, not to mention hundreds of other prophecies fulfilled. They have no explanation for how the book of Isaiah could be found among the Dead Sea Scrolls dated to at least 125 BC describing Jesus as Messiah, especially in chapters like Isaiah 52 and 53. The only option available to them is to say that the written accounts of the life of Jesus were manufactured to match the earlier prophecies.

This belief that miracles are impossible seems to me then to be the foundation on which the house of humanism is built. This leads to many of the philosophical assumptions that go along with the work in the natural sciences. This is in no way meant to impugn the motives and beliefs of scientists in general. There are many born-again Christians scientists. And many agnostic or atheist ones that could not care less about metaphysical claims or are honest enough to say there may be more to reality than we can measure.

These philosophical underpinnings work their way out in various areas of science, especially when throwing around words like *random* and *chance* to explain the processes of nature. Calling something random is often another way to short circuit the study of science and impede its progress. Randomness is best defined as something which does not follow an ordered pattern or something we cannot predict with accuracy. But there may just be mechanisms involved that we are ignorant of and therefore cannot make good predictions of effects.

One of the biggest oversights that I think is commonly made is ascribing causative power to chance. We may say, "It happened by chance." But chance is not a thing. It has no causative power. It can affect nothing. Nothing in the history of the universe has ever been caused by chance because chance is not a thing. That's not to say that chance isn't a useful word in our language. It is very useful in describing likelihood in probability. Like, what are the chances that something happens? It is often used as a synonym for random. Let's take an example to see if it helps us clarify these thoughts. Let's say someone goes down to the convenience store and buys a lottery ticket. They either pick the numbers based on ones they always play, or they let the machine choose them. If they let the computer pick the numbers, a software algorithm is used to generate the numbers and they are printed on their ticket. Neither one of these scenarios was caused by chance. One was caused by them thinking about which numbers they wanted to play, and the other was caused by a computer program with a "random" number generator following the code of a software program. Now they have their ticket. So, they tune into the tv show airing the lottery drawing. The balls are swirling around in the metal cage and a vacuum sucks up each of the six balls into the tray where they are displayed. Now the balls were not selected by some force called chance. The position and momentum of the balls, the speed of the rotating cage, the interaction and forces between the balls, and when the machine is programmed or triggered to suck up each ball determines which ball will be a part of the six-number winning combination. So, we can see that neither the numbers on their ticket nor the determination of the balls was random or caused by chance. If they knew the initial conditions and all the information needed to calculate the outcome, they could choose the right numbers. But even here we cannot escape the providence of God.

Proverbs 16:33

The lot is cast into the lap,

But its every decision is from the Lord.

Even things that physicists try to predict based on statistical mechanics is not immune from God's government. What I believe can properly be described by chance is the likelihood that the numbers on their lottery ticket do in fact match the balls that were chosen. There is no force, causation, or communication of information involved. Instead, just the odds of 1 in 260,000,000 that their numbers would match the balls. Thus, we can properly say that their chances of winning were extremely small. Now the law of large numbers from probability says that if enough people play, someone is almost sure to win. The improbable gets likely when the numbers are large enough.

I believe that this is an apt analogy to finding ourselves in a universe that is friendly to life. This is the main thrust of the arguments for the fine-tuning of the universe and in many books describing the negligible probability that the laws of physics, the initial conditions, and all the free parameters are just right to allow for a universe that can support life. And not only life, but life that can think in logical and mathematical categories and be motivated to discern the natural world that they find themselves in.

When people say that things are random and there is no purpose or goal, they are espousing a worldview and not just reporting empirical data. They are making philosophical assumptions and not physical ones. What often gets Christians in trouble is saying that God is responsible for processes that we do not yet understand. This is often referred to as the god-of-the-gaps argument. We place a supernatural explanation in place of something science is still ignorant of. I think that this is a misunderstanding of ordinary providence. We believe that God in His providence controls all things, including secondary causes like natural laws of physics and chemistry. We may be ignorant of the most fundamental ways in which God uses other secondary causes for His ends, but that is okay. The secret things belong to the Lord. We are held responsible for the things revealed to us. When science has not fully understood how something works, some are tempted to say that God must have done it. Which is fine if God has clearly revealed that truth in His word. But if not, then these types of hypotheses have a generally poor track record. Science usually comes around to fill in the gaps of knowledge and offer a solution. But it does not follow that making god-of-the-gap arguments gives materialism more credence. Providence and design are legitimate arguments to infer the truth of Christianity.

One supposed example is what some claim to be a failed watchmaker analogy made famous by William Paley. This is an inference from design argument that says if you look around the world at the complexity of life and how the universe functions you can infer a designer. Many biologists would say that random mutations, plus natural selection, sexual selection, genetic drift, other evolutionary mechanisms, plus time can account for the *seemingly* apparent design. Paley's argument from design has been getting a renewed interest over the last few decades as the probability of random mutations producing beneficial adaptations is put under the microscope to see if it really has explanatory power. With the advancement of molecular biology and genetics, any explanation for adaptation and the illusion of design using simplistic examples and reasoning often no longer looks persuasive.

But look closely at what is going on with the argument of the illusion of design. These biologists are saying that it appears to us that evolution is a fantastically wasteful way to develop something. It appears to them that the mutations are random and not goal seeking, therefore they feel confident they can conclude that God is not involved in the process in any shape or form. This is not an argument from the data, but from their own assumptions. As we have seen, the way the Bible speaks about God working in His natural world apart from miracles is presumably using His own natural laws and processes.

Next time someone tells you how inefficient and wasteful "nature" is at coming up with a design, ask them if biologists in the 21st century can make a simple cell? The unstated premises in these arguments is that if they were to design a universe, they would do it differently and better. They fail to see that God will be glorified in all His attributes, including His mercy, grace, and justice. They fail to see that one creature differs in glory from another creature, and one star and galaxy differs in glory from another. And that our resurrected bodies will differ from our earthly bodies (1 Corinthians 15:35-49). They imply that they would expect the processes of nature to be different if God exists and created everything. But as we will see in later chapters, this may not be the case from the

Bible's own teaching. The critics of God being involved in biology seem to be saying that they know of a higher standard by which to measure perfection. Why are they critiquing the idea of God based on standards they don't believe exist?

Their arguments also have a lot of assumptions to produce dogmatic conclusions about metaphysical realities. And more importantly for our thesis, it does not conflict with how the Bible portrays the providence of God over the natural world. Just because we can explain how grass grows or the clouds form to make rain does not conflict with the fact that God is behind all things. We know that when He works miracles, He might use different chains of causation compared to the ordinary second causes and natural laws operating in the world. That is how you define a miracle, something that would not have happened naturally in the normal chain of events. The walls of Jericho may have been flattened by an earthquake, but no earthquake happens in the normal chain of events when an army marches around a city for the 7th time and blows the trumpets. That is a miracle. This is just one of hundreds of miracles spoken about in scripture. But it helps to bear in mind that God uses miracles in the Bible, usually called signs and wonders, in very specific ways. They are often to show His power and majesty to His covenant people (Exodus from Egypt), during certain important periods in redemptive history (ministry of Jesus), to demonstrate His superiority over so-called gods (1 Kings 18:20-40), or to verify the veracity of His message through His chosen prophets and apostles (Book of Acts). Whether the sign gifts in the New Testament are still in operation today is a matter of debate within orthodox Christianity and beyond the scope of this book.

I think miracles that happen in space and time is where Christianity must put its foot down when describing many events in the Bible. Christianity is a religion that is based on a God who has done miracles along the way from the beginning of creation up until the present day. From missionaries' stories, it seems God often does miracles when the gospel is brought to a people who have previously been without the gospel. Many Muslims are having dreams and visions of the truths of Christianity prior to someone coming to them to tell them the good news. Often you will hear about blind men being healed when the gospel is first brought to a people. To me this seems consistent with what happened in the New Testament, as miracles were being performed among the first-time listeners of the gospel as a way for God to verify the authority of His message.

There is always a purpose behind miracles in the Bible. God does not do them just to wow people like a magician. Let me be clear; Christianity without miracles and the supernatural is dead. Jesus was either born of a virgin and rose bodily from the dead, or He didn't. It is not some metaphor as to how He becomes alive in our hearts. If Jesus did not rise from the dead around Jerusalem in 30-33 AD, we are of all men most to be pitied (1 Corinthians 15:19). Our triune God, who dwells in heaven with a myriad of angels, who has created the entire universe by His Word, and upholds it by His power, can change the course of causation and introduce energy, force, and matter into any chain of events. God always has a purpose when He performs a miracle, and science has not shown miracles are impossible. But a proper biblical understanding of providence will guard us from thinking that God is only working when He is performing miracles.

Now that we have explored what I believe is the foundation for modern skepticism and the clear biblical teaching that God's providence is ultimately behind all events, ordinary and miraculous, let's look at another very common way that Christianity is attacked. By comparing it to ancient near eastern myths.

Chapter 4

Ancient Near Eastern religion and myth

Since the middle of the 19th century many ancient near eastern texts have been unearthed from different Middle Eastern locations, but especially from the library at ancient Nineveh (present day Mosul, Iraq). We also have plenty of documentation from Egypt that have been used by scholars to reconstruct creation myths and theology from there. This led to a flurry of activity when they were first discovered and translated. Some of these stories had a familiar ring to them. The ancient accounts of creation, humanity, and the flood from Egypt and Mesopotamia are going to be the ones that garner most of our attention, as this was the world and culture around the time of Moses and the writing of the Pentateuch (first five books of the Bible).

Scholars claimed that some of them sounded like the first chapter of Genesis (or the first 9 to 11 chapters), while others were more particular to Noah's flood. The search was on to find the similarities and try to construct theories of where the ideas came from and who borrowed from whom, if any. What I will warn the reader about regarding the ancient near eastern myths is that some of the people who want to use them to claim that Genesis was written using them as a template often distort what the texts actually say. Said a different way, scholars often take great liberty with the myths to try to make them match something from the Bible. When we read these stories in the best translations accepted by scholars of ancient languages, we often find the similarities are skin deep. Many of the ancient near eastern stories have superficial similarities to Genesis, but you would often not know that if you just listen to some professors or scholars who so desperately want to find borrowing, adaptation, or to assign late dates of authorship to books in the Old Testament.

But I believe we need to get the following idea in our minds from the start. It is established in Old Testament scholarship that God used cultural forms of communication to the Hebrews that were common from surrounding cultures. Kenneth Kitchen has done a wonderful service to Old Testament scholarship and Christianity by demonstrating the historicity of the Old Testament over his distinguished career, in particular with his fine book that I am familiar with, *On the Reliability of the Old Testament*. Professor Kitchen shows, among other evidence, that many of the forms and compositions of treaty, law, and covenant in the Bible were used by nearby cultures. He then goes on to show that the times in history when certain forms were common undermine the arguments made by critical scholars that many of the books of the Old Testament were written later than traditionally believed. (Kitchen, 2003) It is important to understand that there are scholars out there like Professor Kitchen who have answers to critical scholarship, as this is the place many students get pummeled by professors at college and seminary. But this idea that God communicated to the Hebrews in forms of literature, covenant, and law that they were familiar with will give us some insight.

If we have taken a comparative religion class, or any Old Testament class in a college, university, or even liberal seminary, we undoubtedly have come across the line of argumentation that says that the monotheistic religions of the world like Judaism, Christianity, and Islam are ultimately just the evolution (or revolution) of older polytheistic religions. From religions mainly originating in Sumer, Assyria, Babylonia, Egypt, and Canaan.

Every culture, whether it is literate or not, has a story about origins. How the world was formed, where the sun, moon, stars, sky, land, and the sea came from. Where the people and animals came from, how they got there, and their purpose in the world. Regarding Christianity, some of these professors say that the Old Testament stories of creation, human origins, and the flood are just adaptions of older stories from these pagan cultures which practiced polytheism (many gods) or animism (ascribing spirits to natural phenomena). We listen to these professors strain the comparisons between some of the ancient near eastern creation accounts and Genesis, trying to show that the Israelites simply reworked the stories for their monotheistic theology. Many of these same professors will adopt the documentary origin hypothesis of the Bible and say that the final version of Genesis was constructed when the Jews were exiled in Babylon during the 6th century BC or later. They will assert that the Jews living in

Babylon got familiar with the pagan texts and then adapted them to make their own story of creation, where mankind came from, and how their early history unfolded to give them an identity. Apart from the evidence to the contrary argued by Professor Kitchen and many other able biblical scholars who recognize that intelligent authors can use the different names of God or literary devices to make their points, we will find that most of these stories are very unlike the early chapters of Genesis. Although they will shed some light on some of our interpretations of these same chapters. And yet other stories, like those about Noah's flood, are too alike to brush aside and not examine critically.

We will see that some of these myths do not share as much in common as some would have us believe, while others point to more of a collective memory of actual people and events. But before we begin to look at the other accounts of creation, the origin of mankind, and the flood from the ancient near east, keep the logic of the skeptic's argument in mind. They are saying that because literature has been discovered from a culture pre-dating our understanding of the dating of the Genesis account, this means that the story, as told in Genesis, is just an adaptation of foreign myths and beliefs. The assumption seems to be that history or stories cannot be accurately known unless written down by contemporary sources. They generally distrust the ideas that God can reveal accurate history through revelation, that older sources may have existed, or that oral tradition can maintain accuracy. But I will argue that God did intend to communicate things to the Hebrews by referencing some of these creation accounts, just not in the way the secular scholars usually believe.

The Bible tells us that God placed Adam in a garden in Eden. This garden was in southern Mesopotamia (current day southern Iraq). I will try to show later in the book that there are very good reasons to believe that the location of the Garden of Eden is now under the waters of the Persian Gulf. Eve was created there, Noah lived in this same geographical area, along with Noah's sons and their wives. Abram was called much later from the same geographic region and given covenant promises that apply to you and me as Christian believers. The Old Testament in Genesis refers to what the Greeks would later call Mesopotamia (meaning between rivers) as the land of Shinar. Other nations contemporary with the Hebrews also used this term for Mesopotamia. So, from the creation of Adam and Eve to the descendants of Noah, to Abram and Sarai, thousands of years of history recorded in Genesis take place in Mesopotamia.

Furthermore, as I will expound further in later chapters, as Moses was writing Genesis, he had knowledge of ancient geography that pre-dates the Epic of Gilgamesh, one of the earliest known literary writings, by thousands of years. And even before the enigmatic Ubaid culture that predated the Sumerians in Mesopotamia. The mention of the rivers Pishon and Gihon, and the related geography of the four rivers coming together as one, are ***only*** mentioned in Genesis. Not in any of the Sumerian or Akkadian writings that have been found. The knowledge of the geography of the Garden of Eden could only have been known prior to about 7000 BC. But we will tease this out in much more detail in later chapters.

I think it is incorrect to say that the Hebrews simply adapted the Mesopotamian myths that they may or may not have been familiar with. I find it more plausible that the collective memory of some of these events (i.e. the flood) passed down through Noah and his sons found their way into the religion and culture of the Mesopotamians. The Sumerians, Akkadians, and Babylonians had a pattern of taking actual historical events and mythologizing them to make their rulers the heroes of the stories and to place the actual historical events in the context of their city-states. In other words, the ancient Mesopotamians, and cultures around them, did not create their stories out of whole cloth but took actual events and exaggerated and mythologized them to fit their political purposes. Professor Kitchen argues that the Mesopotamians did not historicize mythology, they mythologized history. (Kitchen, 2003)

We will see that some of the accounts are very different while others are similar. The account of Noah's flood from the Bible shares the most similarities with some ancient near eastern myths, but I propose that the other flood myths are an adaptation of what happened as a real, historic event as recorded in Genesis. I believe that this is given further support by the fact that the Mesopotamian flood stories changed with time and different elements were added to conform to the bigger stories they were inserted into. The context of all of them is within the city-states of Sumer and make an abrupt change in the kingship of the warring city-states. Put differently, the Mesopotamian myths reference a time in history that corresponds to after 4000 BC, while the context of Noah's flood in Genesis 6-9 puts it in the timeframe of 6500-6000 BC, long before the establishment of cities in Sumer. Another piece of evidence is the inconsistency of practices of worship from the ancient near eastern flood stories. Some have the flood hero landing on a "mountain" with a ziggurat already built to offer sacrifices, others infer that it was a river flood, others that the boat came to rest on the flat land, while others have them offering sacrifices outside the boat. I propose this is inconsistent with what we know about their religious practices, their view of deities housed in temples, and their view that the city-states were founded by the deities and then the city developed around the temple. But above all those considerations, are vast differences in how realistic Noah's flood account is in comparison to these myths. But we will get into all those details.

Ancient Near Eastern creation myths

Mesopotamian creation myths

Enuma Elish (Epic of Creation)

Let's start with the most popular account that many scholars try to make connections with Genesis. The first is the creation account from the Enuma Elish (which means "when on high", based on the opening lines). Different copies and fragments of this creation account have been found at Ashur, Kish, Nineveh, Sultantepe, and other sites. This myth is dated to the 12th century BC by most scholars, with some arguing for an earlier date. Many scholars argue that Enlil was still the chief deity before the 12th century BC, but with the defeat of the Elamites, Nebuchadnezzar I raised Marduk to be king of the gods. But Stephanie Dalley thinks there is good evidence for some or most of the epic being composed earlier than the reign of Nebuchadnezzar I. (Dalley, 2008) Some scholars think that the theme of a storm god defeating a sea god originated north and west of Mesopotamia (Hurrian, Mittani, Hittite, Ugaritic) and was known long before the Enuma Elish was composed, but this is debated.

The primary purpose of the account appears to exalt the god of Babylon, Marduk (sometimes also referred to as Bel, meaning lord), above the rest of the deities. Sumer is the birthplace of written language, although some scholars think it developed around the same time in parts of Egypt. We are told by archaeologists that the first known writing in the world dates to 3500-3000 BC from Sumer in southern Mesopotamia and the script is known as cuneiform for its appearance as wedge-shaped writing. The story of Enuma Elish is essentially about numerous gods warring with each other in cosmic battle, and the creation of the heavens, earth, and humanity. But its primary purpose is about exalting Marduk as the king of the gods.

The summary of the Enuma Elish:

Before the heavens and earth, two gods existed. Apsu (the male god of fresh water) and Tiamat (the female god of salt water). They mixed their waters together and created a host of great gods before anything else existed. These gods grew and matured and produced other gods.

These created gods caused so much noise and disorder that Tiamat and Apsu were irritated and could not rest. Apsu wanted to destroy them all, conferring with his advisor Mummu, but Tiamat was against it. The wise god Ea (Sumerian Enki) becomes aware of the plan, he puts Apsu to sleep, and then kills him. He then either kills or binds the advisor Mummu. Ea then builds his temple on Apsu (which was also the name of Ea's temple and the name of the freshwater) and he and his goddess wife Damkina have a son named Marduk. Marduk is praised for his attributes.

One of the third-generation gods, Anu, (god of the skies and wind, Ea's father, and Marduk's grandfather) creates the four winds which whip up the sea (Tiamat) and upset the gods in Tiamat's belly so they cannot rest. They petition Tiamat to take revenge for the death of Apsu and those gods living in him. She agrees, changes her mind about the great gods that she and Apsu had created, and she now wants revenge for the death of Apsu. Tiamat creates monsters with invincible and terrible weapons to battle with Ea and the gods that sided with him.

She appoints her husband, the god Kingu (Qingu), to be the leader of the gods on her side and the monsters she created. She legitimizes his position by affixing the Tablet of Destinies (which gave the god who possessed it power to decree destinies and fates) on his chest.

Ea learns of the plans of Tiamat and is terrified. Ea goes to his grandfather Anshar and tells him about Tiamat's plans of vengeance, and they are both fearful at the prospect of going to battle with Tiamat, the gods with her, and the monsters. Anshar then encourages his grandson Ea to try to pacify Tiamat by words or a spell. Ea goes to seek out Tiamat but turns back after perceiving her power. Anshar then sends his son, the god Ann (or Anu), to try to pacify Tiamat, but he turns back after perceiving the power of Tiamat. Anshar, their father, is dismayed and falls silent.

Ea summons his son Marduk to approach his great grandfather Anshar to be their champion. Marduk agrees to be the avenger and fight Tiamat on the condition that the great gods give him preeminence among the gods if they let him be their champion to defeat Tiamat. Anshar is overjoyed to have a possible solution to the war with Tiamat. Anshar then assembles all the gods, including his parent gods, and they have a party of eating and drinking in which they get drunk, their spirits are light and merry, and they decree Marduk to be their champion and king of the gods. The assembly of the gods pour out praise for Marduk their leader and promise loyalty to him and to obey his commands. The other gods then robe him with the kingly garments and ornaments and seat him on the throne. They decree his sovereignty and destiny and make him king of the gods.

Marduk is girded with weaponry and given the powers of the wind from the god Anu. He hooks up four terrible horses to his storm chariot and sets off for Tiamat wielding a net and his terrible weapons. The two of them meet in battle and Marduk captures Tiamat in a net while using the wind powers to subdue her. Marduk then inflates Tiamat, pierces her through with an arrow, and slits her insides. He gathers up all the gods that were arrayed against him on Tiamat's side, along with the monsters, and binds them. Marduk takes the Tablet of Destinies from Kingu and fastens it to his own chest to legitimize his rule and dominion.

He smashes the head of Tiamat with a club and splits Tiamat's corpse in two. From one of the halves, he creates the heavens by stretching out her skin. Marduk assigns a role so that the waters of Tiamat do not escape. He surveys the heavens, prepares the Apsu, and builds temples for Anu, Ellil (Enlil), and Ea.

Marduk makes appointments for the gods of the heavens, and he creates the constellations in the stars which correspond to different gods. He sets the timing of the year by the stars and the constellations, who are themselves gods. He creates a moon god and gives him guidance about the waxing and waning of his cycles. He gives commands to the moon god with how to act in conjunction with the sun god.

Marduk makes the rain and springs from Tiamat's spit and the Euphrates and Tigris flow from her eyes. He uses various body parts of Tiamat's corpse to create the earth, mountains, and ways of separating the parts and functions of creation. He brings the captive defeated gods into the divine assembly and fashions statues of the eleven monsters to be at the gate of a temple for remembrance. The gods honor Marduk for his achievements, give him lofty praise, and officially inaugurate him as king of the gods. Marduk purposes to create a temple in Babylon for himself and make that city the center of religion. All the gods of the sky and fresh water can find their abode there. Marduk then divides the gods between heaven and earth.

Marduk conceives of creating mankind of blood and bones to relieve the work of the gods by sacrificing the god Qingu who instigated the rebellion of Tiamat. They slay him and Ea uses his blood to create mankind to relieve the work of the gods. The other gods are grateful and ask Marduk what they can do for him in return. Marduk asks them to build Babylon and its temples and shrines. The gods do the work and construct a temple for Marduk in Babylon, along with its ziggurat. They finish constructing the temples and their own shrines, at which point they gather for a grand banquet and feast to celebrate Marduk and all the destinies being fixed.

The epic ends with the fifty names of Marduk pronounced by the other gods which extol him, give his attributes and functions, and reconfirm his role as king of the gods. (King, 1902) (Lambert & Parker, 1966) (Dalley, 2008)

I think the story speaks for itself and is easily distinguished from the account we are given in Genesis. With the possible exceptions of the world beginning with water, then being differentiated, the implied waters above the heavens, and different functions for the elements of creation.

But Genesis is not the story of gods battling one another vying for power, it is about one sovereign God creating from nothing. Some say that the creation account in Genesis begins with matter already existing (the Earth and the deep), but Genesis 1:1 is best understood as not being a title for the rest of the chapter but a declaration that God has created everything. The Hebrew word bara' is only used in relation to God being the subject of creation and never with preexisting matter. Heavens and earth is called a merism and describes everything, like saying "the land and the sky" implying everything. This figure of speech was common in the ancient near east, and the ancient Hebrews had no word for universe.

The point in Genesis 1:1 is that all things have been created by God and this point is further elaborated in the New Testament in John 1:3 and Hebrews 11:3. The idea is that God created the universe, including the Earth, prior to him developing it into a habitable place for man and beast. Some want to equate the Hebrew word tehom in 1:2, translated as deep, with the god Tiamat. Although they *may* be derived from a similar root word for ocean or deep, the idea of Genesis 1 can be seen as a polemic (a strong argument against an opposing view) counter to the polytheistic cultures and their mythology deifying elements of nature. But we will explore this further when we discuss Genesis 1.

Atrahasis Epic – "Creation" account

This story dates from the 17th century BC in Babylonia, was written in the Akkadian language, and is almost complete.

The summary of Atrahasis:

The "creation" portion of the tale begins by already assuming the existence of two categories of gods. After the lots were cast, some great gods went into the sky and the others into the fresh water of the Apsu. The lower gods were tasked with doing the labor on the earth under the supervision and authority of the gods of the sky. These lesser gods were under the burden of heavy manual labor digging ditches, canals, the Euphrates and Tigris rivers, and building the mountains. These lesser gods were under the supervision of god Enlil (Ellil) and after thousands of years of hard labor, toil, and complaining day and night, the lesser gods rebelled and conspired to storm the dwelling place of Enlil and protest the undue burden of their labor.

The laboring gods took up weapons and fire and surrounded Enlil's mountain temple. Enlil is awakened from his bed by his servant, and they summon the sky god Anu and the god of freshwater Enki. The servant asks the lesser gods what their complaint is, and they tell him that their labor is too great. Anu goes back to the sky to convene with the great gods, and they seem to agree that the burden of manual labor is indeed too great for these gods.

Enki advises the gods on creating human beings to take up the work in place of the lesser gods. The gods sacrifice one god after some purification rituals and then the god Nintu mixes clay provided by Enki with the flesh and blood of the slain god. Somehow human beings would remember the slain god because the spirit of the slain god was now part of them. (Maybe a reference to the heartbeat?) All the other gods spit into this mixture and the midwife god Mami calls on the womb goddesses and out of this clay seven male and seven female humans are created. They decree certain fates for the humans, including a 10-month pregnancy, and require certain rituals when a human woman gives birth. These human beings begin to do the labor that the gods used to do, grow food, and provide nourishment for the gods themselves.

All seemed well and good until the humans multiplied and created too much noise for the gods and thus keeping them perpetually irritated. Around 600 years (the Mesopotamians used a base 60 number system) after their creation, the population of people grew large and their noise was irritating to the gods, especially Enlil. So, he ordered a disease to break out among the people.

Atrahasis (meaning exceedingly wise) is portrayed as a pious and devout man to his patron god Enki (who is usually the compassionate god in these myths). Enki tells Atrahasis to tell the people to stop praying and offering sacrifices to all the gods and goddesses except the one responsible for the disease. They do this and the god is shamed into relenting by removing the disease. This pattern continues every 600 years or so with drought, famine, disease, and infertility. And every time Enki gives counsel to Atrahasis, the god(s) responsible for the calamity relent when they are the sole focus and recipient of the people's prayers and offerings.

Eventually Enlil has had enough and convenes a council with the gods, and they decide to flood the earth. He and Enki seem to have an argument and Enlil is furious that although people were created to do the work of the gods, produce food, and take care of the god's needs and temples, they are far too noisy and must be destroyed.

(We will return to the flood portion in the next section.)

After the flood, the gods once again decide to limit human population. It is unclear what some of the mechanisms for limiting the population are, as the extant tablets are missing lines. But from what is legible, some women are to lose their babies at childbirth and some women will have roles that make it taboo for them to be mothers. (Clay, 1922) (Lambert & Millard, 1969) (Moran, 1987) (Dalley, 2008)

Those who would wish to compare the "creation" part of this epic with Genesis I think are reaching at best. Like Enuma Elish, it is describing the conflict among the gods and the purpose for creating mankind is to lessen the burden on the gods and to work for them. Let the story speak for itself and I think we will agree that it has little similarity in the "creation" part. The flood part is a different matter, which we will return to.

Adapa myth

Some have suggested a connection between Adapa, Adam of Genesis, and the Fall of mankind in this tale from the 14th century BC, with its earliest tablets attested from Amarna in Egypt.

The summary of Adapa:

This story describes the man Adapa being created as a wise but mortal sage created by his god Ea (Enki), although the myth doesn't say how he was created. Stephanie Dalley points out, based on other Mesopotamian literature, that Adapa was the first of the seven sages before the flood to bring civilization to mankind. In his case, Adapa was to bring proper religious observances and rituals as the first priest of the first temple in Eridu. (Dalley, 2008)

He was a wise, pious, and intelligent man of the city of Eridu (in southern Mesopotamia, regarded as the first Sumerian city), was chief among men, and no one questioned his judgements. He conducted priestly duties like setting and clearing the offering table, baking the bread, bringing the water, and closing down the temple for the night. He also did the fishing for Eridu.

Upon having a fishing mishap while out sailing on the sea because of the south wind, he cursed the wind and broke the wing of the south wind, rendering it impotent to blow for seven days. The god Anu, god of the sky and winds, heard about this and summoned Adapa to appear before him in heaven to give an account of his actions against the south wind.

Ea was aware of this summons by Anu, so Ea advises Adapa to mess up his hair and show up in mourning clothes. Ea tells Adapa how to act when he gets to heaven. He advises Adapa to tell the two deities at the gate that he is mourning because these same two gods, Tammuz (Dumuzi) and Gisbzida (Gizzida), have vanished from the land. And because of this they will show favor on behalf of Adapa to put Anu in a good mood and favorable disposition towards Adapa. Ea tells Adapa not to eat the bread and water that are offered to him by Anu as they will bring him death, when in fact they are the bread and water of eternal life. But he is to accept the clothing and be anointed with the oil.

Adapa explains his actions for breaking the wing of the south wind to Anu, which is that he was fishing for the temple and when the south wind came out of nowhere and capsized his boat he lashed out in anger. And then Tammuz and Gisbzida seem to advocate for Adapa, Anu's anger is calmed, and he seems satisfied with the answer. Anu offers him the bread and water of eternal life, but Adapa heeds the advice of Ea and does not eat nor drink. Adapa does accept the clothing and anointing oil. Anu seems surprised that Adapa did not eat and drink the food and drink of eternal life. Adapa explains that Ea gave him the advice, and then Anu sends him back to earth still a mortal man. And then Anu seems to be angry with Ea for giving advice to Adapa. (Rogers, 1912) (Dalley, 2008)

Some have been tempted to infer the concepts of temptation and the Fall in this myth to make a connection with Genesis 3. They make the comparison of Anu clothing Adapa with a garment with Adam being clothed by God with animal skins. They also say that both Adam and Adapa had to give an account of what they did before God/god.

In my view the comparison is a stretch. Adapa is never presented as the first human being or head of humanity. He is presented as a sage and priest among the men in the city of Eridu in southern Mesopotamia. Stephanie Dalley gives us further insight from other Mesopotamian literature in that she argues Adapa was the first of the seven sages before the flood, and that all the sages lost the chance at immortality from the sky gods and were banished from the earth to live in the Apsu with Ea. (Dalley, 2008)

Adapa's breaking the wing of the south wind is never conveyed as his first act of disobedience or a falling away from fellowship with his father god Ea or the god Anu. Adapa is offered eternal life *after* making Anu angry, and Adapa is deceived from achieving immortality by Ea, his patron god of Eridu, although it is freely offered to him by Anu, a god of higher status. And Adapa's breaking the wing of the south wind angered the god Anu because he was the supreme god of the sky and in charge of keeping order in southern Mesopotamia. We could say a couple similarities would be that Adapa and Adam are both portrayed as righteous and holy men, and that both can be viewed as priests. More on that later.

I am not saying that Adapa could not be a remnant in the minds of the Mesopotamians of the true Adam of Genesis passed down by tradition. As he is described as holy, pure of hands, and a priest. But if breaking the wing of the south wind is conceived as sin, then being offered eternal life after doing it is anathema to the biblical story.

Some claim that the name in the myth should be translated Adamu instead of Adapa, and that this name would correspond to the second king listed on the Assyrian King list. It is possible, though very unlikely, that this is a memory of Adam passed down through thousands of years. Whatever the conjecture, this story and the one in Genesis have vast differences.

Egyptian Creation myths

Egypt had several different creation myths from different cities and times that appeared to evolve within the long-lived kingdom, but they did share some common themes.

One of these themes is that everything arose out of the chaotic waters known as the god(s) Nu or Nun, and Nunet. This water itself seems to possess the creative power for the initial god(s). A hill of dry ground emerged from this primordial water and was the beginning of all things. The self-begotten god Atum, the sun god Ra or Re, or the composite Atum-Re was found on this initial mound of earth. And this original god led to the creation of further gods.

One of the most ancient, arising in the Old Kingdom, was the Ennead (or nine gods) worshipped in the city later called Heliopolis by the Greeks and Romans.

Ennead

The Ennead (nine gods and goddesses) creation myth begins by the self-begotten creator god Atum or Atum-Re arising from the primordial waters (god Nun) on a mound (god Tatenen). Atum, Re, and Atum-Re were solar deities. The god masturbated or spat and created two other gods, Shu and Tefnut. Shu represents air and Tefnut moisture. Shu separated the earth from the sky and his domain was in the clouds. Tefnut was the personification of rain and dew. These two siblings have intercourse and produce Geb and Nut. Geb represents the earth and Nut the sky. Geb personified the green plants and vegetation, and Nut was often show as a woman arching across the sky with stars covering her body. She was thought to swallow the sun each night and give birth to it the next morning. Geb and Nut had relations prior to being separated by their father Shu (air between earth and sky) and produced four children. Osiris, Isis, Seth, and Nephthys. These four gods were more in touch with the everyday affairs and destinies of men.

Osiris was a benevolent and good god, but according to some later traditions was murdered, chopped up, and scattered throughout Egypt by his brother Seth because of jealousy. Seth was an evil god of chaos who was associated with war and storms. (Interestingly, Seth was worshipped through much of Egyptian history and some pharaohs used his name in their throne names.) Osiris's wife/sister Isis and other sister Nephthys gather up the pieces of Osiris to put him back together and Isis gives him life. Nephthys was technically Seth's wife, but he was evil, and she had a child with her other brother Osiris and named him Anubis. Anubis in later Egyptian religion was considered a god of funerary rights, mummification, a guide to the afterlife, and participated in the weighing of the heart along with Osiris and others. It is likely that he was one of the main gods of the dead and underworld prior to the rise of Osiris in Egyptian religion, although Osiris is also attested quite early in Egypt.

After coming back to life, Osiris and Isis conceive a son (who is named Horus at his birth) and then Osiris is appointed the judge of the underworld where he rules the Duat and judges the hearts of people to see if they are worthy of the afterlife. Isis is also presented as a benevolent and compassionate goddess. Horus later avenges the death of his father Osiris by defeating Seth, although there are a variety of traditions as to the nature of the conflict and ending. But in my view, it seems that most end with Horus regaining the kingship of Egypt and is used as the analogy and justification for Pharaoh ruling both upper and lower Egypt as a unified kingdom, subduing the chaos personified by Seth, and bringing Maat (justice, order, and goodness) back to the kingdom.

The Ennead seems to have been combined with the myth of Osiris to make a coherent story and to explain why Pharaoh must rule as Horus on earth to maintain Maat, because the Pharaoh instantly becomes Osiris upon his death, and his living son is now Horus on earth. All Pharaohs were also called the sons of Re. But with most Egyptian mythology, the texts, traditions, and religious significance are fragmentary, fluid, and flexible. And the gods take on different attributes and roles throughout history, so there is debate among scholars.

One of the next oldest is known as the Ogdoad (or eight gods) worshipped in the city later called Hermopolis. Mentioned in the Coffin Texts of the Middle Kingdom, it is thought to go back to the Old Kingdom.

Ogdoad

The Ogdoad (eight gods and goddesses) starts by describing eight deities who existed in the primeval, chaotic waters of pre-creation described with negative attributes. There were four male gods with frog heads and their four female wife goddesses with serpent heads. They are known as (sometimes spelled differently) Nun and Nunet, Hok and Hoket, Kuk and Kuket, and Amun and Amunet (Egyptian female goddesses can usually be identified by their translated English names ending in t). There doesn't seem to be a total consensus on the attributes of these gods among scholars, but there is general agreement that Nun/Nunet represent the primordial waters, Hok/Hoket timelessness, infinity, or formlessness, Kuk/Kuket darkness, and Amun/Amunet hiddenness or invisibility.

There are different versions of this creation story. One version has these eight deities creating an egg which is either laid by an ibis bird, representing the god Thoth, patron deity of the city of Hermopolis (from which this creation myth originated), or a goose laying the egg. This egg then hatched a deity (usually the sun god Re, or later the composite god Amun-Re of Thebes) who it is assumed was responsible for creation and everything that followed. Sometimes from the primordial waters a mound of land (called the benben) arises, which is personified in the god Tatenen (and most likely signified by obelisks and on the tops of pyramids) and a lotus flower springs up from it. Inside the lotus flower, the sun god Re or a scarab beetle personified by the sun god Atum (sometimes called Atum-beetle) is produced. Traditions from other parts of Egypt, particularly Heliopolis, used the background of the Ogdoad to form a creation narrative with Atum creating himself on the benben and then the story of the Ennead follows. This synthesis of different creation narratives was common in Egypt as the religious and political power waxed and waned among the Egyptians cities through the millennia.

Although it's not really accurate to say that these are creation narratives. Unlike Mesopotamia, scholars in Egypt don't have the benefit of translating and interpreting generally coherent and linear literary works of creation. They must piece together myths and traditions from various sources like the Pyramid texts, coffin texts, and the book of the dead. But I personally think it's safe to say that most Egyptians would have thought of the Ogdoad as what existed prior to the ordering of the cosmos and the generations of the gods, which in the Egyptian mind were one and the same process. Nature was made up of gods and goddesses, personifying aspects of nature.

Ptah

A third god (with his Memphite creation myth) that scholars like to try to make connections to Genesis and Jewish and Christian theology, involves the god Ptah, who was initially a local deity of the capital city of Memphis. Memphis was the first capital of the united kingdom of upper and lower Egypt, so their deities had important roles for the kingdom. It seems quite clear that this god was important in Memphis from the very early times in the Old Kingdom, as he had a temple constructed in the city from the earliest times. But his role and the theology surrounding him seemed to evolve over time.

Ptah's identity seemed to change (or just added new roles and functions) from being a patron god of craftsmen, blacksmiths, architects, and skilled labor, to a god of mummification and funerary rights (including the opening of the mouth ritual), to the association with the primeval mound (Tatenen) arising from the primeval waters (Nun), into absorbing the roles of other gods, to finally being the one that created them and imbibed them with life as his status in the religion of Egypt was elevated (or the city of Memphis wanted its own creator deity exalted above Heliopolis, Hermopolis, and Thebes). He seemed to be linked in the Old Kingdom to the god Tatenen, which could have been his initial association with the primordial mound or benben.

As far as I can tell, his role as a creator god who is responsible for the creation of all things (although this language is not so straightforward from the Memphite theology as we will see) is not attested prior to an artifact known as the Shabaka Stone dated to around 700 BC. He still seems to be known as a creator god, along with Re and Atum, around 1200 BC in the New Kingdom. But not as a god who creates all other gods from his heart and tongue. Scholars debate when this concept of Ptah as the chief god who creates all other gods by heart and tongue took form, with many scholars dating the concept to the late second millennium BC (late New Kingdom).

The text of the Shabaka stone in which this creator god Ptah allegedly spoke all things into existence is summarized as follows:

It starts with a section giving names and titles to the Nubian Pharaoh of Dynasty 25 (around 700 BC), whose throne (royal) name is Neferkare, and whose given name is Shabaka or Shabaqo. It then goes on to explain that this Pharaoh made the stone inscription because he found another copy that was worm-eaten and in bad disrepair and he wanted to make sure the name of Ptah-Tatenen endured forever. Another section of the stone states that Ptah was the unifier of Upper and Lower Egypt and the creator of Atum and the Ennead.

The text then goes on to discuss the conflict between the gods Horus and Seth and how the Ennead gave Seth kingship over Upper Egypt and Horus kingship over Lower Egypt. The god Geb (leading the council of the Ennead) then changes his tune and gives the entire inheritance of Upper and Lower Egypt to his firstborn son Horus. And so, Egypt is unified under Horus. This feat is then codified by signs adorning the temple of Ptah in Memphis as the place and deity of Egypt's unification. Then there is a section which talks about the god Osiris and appears to relate how Osiris's body was rescued from the waters after he drowned (one of the versions of the myth about how Osiris was killed) and how he was apparently buried in Memphis and his son Horus was uniter of Egypt.

And then the section that has been termed the Memphite theology (the theology of Ptah coming from the city of Memphis):

Ptah is given various titles, including composite names with Nun and Nunet, two of the gods from the Ogdoad representing the primordial waters and seen as the gods from whom Atum arises in the Ennead. Ptah is then presented as the father and mother of Atum through Nun/Nunet and this appears to validate him as the creator of all subsequent gods. He is called the heart and tongue of the Ennead and Nefertem (who was Ptah's son in the triad of deities in Memphis with his wife Sekhmet, and is sometimes thought of as the young Atum himself) closely associated with the sun god Re. Ptah is then declared as the one who gives life and a component of the soul (the ka) to Atum, Horus, Thoth, and all the other gods. It then seems to explain that Ptah is the source of the heart and tongue for all the gods and gives them their animate power to think and create all that exists. It still acknowledges the creation of the other eight gods of the Ennead from Atum's semen, but seems to credit Ptah as the source of the will and power for the Ennead to do their creative work, including humans and their actions, arts, and civilization, etc. In other words, Ptah is the creator of the gods of the Ennead who then have the power and ability to form the cosmos by the gods of those cosmos. Ptah is once again associated with Tatenen and the benben arising from the primordial waters. Ptah is said to have created the temples, shrines, and statues of the gods and then all the kas (part of the soul) of the various deities were united with their statues in their temples under the uniter of the two lands (and all is as it should be).

The inscription ends with a praise for Memphis and the temple of Ptah as seemingly the burial place of Osiris and where his son Horus united the kingdom of Egypt. (Lichtheim, 1973) (Dungen, 2004) (Bodine, 2009)

So, it is not at all like Ptah spoke the cosmos into existence and then separated them like the God of the Bible, but because Ptah existed first in the primeval waters and the benben, he gave all the other gods life and everything that followed from that. There is no mechanism or timing in the story about creating human beings, as it is a more abstract concept of the role of Ptah in enlivening and empowering Atum and the other deities to create the other gods, and by extension, the world, humanity, and all that exists.

When it comes to the creation of mankind, ancient Egypt had various myths to account for them. Some records tell of the god Khnum, also a craftsman god, as the one who fashioned people and animals on his potter's wheel, not Ptah. Later texts included the creation of the gods as well coming from the potter's wheel of Khnum. Or people come from the tears of the young sun god birthed from the lotus flower. Or from Atum's tears as he is waiting for his children to come back. But overall, it doesn't seem that Egypt was too concerned about myths explaining the creation of mankind.

In actual history, the first Pharaoh of Egypt to unite Upper and Lower Egypt is known as Menes or Narmer. Ptah is usually found in a triad of deities in the city of Memphis, including his wife Sekhmet and their son Nefertem. Nefertem is sometimes described as a younger Atum. Later Egyptian religious beliefs seem to conceive of the presence of Atum in the sacred mounds on which the holy sites were built and the presence of Ptah in the gods and other animate life. But it seems that this abstract concept of Ptah giving the other gods power through their hearts, tongues, and kas did not meet with widespread acceptance in Egypt, even after the Shabaka Stone was created.

Confused yet? Me too. I don't think even scholars of Egyptian religion would claim that the Egyptians had one unifying idea about creation and the theology surrounding it. In fact, the Egyptians seemed quite comfortable (as far as I can tell) having various creation stories and theologies that they didn't seem anxious to make consistent. What seems to be going on is that each major city in Egypt had a triad of deities they worshipped, among the numerous other less important ones to them. One male, one female, and then a son or daughter. The males in these triads would often assume or evolve into the first creator god. It could be said that Egyptians at different times and cities (this was one very long-lived kingdom) exalted numerous deities as the first creator god, including Atum, Re, Atum-Re, Amun, Amun-Re, Khnum, Min, Tatenen, Aten, Ptah, and probably some I am missing. But none of these gods exist eternally and then speak material creation into being.

But before we scoff too quickly at the Egyptian beliefs, we must remember that many of these same scholars would say the same things about Judaism and Christianity. Many of them would claim that we have two distinct creation stories in Genesis 1 and 2 and that the theology of creation and God's attributes in Judaism and Christianity evolved over time. I hope to show that this skepticism is unwarranted.

It is sometimes asserted that the Jewish and Christian idea of a self-existent Creator God originated from this idea of Ptah from Egypt, but we can see from the above study of Ptah that his identity was morphing through time to absorb the other local deities and reconcile them into an idea of a first creator god. Furthermore, the theology that he thought and spoke the gods and creation into existence is not even an accurate way to render and understand the text. Ptah being identified with the preexistent waters of the gods Nun and Nunet, and with the benben of the god Tatenen, is far different than the theology of Genesis 1, and is only documented long after we assert Moses had written the Pentateuch. Ptah also spawns many other gods, so it was definitely still polytheism at heart. I think what is going on with Ptah is that he was initially a craftsman god, then associated with the god of the benben, and later a god of the dead and funerary rituals. Which included the opening of the mouth ritual to give the mummified bodies their kas so they could navigate to the afterlife. This opening of the mouth ritual was also associated with animating the idols (statues) in temples by giving them their kas so they could be a manifestation of the god or goddess in their temples. (Walls, 2005) It seems to me that in the Memphite theology this same power of instilling the kas is why Ptah is thought to give animating life and energy to all the other gods. It was only a short jump to associate him with the primordial waters in the gods Nun and Nunet and therefore claim that he was at the beginning and gave all the other gods life.

What is interesting and I think worthy of note is the idea of primordial waters at the beginning of creation and the ideas of initial disorder or negative attributes that seemed to be shared among Genesis, Mesopotamia, and Egypt. We will come back to this.

One theme to tie these pagan creation myths together (apart from the obvious polytheism) is the fact that they all share the idea of a realm in which the gods themselves spring out of. They are seen rising from primordial waters, a mound of land, and/or a cosmic egg in the case of Egypt, all of which are gods themselves. Or from salt and fresh water who are gods in the case of Mesopotamia. And they all produce generations of other gods, usually through sexual reproduction, or euphemisms to imply as much. But the myths that contain a coherent beginning to explain where the gods and creation comes from all have deified waters and these deified waters then produce generations of gods.

And the main and significant difference with the Bible (again, apart from the explicit polytheism) is that all these gods are made in the image of man. What I mean by this is that all these so-called gods act like fallen human beings. They have passions and emotions, they are in conflict with one another, they plot and scheme, they have sexual desires and procreate, they work and labor, they need people to meet their needs, and they need to control chaos and disorder in the realm of the gods to reflect the ideal and balance in the world of men. They act like people act.

You will often hear from the teachers of some of these courses on comparative religion, or scholars of ancient religion, that the idea of God creating the universe out of nothing is solely a product of later Christian theology. But this is simply not true. It is true that a great deal of discussion about the Hebrew word bara' (rendered created) in Genesis 1:1 has gone on throughout the centuries. It was being discussed for sure in the 2nd century BC. But the verb bara' is only used in the Old Testament with God as the subject creating something and it always refers to something new being created, with no mention of pre-existing matter from which it came. In the context of Genesis 1 where God is creating everything using the figure of speech known as a merism "heavens and earth", this verb bara' could only be understood as creation from nothing. (Sailhamer, 1996) The rest of the Old Testament clearly teaches that God Himself is eternal and has no beginning nor end, not to mention that the New Testament makes creation from nothing even more clear.

In all fairness, there are a good number of creation myths from Egypt and Mesopotamia, and much later in Greece, China, and North and South America, that describe men created from clay (unlike the dust in Genesis 2), but they all have to do with creating many humans and even humanity in general from the clay. As clay was used for pottery, to make and fashion figures and vessels, this would be the natural substance to describe gods "molding" humanity. We can read these creation myths ourselves and we will see that they are very fanciful accounts that I believe were not meant to be based in actual historical chronology. That is not to say that the ancients didn't believe them. But unlike some scholars, I like to give ancient people more credit for their reasoning abilities. They really were just like us mentally. I believe that they were sincere in their religious devotion and thought that angering the various gods led to disorder and suffering, while placating them made life run more orderly. I am just not so sure that when a normal Egyptian adult looked up into the sky, they thought that a woman deity was contorted and had stars in her body.

But this could just be my Christian and modernist assumptions showing through. Most Egyptian and Mesopotamian scholars do seem to think that the worldview of these two peoples was a view where deities made up reality and it was inseparable from material reasoning. Tom Holland has argued that a concept like religion would not have made sense to pre-Christian cultures, like Greece and early Rome, and that the distinction between the religious and secular is almost solely a product of Christianity and Christian theology. (Holland, 2019) And I think he is probably right. We can probably apply this same logic to Egypt and Mesopotamia and say that if we asked them what their religion was, they would have just looked at us funny. There was no distinction between life and religion. What these nations have done with their false worship is exactly what Paul said they did and do; they exchanged the truth for a lie and the glory of the one true God for creatures and created things. (Romans 1:18-23)

Commonalities and differences with biblical creation

Both the Mesopotamians and Egyptians had a pantheon of gods to worship. This defined their polytheistic religions from start to finish in the history of their empires, kingdoms, and city-states. I think there is stronger evidence coming from Mesopotamia compared to Egypt for the idea that they believed that there really were no false gods. If they encountered another culture with different gods, their impulse was to assimilate and adopt those gods into their pantheon or to see gods with different names as corresponding to the same gods. Sometimes Christians tend to think that Nebuchadnezzar II abandoned his polytheism to worship the God of the Israelites, but I think this is very unlikely. In the book of Daniel, we see him exalting the God of the Israelites and making decrees about worshipping Him. But it would be very unlikely that he would denounce the other gods that he and his people had spent generations worshipping. And we see no evidence of him outlawing Mesopotamian worship and religion. As is common to both Mesopotamia and other ancient near eastern polytheistic cultures, these people believed that maybe the calamities that had befallen them was a result of ignorance or displeasing a god they had neglected. But I think it is clear that the Mesopotamians did not view the gods of other cultures as unreal or false.

Shalmaneser, king of Assyria, let the people left in Israel and Samaria worship their God or gods and even sent priests back to the land after deporting many to Assyria. (2 Kings 17) Hezekiah was king in Jerusalem and Judah during this time and did right in the eyes of the Lord. Eight years after Shalmaneser had captured Samaria and deported the Israelites, the next Assyrian king Sennacherib came against the southern kingdom of Judah and seized many of its cities. The king sent generals and an army against Jerusalem which included a man named or titled Rabshakeh. He mocks Hezekiah and the people of Jerusalem, and showing his ignorance of God's covenant with Israel, asks why they have torn down all the high places and displaced the priests that ministered there. And then he asks why they would trust in the Lord to deliver them as he relates all the other gods of the lands and peoples that Assyria has already conquered. (2 Kings 18) The Assyrians, like other empires who temporarily possessed power and dominion, thought that their gods were superior. And the evidence was in the fact that they conquered peoples and lands with other gods. But to say that they believed that the other gods were false is I believe against the evidence from their own sources and history. As far as the Egyptians, Pharaoh wonders who the God of the Hebrews is and why he should obey Him. But Egypt had a vast pantheon of gods, so I would guess their mindset about other gods was the same.

In contrast, the Israelites believed that there was only one true living God and that all other so-called gods were false. Even though many secular scholars will try to argue that the Israelites initially believed that many gods existed but only one God, YHWH, was to be worshipped. And that this idea evolved and culminated in the strict monotheism during the time of a prophet like Isaiah. This idea of there being many gods but one greater god to be worshipped is generally called henotheism. But I think this can be shown to be incorrect by the earliest books of the Bible and God's condemnation and judgement of the Israelites' idolatry. Furthermore, as we will examine, the idea that there is only one true living God is right in the theology from the beginning in Genesis 1. Of course, this has little persuasive power over the critics who are convinced that Genesis 1 is a late writing composed by the priestly source, despite evidence to the contrary, including the fact that there are tons of Old Testament references to passages from Genesis 1.

Another striking difference between the Bible and the ancient near east is that the gods of the near east personified or embodied natural phenomena. The sky was a god, the earth was a god, the air and moisture were gods, storms were gods, stars and constellations were gods, the sun and moon were gods, etc. In stark contrast, the God of the Bible creates the material world and He Himself is separate from and not of the same substance as that creation.

A third difference that we touched on earlier is the anthropomorphic nature of the gods in Egypt and Mesopotamia. They were controlled by passions and desires like human beings. They may be more powerful than humans and live indefinitely (although all of them except the first water gods have a clear beginning through birth and sometimes they are portrayed as aging), but they exhibit the same psychology and behavior as mankind. They work, sleep, lust, have sex, produce children, get irritated, have meetings, socialize, plot, scheme, and get even. The God of the Bible is eternal, spirit, without passions that influence Him and cause Him to change, morally perfect, holy, separate, and exalted above His creation. We should expect a God who exists to be different than us, not gods created in our own image. But the Bible does say that man is made in the image of God, so we can see something of what God is like by looking at man. Theologians say that there are communicable attributes that are similar in God and man, and we share these because we are made in His image. But that God has those communicable attributes in divine perfection, while ours are but an imperfect shadow. But we will go into the image of God much more in depth in a later chapter.

The Mesopotamians, Egyptians, and Israelites all had temples, priests, sacrifices, and offerings in their religions. The Mesopotamians, Egyptians, and Israelites all had a priestly class. The Israelites also needed to make God propitious through repentance and atoning for their sins through sacrifice. But God communicated to His people by the mouth of His prophets and gave the people His law and commandments. The Israelites had a clear revelation and idea of what breaking the law meant and didn't have to guess which god was causing their calamity, wondering if the gods were irritated, at odds with one another, or capricious in dealing out suffering. The pagans did not have clear moral standards with which to expect blessings and curses, and would often use divination of some kind, like reading the organs and intestines of animals, to discern the will of the gods.

Egypt and Mesopotamia shared the idea of making idols to reside in the holiest place of their temples and they clothed, perfumed, applied makeup, and fed these idols. The gods needed the people. The gods that were incarnated in the idols were often thought to be rather arbitrary. In contrast, the God of the Bible does not need humanity. He does not need to be housed in a temple, for all creation is His and he made the cattle on a thousand hills. He is not passively led by emotions and exists eternally without a wife or consort. His presence is said to dwell in the mercy seat above the ark in the tabernacle and temple, and he required a mediator to bridge the gap between Himself and sinful people. But He does not need a house to dwell in, for heaven is His abode and the Earth His footstool (Acts 17:24-25). He is not a god who changes and is the same yesterday, today, and forever.

In their creation myths, the Mesopotamians and Egyptians had multiple gods producing one another and there was always preexisting matter, even if it was personified in water deities. The biblical God is presupposed from the beginning to be eternal and created all matter and energy Himself. They all have accounts of turning the dark, formless deep into an orderly creation, but the other two kingdoms deify nature and its phenomena, whereas God works with only matter and energy. The Bible demystifies and de-gods nature. The other two kingdoms used gods both as actors and as the raw "material" to separate and differentiate the creation, whereas God wills and speaks it. The others use parts of slain gods and bodily fluids to create man, mainly to relieve the work of the gods, produce food, and to serve in temples. The Bible elevates man to the pinnacle of creation by the will of God. We will expand these ideas when we look at Genesis 1. Let's now turn to the flood stories from Mesopotamia.

Mesopotamian flood stories

Epic of Gilgamesh (flood hero Utnapishtim)

Although the Mesopotamian Epic of Gilgamesh does not contain a cosmogony (an account of the origin of the universe), it is often employed as another comparison with Genesis. It will be the first story we compare, as many people claim it contains the storyline of the early chapters of Genesis, particularly the account of Noah's flood.

The Epic of Gilgamesh is dated to around 2100 BC for its literary beginnings. Various poems in Sumerian seem to make up the heart of the tale from either before or around the time of the Akkadian empire, which first began under Sargon the Great. From these an Akkadian version was constructed around the time of Hammurabi, the famous Amorite king who is known for his law code (even though the Mesopotamian king Ur-Namma had one that survived three centuries earlier). The composition of the epic from this time did not have the prologue and probably did not contain the flood story, although the tablets may just be missing. In the Old Babylonian version, Gilgamesh is still searching for Utnapishtim and eternal life after the death of his friend Enkidu, but many scholars think that the flood story in the standard version shows evidence of being a later insertion in the epic.

The standard version that included the prologue, epilogue, and flood story appears to have been written and edited by Sin-leqe-unnini, a poet scholar, around 1200 BC. It is feasible that he put the prologue, epilogue, and flood narrative in the story of Gilgamesh and gave it its final form. But the only attested tablets that we have that contain the flood story are from the library at Nineveh dated to the 7th century BC. The flood story itself existed independently and was much older. It is known from other myths like the Sumerian Eridu Genesis, and the Babylonian Atrahasis epic from around the 17th century BC. Both of which we will look at in this section. The standard version is transformed and added to substantially compared to the Old Babylonian version as far as we can tell, but again this may be due to the Old Babylonian version being incomplete. The version found on tablets from the library at Nineveh is now referred to as the standard version. It consists of the twelve-tablet epic. Dalley argues that while the eleven tablets probably go back to at least 1200 BC with the composition of Sin-leqe-unnini, the twelfth tablet seems to have been a much later addition. (Dalley, 2008)

The standard version of the Epic of Gilgamesh can be summarized as follows:

The epic of Gilgamesh starts with a prologue extolling Gilgamesh for his characteristics and summarizing his adventures. Gilgamesh is the king of Uruk. He is 2/3 god and 1/3 man, the son of Lugalbanda, a former king of Uruk, and Ninsun, a female goddess. Gilgamesh is a despotic, harsh king who is overbearing and harasses the men and young women by not leaving the young women to their husbands or mothers. The gods of heaven, including Anu, the sky god of Uruk, hears the pleas and complaints of the citizen of Uruk, and they call upon a mother goddess to create a man equal to Gilgamesh to rival him. The mother goddess takes a piece of clay and throws it into the countryside and the valiant warrior Enkidu is created.

Enkidu is uncivilized and lives among the animals in the wilderness. An animal trapper sees Enkidu living with the animals and observes him removing traps and other things to thwart the trapper from getting his animals. The trapper's father suggests telling Gilgamesh from Uruk, who will send a prostitute named Shamhat out to seduce Enkidu and alienate him from the animals. The harlot seduces Enkidu, and they have a passionate excursion for seven days, after which the animals flee from him, and he seems to lose the wild abilities to keep up and have camaraderie with them. Enkidu's mind seems to have changed by the experience with Shamhat and she proclaims that he has become like a god. She implores him to go into Uruk to live a civilized life and enjoy its fruits. Enkidu wants to challenge Gilgamesh for his arrogance over his people, but Shamhat tries to persuade him otherwise. The harlot then relates some dreams Gilgamesh had that were interpreted to mean that Gilgamesh would love Enkidu as a brother and friend.

Shamhat brings Enkidu to some shepherds, and they offer him food and beer. He eats and drinks, washes himself with water, and puts on clothing. He is now civilized, and he takes up weapons with which he fends off wolves and lions so the shepherds can relax. She brings Enkidu into the city of Uruk, and he challenges Gilgamesh to a match of physical prowess when he learns that Gilgamesh is going to forcibly sleep with a bride-to-be at a local wedding, as is his usual evil habit. Gilgamesh and Enkidu seem to be evenly matched in the battle, exchange pleasantries, and the two mutually respecting men become friends.

For some reason (the text is missing or fragmentary, maybe to make a name for himself, to get fame and renown, or to please the sun god Shamash?) Gilgamesh wants to go the cedar or pine forest to kill the monster Humbaba, who guards the forest. Enkidu is against it, but Gilgamesh has the craftsmen forge weapons for them both and then tells the men of Uruk of his plan. Enkidu asks the elders of the city to talk Gilgamesh out of it, but he seems unpersuaded by their initial counsel and then they seem to relent and grudgingly give their blessing if Enkidu leads the way. He then seeks out the advice of his mother, the goddess Ninsun, and she is grieved by the foolhardiness of her son. But she offers a sacrifice to the sun god Shamash, who seems to want the monster Humbaba killed, and concludes that they are to go to the forest and asks Enkidu to keep Gilgamesh safe.

The two of them travel to the forest of Humbaba and eventually kill the monster. The goddess Ishtar is impressed with Gilgamesh after he washes and clothes himself and she tries to seduce him, but he resists her advances. Gilgamesh insults her repeatedly and recalls how all of Ishtar's lovers ended up dead or in a pitiable state. She is enraged and asks her father Anu, god of the sky, to send the bull of heaven to punish Gilgamesh or she will ruin the underworld and cause the dead to come up and outnumber the living. Anu learns that Ishtar has filled the granaries of Uruk and made the pasturelands productive, thereby Uruk will avoid famine as a result of sending the bull of heaven. Anu therefore gives his daughter Ishtar the bull of heaven, which she sends down to Uruk.

Enkidu and Gilgamesh manage to kill the bull of heaven and then celebrate in the city of Uruk.

The gods confer and decide that one of the two must die for their actions and it is determined that Enkidu will be the one to suffer this fate. Enkidu curses the trapper that first saw him and the harlot Shamhat that civilized him, but then blesses them after hearing from the sun god Shamash. Enkidu has a dream where he will go to the dark underworld to be clothed with feathers, drink dirt, and eat clay (which seems to be the dreary conception of the underworld and life after death for the Mesopotamians). Gilgamesh eulogizes and mourns bitterly for his friend.

This event gets Gilgamesh thinking about his own mortality. He then sets off on a journey to find the secret to eternal life. After much trekking through perilous terrain, including mountain passes in darkness, Gilgamesh arrives in the sunlight and a brilliant environment by the seashore where plants produce gemstones. Along this seashore a bar maid and tavern keeper converses with him and learns that he seeks Utnapishtim (whose name means "he found life") because he fears death, but she tells Gilgamesh that although no mortal has made the sea crossing, there is a ferryman who can possibly carry him across.

Gilgamesh finds the boatman and somehow ruins the mechanism that is used to launch and navigate the boat. The boatman tells him to go into the forest and cut a bunch of timber and they will fashion something to make the crossing feasible. They pass the treacherous waters and reach the man named Utnapishtim. This man survived the great flood and was granted eternal life. Gilgamesh wants to know the story and if he too can become immortal. Utnapishtim chastises him for wearing himself out when the fate of man has already been fixed by the gods, no mortal can see death coming, and life is often vain despite man's efforts.

Gilgamesh observes that Utnapishtim is just like him and demands that he tell him the story of how he obtained eternal life from the gods. (Many scholars recognize the different literary structure and excursus of this portion of the epic as being out of place with the rest of the epic, lending further evidence that it was probably absent in the Old Babylonian epic of Gilgamesh and later inserted from flood stories common to Eridu Genesis and Atrahasis, but this is debated).

Utnapishtim, a man of the city of Shuruppak, and son of Ubara-Tutu (listed as the king of Shuruppak just before the flood swept over in some of the Sumerian king lists), was told by the god Ea (Sumerian Enki) (in a roundabout manner to preserve his oath to the gods) to dismantle his house, leave everything behind, to build a giant ship with equal length and width, and to board all living things upon the boat. He was to tell the inhabitants of the city of Shuruppak that Enlil was angry with him, so he had to leave Enlil's earth and live with Ea in the waters. Once Utnapishtim did this, Ea (or maybe Enlil?) would rain down grain, bread, meat, and wealth upon them. This building project was done with the help of many skilled and unskilled laborers from the city after Utnapishtim told them this lie so they would help and not panic. It was built either like a giant cube or a very large circular boat (Although the text to me seems to clearly describe a cube shaped vessel, see section below in Atrahasis for further clarification)

The boat had seven levels and nine compartments within it. It was then waterproofed with pitch and bitumen. Utnapishtim fed the laborers a giant feast, finished up the waterproofing, and stored oil away on the ship. (The timeframe of total construction seems to be within a week or so because the frame was laid out on day 5 and then it says it was completed in the evening a few verses later).

The boat was filled with gold, silver, animals, craftsmen, and his own family. Launching the ship was very difficult. The sun god had set a time for the flood to commence and when the time had come and the skies grew threatening, Utnapishtim sealed up the entry to the boat. The various gods thundered and unleashed the flood upon the earth. The gods themselves were frightened by the tumult and fled into the sky, where they cowered like dogs. Ishtar bemoans what they have done and the other gods weep with her and are parched with thirst.

After seven days of storms raging and flooding, the winds calm, the sea is still, and all the people have turned to clay. Utnapishtim makes an opening in the boat, the sun falls on his face, and he breaks down weeping and lamenting. He sees a bit of land emerging and the boat grounds on Mt. Nimush, thought to be a reference to Mt. Nisir (Pir Omar Gudrun), an 8500 ft. mountain in northeastern Iraq. Although some have suggested that the text can be translated differently, referring to a hill instead of a mountain. (Best, 1999) Stephanie Dalley notes that in the Sumerian version of the flood with Ziusudra, despite gaps in the text, it seems to present the boat as coming to rest on flat land, not on any mountain. And that the only attested texts in which the boat lands on Mt. Nimush are from the 7th century BC. But she concedes it could have been an earlier tradition. (Dalley, 2008) This discrepancy of where the boat comes to rest will be germane when we look at Noah's flood.

The boat stays fixed on this mountain or hill for seven days and then Utnapishtim sends out a dove, but it comes back to him. Then he sends out a swallow and it returns. He lastly sends out a raven, but it does not return to him. Utnapishtim offers a sacrifice and then an incense offering either at a "mountain" ziggurat or at the "mountain" peak. The gods smell the incense offering and gather over it like flies. The great mother goddess appears and says that the gods should enjoy the sacrifice, but Enlil should not partake of it because he caused the flood.

Enlil arrives and is angry to find the boat and people who have survived. The gods implicate Ea in helping people escape and he responds by saying Enlil should not punish everyone indiscriminately and that he should have sent wild beasts, famine, and disease to control the human population instead of the flood. Ea further defends himself by saying that he didn't tell the secret of the coming flood but gave Atrahasis (used here instead of Utnapishtim) a dream and he figured it out. Enlil goes into the boat and takes Utnapishtim by the hand and blesses him and his wife and says they should be like the gods, meaning that are made immortal and have everlasting life. The gods take them away to settle at the mouth of the rivers.

The excursus of the story of the flood then comes back into dialogue with Gilgamesh. Utnapishtim asks Gilgamesh a rhetorical question: How can they convene the gods to give him a similar fate? He tells Gilgamesh that first he must try to stay awake for seven days, upon which he fails almost instantly and falls asleep. The moral seems to be if this man cannot even stay awake, how can he have immortality? Utnapishtim and his wife tell Gilgamesh to go, wash, and put on new clothes and anoint himself with oil. The ferryman assists Gilgamesh to clean, anoint, and clothe himself and they get ready to disembark. Utnapishtim's wife implores him to give Gilgamesh something for his troubles and long journey, so Utnapishtim tells Gilgamesh about a plant that will make him young again.

Gilgamesh puts heavy stones on his feet and sinks to the bottom of the fresh water and retrieves the plant. He plans on taking it back to Uruk for an old man to try and then he himself will eat it and be restored to his youth. Upon their travels back to Uruk, Gilgamesh stops to bathe, and a snake senses the plant of youth. The snake snatches it, carries it away, and while doing so sheds its skin (signifying its youthful regeneration). Gilgamesh is distraught and breaks down weeping, seeming to lament about the vanity of his quest and his own inevitable mortality. Gilgamesh and the ferryman continue their long voyage back to Uruk and when they finally arrive Gilgamesh tells the ferryman to look and marvel at the majesty of the city of Uruk, (in a condensed version of the prologue, possibly coming to terms with life and what he has accomplished and is grateful for).

The 12th tablet of the epic is not included in many translations, as it is a late addition to the epic, and is rather discordant with the storyline. Like the epic of Gilgamesh itself, the 12th tablet incorporates material from various older Sumerian tales. Basically, it picks up with Enkidu (who we know died earlier in the epic) volunteering to retrieve two objects that fell into the earth for Gilgamesh (these two objects are from a separate myth usually called Gilgamesh, Enkidu, and the Netherworld). Enkidu goes down into the netherworld but is not allowed to return to the surface. Gilgamesh digs a hole and releases Enkidu's spirit and then Enkidu relates the conditions of the netherworld to Gilgamesh. (Thompson, 1928) (Tigay, 1982) (Kovacs, 1989) (Dalley, 2008)

I think we can see that the epic of Gilgamesh is a far cry from Genesis. It doesn't even deal with creation, the origin of mankind, or a fall from grace. But some people still want to try to make comparisons. For the sake of argument, let's first talk about the Garden of Eden. They claim that Enkidu is Adam because they were both created from the earth by the gods and that the prostitute Shamhat has evolved into Eve, because Enkidu's "eyes were opened" after his episode with Shamhat, Enkidu became wise and thereby "could discern good and evil", and Shamhat proclaims that Enkidu has become like the gods, making comparisons with Adam and Eve after they ate the fruit.

They think that Adam living in the garden was either likened to Enkidu in the woods or more likely to Gilgamesh in the land of the mouth of the rivers when searching for the youth-giving plant. They say Eve is like Shamhat because she tempted the man, he covered his nakedness, and then returned to civilized life. The serpent stealing the plant from Gilgamesh represents the serpent in the Garden of Eden.

Let's take these claims in order.

First, Enkidu is created from a piece of clay (like the rest of mankind in Mesopotamian creation myths) thrown into the wilderness. Adam is created directly by God from the dust. Enkidu is created as a savage who frolics with the beasts. Adam is created upright and righteous. Enkidu is never presented as righteous and sinless, and his subsequent civilizing is the opposite of Adam's trajectory after the Fall. Enkidu's purpose for being created in the myth is to rival Gilgamesh and provide relief for the citizens of Uruk, not to be made in the image of God and a covenant head of humanity with repercussions for his sin and the rest of humankind, like Adam.

Second, Shamhat is a harlot and Eve is holy and blameless. Adam falls from grace because he sits idly by and partakes in the fruit as Eve is tempted by the serpent. And sex between the pair is only presented as good and right in the context. Shamhat makes Enkidu "wise and like the gods" through a seven-day sexual encounter and this plan seemed to be the intention of the gods for Enkidu from the start. Adam is not tempted by Eve; both are tempted by the serpent. And God clothes both Adam and Eve in grace because they cower in guilt and shame.

Third, the myth does not explain why or how Enkidu has become wise or like the gods. It seems that the explanation is in how Enkidu goes from beast to civilized. This is in far contrast to Adam and Eve, who do not become civilized but lose an aspect of their true humanity made in the image of God when they disobey. Nothing about the context of Gilgamesh gives us any indication that what Enkidu was doing is negative or frowned upon by the gods.

Fourth, trying to make comparisons about the garden of Eden from the myth is reaching at best. The forest of Humbaba is the realm of many gods and involves the two demi-gods, Gilgamesh and Enkidu, killing the monster that guards it. The mouth of the rivers where Utnapishtim lives is the place the gods placed Utnapishtim and his wife (not sure what happened to their family and the craftsmen?) and has no bearing or relation to the garden from Genesis.

Lastly, the youth-giving plant that is stolen by the serpent is more interesting and intriguing (I will touch on it again later in the book), although it bears little resemblance to Genesis. Serpents generally play a mischievous role in ancient myth. And little wonder because people are naturally scared of them. The serpent in the garden is the mouthpiece of Satan and his temptation brings misery and consequences for Adam, Eve, himself, and mankind. The story in the Gilgamesh myth seems to be an explanation for why snakes shed their skin and it adds to the drama of the epic, finally pushing Gilgamesh to accept his mortality and be grateful for his accomplishments as the king of Uruk.

As far as the garden, I believe the evidence is very strong for the Garden of Eden now being under the waters of the Persian Gulf, which we will expound later. The geography explained in Genesis 2 for the Garden of Eden explains a landscape that would have dated to around 8000 BC or before. And by the time of the Epic of Gilgamesh, it would have been under the waters of the Persian Gulf for many thousands of years. With all those things said, and simply for the sake of argument, the storylines are completely different. The similarities of Adam and Enkidu being created from the earth are not even accurate. Adam is created from the dust of the ground and Enkidu is created from a piece of clay thrown into the wilderness. Many ancient myths of creation use clay (think about the ancients making pottery and figurines), and the theological point of Genesis is Adam is earthy and to the earth he shall return.

People are free to take some time to read good translations of the Epic of Gilgamesh for themselves and see how fanciful the story is. We can see that it is not a story of origins. But we cannot simply dismiss the flood portion in the standard version. Let's look at a couple other flood myths from Mesopotamia that most scholars agree are the foundational flood stories inserted into the standard version of the epic of Gilgamesh.

"Eridu Genesis" (flood hero Ziusudra)

The story of Ziusudra (whose name means "life of long days"), which some scholars used to call "Eridu Genesis" but is now sometimes just called the Sumerian flood story, is from a partial Sumerian tablet dated to the 17th century BC (a little later than Atrahasis). But some scholars seem to think it is probably much older and is therefore usually believed to be the oldest story of the flood from Mesopotamia. Although I haven't found much evidence to validate that claim, I could just be ignorant of the arguments. The only circumstantial evidence I could find was from a Sumerian king list dated to around 1750 BC that matches the first five cities from this myth before the flood swept over. Whatever the truth about the oldest tale, the hero Ziusudra is listed as the king of Shuruppak (the same city as Utnapishtim) in one of the Sumerian king lists prior to the flood.

The summary of Eridu Genesis is as follows:

This highly fragmentary myth begins after a missing section with a mother goddess (?) wanting to gather up people again so they can build cities and temples and institute religious rituals to appease the gods and bring peace to the land. Four gods, An, Enlil, Enki, and Ninhursag, are said to have created the black-headed people (the Sumerians), animals in abundance, and four-footed animals to roam the plains. After the cities and temples have been completed, kingship descends from heaven and apparently (hard to discern from the texts) the mother goddess (?) appoints a god for each of five cities. The first Eridu, then Bad-Tibira, Larak, Sippar, and finally Suruppak. After a missing section, the mother goddess and Inanna are lamenting because the gods have decided to send a flood on mankind. The same four gods who created the Sumerians make the other gods swear an oath, but Enki deliberates within himself. Enki again appears as the compassionate god, and he warns the king and priest Ziusudra of the impending flood in a roundabout way by giving the revelation of the flood to a wall. Several lines of text are missing, but when it resumes the flood and storms are raging and continue for seven days and nights, accosting the large boat. The storm breaks and the sun god shines daylight upon the heavens and earth. Ziusudra makes an opening in the boat and the rays of the sun god stream into the boat. He then sacrifices sheep and oxen, and animals disembark from the boat. An and Enlil reward Ziusudra with eternal life like the gods have. They then send Ziusudra to the land or "mountains" of Dilmun and call him the preserver of the seed of mankind and animals. (Jacobsen, 1987) (Black, Cunningham, Robson, & Zolyomi, 2006)

Babylonian version (flood hero Atrahasis)

We have already looked at the "creation" portion of this tale. Next, we will examine the flood portion. The most complete version of this Babylonian tale is written in the Akkadian language and is dated to around 1700 BC. Later versions have been found at the Library of Ashurbanipal in Nineveh.

The summary of Atrahasis:

After the humans are created to lessen the hard labor of the lower gods, they prove to be an annoyance with their excessive noise, and the supreme god Enlil decides to destroy them by bringing a flood upon the earth after many other attempts to limit human population. He wants the god Enki to carry it out, as he blames him for creating mankind, but Enki steps in and sends the flood hero Atrahasis a dream and interprets the dream for Atrahasis by speaking to the wall and house. He tells Atrahasis (his name means "exceedingly wise") to tear down his house, forsake possessions, save living beings, and build a boat to escape the flood. The section on how to build the boat is missing, but British Museum scholar Irving Finkel translated an Old Babylonian fragment (dated to 1900-1700 BC) in recent years in which Enki tells Atrahasis how to construct the ark. And it turns out to be a very large round boat. The boat was around an acre in area and was a type of boat called a coracle, a vessel used in ancient Mesopotamia right up until the recent past. The boat was made of plant fiber rope, wooden ribs for structure, and waterproofed with bitumen. (Finkel, 2014)

Our text resumes with Enki telling Atrahasis how to roof the boat, to make it with two decks, and to cover it with pitch (bitumen) to make it strong. He then tells Atrahasis that a seven-day flood is coming. Atrahasis assembles the elders of the city and tells them that Enlil is angry with Enki and that because he is loyal to Enki, Atrahasis cannot no longer live on dry land. Atrahasis has certain skilled and unskilled labor help him collect materials and construct the boat. He brings on board cattle, birds, and various creatures. On the same Old Babylonian tablet fragment describing the construction of the round boat, Irving Finkel translated that the animals were to be loaded two by two. (Finkel, 2014) Back to the text.

Atrahasis invites the people to a feast where they are eating and drinking, and he boards his family on the boat. But he is too distressed and sick from the coming calamity. The storm gods begin to roar in the clouds and Atrahasis seals the door with pitch. The flood commences and Atrahasis releases the moorings and launches the boat. The storm gods rage and no one can even see each other through the calamity. The gods are distressed, famished, and parched. The womb and midwife goddesses weep and lament the destruction of the people they created, for allowing Enlil to carry out such a command, and because Anu approved it. The gods are inconsolable with grief and are hungry and dehydrated as the storm and flood rages on.

There is a large gap in the text, and it does not describe the end of the flood storm, where the boat grounds, if birds are sent out, and what disembarking the boat looked like. When the text resumes, Atrahasis has made a sacrifice and offering, and the gods swarm to it like flies. All the gods blame and chastise Anu and Enlil for their decisions. Enlil comes and is angry to find that humans have survived the flood. Anu and Enlil blame Enki and he defiantly takes responsibility. Atrahasis is neither rewarded, punished, nor sent away to a faraway land in the extant text of the myth. If fact, he is not referenced again after his offering to the gods.

Then the womb goddess and Enki do something to a third of mankind (missing), do something else to another third (missing), and for the final third of mankind they make women lose their babies at childbirth and create a class of women who do not become mothers. (Clay, 1922) (Lambert & Millard, 1969) (Moran, 1987) (Dalley, 2008)

Commonalities and differences with Noah's flood

I think we can see by reading these ancient accounts of the flood that many similarities in the stories emerge. We can see that they all have a person chosen prior to the flood and they are all instructed to build a boat. The setting for all the floods is within ancient Mesopotamia, and the flood is sanctioned by the principal god(s) or the true God. The main actor saves himself, his family, and some animals, while the rest of humanity dies (or at least this is implied). The boats were constructed of either wood or plant fiber rope and wood (Finkel, 2014) and sealed and waterproofed with pitch (bitumen) and sometimes oil. The floods appear to cover the whole land and wipe out all life. In the standard version of the epic of Gilgamesh, the similarities between it and Genesis are the most pronounced, including doves and ravens that are sent out while the flood is diminishing to find dry ground, and the boats come to rest on "mountains". After departing their boats, the survivors sacrificed offerings to their gods or the true God.

Some of the differences are the reasons given for the flood. In Atrahasis, it is clear that humankind was too noisy so the gods were annoyed, they could not rest, and the previous attempts to limit the population proved unsuccessful and did not last. In Gilgamesh, no reason is given prior to the flood except that the great gods decided to do it. But after the flood, the other gods are upset and question why Enlil did not use other means to reduce the human population, so I think it is inferred that the audience understands the background story of trying to limit the human population through other means. In Eridu Genesis, the text is missing that may possibly explain the reason for the flood, but many scholars think it was for the same reason as Atrahasis. That human noise was annoying the gods, especially Enlil and An, the chief of the gods. So, I think it is safe to say that the reason that the gods send the flood in the Mesopotamian myths was the noise of mankind, whereas in the Bible, the cause of Noah's flood was the moral depravity of man.

Noah was chosen because he was a righteous man who found favor in God's eyes, although we could possibly say that the other flood heroes were presented as wise, with no reference to their moral character, apart from the inference of a priest like Ziusudra. Atrahasis's name means extra-wise or exceedingly wise. Although, I think this is stretching the similarities. Ziusudra is not defined in his myth by his character or office, but one of the Sumerian king lists names him as king of Shuruppak, the last king before the flood. Utnapishtim basically lies to the people of his city so they help him construct the boat. Atrahasis tells a similar story to the citizens of Shuruppak, and they also help him build the boat. So, these are not moral beacons.

But they pale in comparison with the capricious behavior of the gods. For the gods create mankind to do the work and meet their needs. And then try over and over again to destroy them and cause them suffering. Although there are clear character distinctions among the gods in the myths, ranging from the hard-hearted king of the gods Enlil (Ellil), and sometimes An (Anu), to the much more compassionate god Enki (Ea). Although he too suggests killing people off with other means than the flood (maybe to appease Enlil and save some people though?). And the female goddesses in Mesopotamia are much more caring than the male gods on average. Many are associated with wombs, birth, and midwife duties and therefore have motherly instincts. The gods truly were made in the image of man.

Some other differences?

It took 120 years for Noah to build the ark while God was giving mankind time to repent, while the others seem to be constructed in about a week. The gods were fearful themselves in the pagan accounts, and were dying of hunger and thirst, while God is always portrayed as in control of nature. Noah's flood lasted over a year from start to finish and the pagan accounts from 1 to 2 weeks.

In Eridu Genesis, Ziusudra is given eternal life by An and Enlil for making it through the flood and sent to dwell in the land or "mountains" of Dilmun (probably the island of Bahrain), although we don't know why. In Atrahasis, Enlil is furious that someone survived the flood, but Atrahasis is not given eternal life nor sent to live anywhere. In Gilgamesh, Enlil is again furious that someone has survived the flood and then for some reason grants Utnapishtim and his wife eternal life and settles them at the mouth of the rivers. In stark contrast, Noah is still seen as mortal and beset with original sin after the flood. God makes a covenant with Noah, his family, his descendants, all the animals with him, and the earth itself not to destroy them again with a flood. The other stories give accounts of how the gods thought of other mechanisms by which to curtail population growth, while God commands them to multiply and fill the earth in a reiteration of the divine blessing and mandate originally given to Adam and Eve.

What are these pagan flood stories about and what did the people of Mesopotamia make of them? Stephanie Dalley helpfully points out that the story of the flood in Gilgamesh reinforces that people now live in a time in which mortals can no longer be given eternal life by the gods. While the flood stories in general teach the lesson that this golden age of sages who brought civilization and its fruits from the gods to the people before the flood is over. (Dalley, 2008)

So, what are we to make of these flood stories and their obvious similarities (albeit far different theology) to the flood account in the Bible? Most would claim that Noah's flood is just an adaptation of the ancient near eastern myths. I think we can see that the similarities between the flood stories do seem too close not to possess a common foundation. But this is not the same as a claim of literary borrowing. While the consensus view among secular scholars seems to be that

Sumerian accounts of the flood existed first, were then incorporated into the Akkadian Atrahasis, which was then adapted for the epic of Gilgamesh, many scholars and scientists still believe that these events do have a grain of truth and correspond to actual catastrophic river flooding in the Mesopotamian river basin.

They point to the Sumerian King Lists and how the names on some of the lists seem to correlate to actual kings of Mesopotamian city-states and how the centers of influence and power change after the flood. There is also documented evidence from geology and archaeology for massive, but localized, flooding at Ur (c. 3500 BC), two at Kish (c.3000 and 2600 BC), and one at Shuruppak (c. 2900 BC). All in the timeframe of the early Dynastic period of Mesopotamia, which would probably correspond to the age of the prediluvian sages in the Mesopotamian mindset. The flood at Shuruppak in 2900 BC may have given rise to the story of Ziusudra and Utnapishtim. It is likely that Ziusudra (Akkadian Utnapishtim) was king of the Mesopotamian city-state of Shuruppak before a great flood on the Euphrates and/or Tigris. The Sumerian king lists are interesting, and I think shed additional light on these flood stories. There are various versions (from many fragments) of the king lists that have been found, which differ in the details recorded. The earliest of these dates from the Ur III period around 2100-2000 BC. The interesting thing about these early versions is they lack any mention of a flood to separate kingship before and after a flood. King lists from the Old Babylonian period, including the best preserved one from about 1750 BC, match the first cities given in Eridu Genesis, the last city exercising kingship before the flood is Shuruppak, and then kingship descends on Kish after the flood. So, we have a possible inspiration stemming from the floods of Kish and Shuruppak around 3000 BC for the Sumerian flood tales which influenced the king lists, Atrahasis, and Gilgamesh. But I think the story is more complicated.

I believe that these Mesopotamians flood myths are better explained by a collective memory and the traditions passed down through the ages of the actual event of Noah's flood. As we mentioned previously, the Mesopotamians did not generally historicize mythology, but instead mythologized actual history. (Kitchen, 2003) The story of Noah's flood assumes a much earlier date than the Mesopotamian stories would convey, and therefore makes it more reasonable and plausible that the Mesopotamians would adapt the knowledge of this event to convey meaning about kingship after their local flooding events.

We have no evidence of direct borrowing between Genesis and the Mesopotamian stories. And the Genesis account is quite unlike the others. The story in Genesis is much more realistic to start with. It involves the construction of a ship that would have taken a long, long time to build, the dimensions of the ship are actually seaworthy and fit for housing animals, the number of days of rain required are more realistic regarding a flood that appeared to cover the whole earth, and the amount of time to drain away all the water is far more realistic in the Bible's account.

Curiously, Josephus in his *Antiquities of the Jews* mentions more than 4 previous historians who had documented where the ark had landed and he himself said that it was still being shown in that region during his day. (Josephus, 1999) I think this rather curious claim should be taken with a grain of salt. Many legends and traditions surrounded people using the wood or bitumen from the ark for various purposes. And one of Josephus's sources, the Greek historian Berossus, actual wrote about the flood hero Xisurthros (Ziusudra), not Noah, and hundreds of thousands of years of genealogies prior to the flood, just like the Sumerian king lists do. (Dalley, 2008) If Noah's flood happened around 6500 BC, which I believe is the approximate date, it would be hard to believe that everyone knew where it was, and that people were still using pieces of it 6500 years later. And then for another 700 years or more according to some traditions. All four ancient historians referenced by Josephus, and he himself, identified modern eastern Turkey or Armenia as the place where the ark came to rest.

But let me be clear here. I don't believe that the Bible demands that the flood covered the entire planet. In fact, I think it is best interpreted as a regional flood in the flood plain of the Euphrates and Tigris rivers. And I don't think this does disservice to the text of scripture when we look at it closely, which we will later. Given the topography of the area, it is quite reasonable that a flood would have *appeared* to cover the entire earth or land. Others like Lorence G. Collins have done the work of showing how this is not just possible but plausible with rain in the mountains of Syria and Turkey as well as in Iraq and Saudia Arabia. (Collins L. G., 2009) And we must keep in mind that the climate and weather of this part of the world was much wetter than it is today. Even what is now the Sahara Desert was a grassland with numerous lakes prior to around 5000 BC. When we add to this analysis that the dating of the event of Noah's flood was much earlier than the Mesopotamian stories would correspond to, then we can also include the fact that the Persian Gulf was essentially dry until about 12,000 BC and began to fill up as sea levels rose. It then continued to fill until the shoreline was about 150 miles further inland compared to what we see today and reached this sea level as of around 5000-4000 BC. The shoreline has since slowly receded south mainly due to silting up from the rivers.

So, river basins with very shallow gradients, possibly even more shallow with the ancient Persian Gulf shoreline, long-lived heavy rains, subterranean waters, and snowmelt from the headwaters of the Euphrates and Tigris and the Zagros mountains to the east could easily have produced what appeared to be a flood that covered the entire land. Not to mention that draining such lakes and massive amounts of water on a slight gradient would indeed have taken many months, as opposed to days in the Mesopotamian accounts. Add to all of this that the Mesopotamian myths talk about kingship descending to a new city-state after the flood without any intervention of the gods to create new humans to populate the lands (in fact, in the case of Atrahasis, new mechanisms to prevent population growth), I believe that we can infer from the context of Noah's flood and from these Mesopotamian myths a description of a local, not global, event.

It seems to me to be an unlikely, but not impossible, position that some parts of the Mesopotamian stories could have borrowed from Genesis. Like possibly some elements in the standard version of Gilgamesh that correspond more with Genesis, like the birds sent out to find dry land and the boat coming to rest on "mountains", as these details are only attested from tablets coming from the 7th century BC, hundreds of years after we assert that Moses wrote Genesis. As conservative Christians we believe that Moses wrote the first five books of the Bible. But even with an early chronology for the Exodus from Egypt, this would put Moses writing the flood account in the 15th century BC. Although I think a later date of 1250 BC for the Exodus is much more supported by the evidence, including that the Hyksos appeared to introduce the horse and chariot into Egypt, that Egypt controlled the Canaanite region up until around 1200 BC, that slavery became more widespread after the expulsion of the Hyksos around 1550 BC, that the hatred and animosity towards Semitic people would have been high after expelling the Hyksos, that the first documented mention of the Israelites as a people is widely accepted to be from around 1210 BC on the Merneptah Stele (although within recent years the discovery and decipherment of hieroglyphic name-rings on a statue pedestal from Egypt dated to around 1400 BC may mention Israel), the names and locations of the storage cities mentioned to be built by the Hebrew slaves, and that Ramesses II appeared to change his behavior and was a different Pharaoh after the death of his firstborn son.

If the later date of c. 1250 BC is the correct one for the Exodus, we would therefore have Moses writing somewhere between 1250-1200 BC. Many biblical scholars (and I agree) are fine with the idea of Moses composing most of the Pentateuch and later editor(s) updating language and place names and putting the text into its final form under the guidance of the Holy Spirit. But Eridu Genesis and Atrahasis predate even an early date for the Exodus, and therefore the composition of the Pentateuch.

But let's think about this further. I think most theologians and biblical scholars are okay with the idea that the doctrine of inspiration and inerrancy do not preclude using source material. Then the idea of Moses having access to source material, whether written or orally past down, does not minimize inspiration or as God being the Author of scripture. Many Bible commentators recognize that Genesis is broken up into ten sections, with the Hebrew word toledot delineating the different sections, possibly referring to source material being used.

We need a little aside here to address the skepticism of scholars who find it doubtful that Moses wrote the first five books of the Bible. And we need this background information and argument because many secular scholars of religion and history think that the early chapters in Genesis took their inspiration and form when the Jews were exiled in Babylonia in the 6th century BC or later. There are many good reasons to maintain early traditional dates for the first five books of the Bible, including the testimony of Jesus, scripture itself, and persuasive ones made by K.A. Kitchen in his book *On the Reliability of the Old Testament*. But we need to demonstrate the reasonableness of Moses writing this material and then tie that into Noah's flood and how it may or may not relate to the Mesopotamian myths.

We must keep in mind that Moses was educated in Pharaoh's court and would have been among the top educated individuals in his culture. Some secular scholars assert that Moses could not have written the Pentateuch because most common people were illiterate and using hieroglyphs or hieratic script to write literature was not very practical, especially for a long fluent document like the first five books of the Bible. But is this so?

Hammurabi, king of Babylonia (1792-1750), was an Amorite king and descended from Semitic people of western Syria. He would have spoken a Northwestern Semitic language in the same language family as Hebrew. Even during his reign, he could have scribes produce his law code of 282 laws using the cuneiform script and the Akkadian language (an eastern Semitic language). And a large corpus of other inscriptions, letters, and literature from Egypt and Mesopotamia predate this law code. So, between the inscriptions and literature of Egypt and Mesopotamia, people had already been writing creative and fluent literary texts. We therefore can see that there was no barrier to writing complex literature by the late 3rd millennium BC or earlier.

The Egyptians used writing almost as early as the Mesopotamians, around 3100 BC. Cuneiform was the first documented script developed for writing in Mesopotamia. But when we find our first documented evidence for hieroglyphic writing in Egypt, it is already based upon a phonetic idea of mapping sounds in language to symbols, although it was still very complex, was far different from an abjad or alphabet, and was mainly reserved for the highly educated scribes. And the hieratic script was developed not long after the more pictorial hieroglyphs. But we already asserted that Moses was highly educated, so that alone is no barrier to him writing. Furthermore, the Proto-Sinaitic script is attested in Egypt as early as the 19th century BC, developed by a person or people who spoke a Northwestern Semitic language. This script, sometimes also called Proto-Canaanite (in case it was actually invented there), was the common ancestor to the Phoenician alphabet and is attested to on an ivory comb from Lachish dated to around 1700 BC, in what would later be the kingdom of southern Israel and then Judah. There are many artifacts from Canaan that confirm this script was used hundreds of years prior to the Phoenician alphabet. Whatever the truth for this invention of the alphabet in Sinai or Canaan, it was being used when Abraham and his descendants would have been in Canaan.

I think that Abraham was born around 1750 BC (although I am not dogmatic about this date), and that Joseph was probably not in Egypt prior to 1650 BC. I think Jacob's family probably arrived in Egypt closer to 1600-1550 BC, and that the 400 years that God said Abraham's descendants would be enslaved and oppressed began with these revelations made to Abraham by God (Genesis 15:12-16). The point for the current discussion is that Abraham, Isaac, and Jacob would have been living in Canaan while the Proto-Sinaitic (or Proto-Canaanite) script was being used and taught. It is much easier to teach children and adults an abjad or alphabet than hieroglyphs or cuneiform. In fact, after the script evolved into what the Phoenicians used and disseminated through the Mediterranean, the Greeks added vowels, and this is pretty much what many of us use today. But the ancient Hebrew script used in the Bible is thought by most scholars to have begun no earlier than the 10th century BC. But it is a direct descendant of the Proto-Canaanite script, and the Hebrews were in Canaan while it was being used. This alphabet could be used by any language but appears to be initially used by people speaking Northwestern Semitic languages, like Canaanite dialects and Hebrew. Therefore, I propose that Jacob and his family spoke ancient Hebrew (or something very close), may have learned the Proto-Canaanite script, and might have continued to pass this knowledge on to their children born in Egypt.

Moses would most likely have spoken both Egyptian and a Northwestern Semitic language similar to Hebrew. And he would probably have been able to write using the Egyptian, Cuneiform, and Proto-Canaanite scripts. My proposal then is that Moses, a highly educated man and called by God, most likely used the Proto-Canaanite script (which was not too far from Paleo-Hebrew script in 1200 BC) to write the Pentateuch and some Israelites could read it and copy it. This Proto-Canaanite script directly evolved into the Paleo-Hebrew script. That is not to say that most of the Israelites were literate. It is quite probable that most of them were illiterate, like the rest of the people in the surrounding cultures. But the law given to Moses was read to the people; they did not read it themselves. And surely there were a number of educated people and scribes among the Israelites who left Egypt. People who could copy the books of Moses. Not every Hebrew in Egypt was a brick making slave from childhood to adulthood. Slaves often did more intellectual work.

Therefore, between the high education of Moses, the Proto-Canaanite script used in Canaan and by others speaking northwestern Semitic languages long before Moses, and the direct lineage of the Proto-Canaanite alphabet into the Paleo-Hebrew one, there is no reason why Moses cannot be the human author of the first five books of the Bible in a Paleo-Canaanite or Paleo-Hebrew script and copied by others. And it is probable that he had access to other source materials, either from the Egyptian libraries, oral transmission, written documents from Abraham, Isaac, and Jacob, or a combination. Even Luke did his research before writing his gospel, yet it is still inspired by the Holy Spirit and are the very words of God. A proper understanding of the doctrine of inspiration does not preclude human authors using their vocabulary, personalities, writing styles, and sources. And I think you will find most conservative biblical scholars and theologians hold such a view.

Furthermore, and here is the important point relating back to the flood myths, the story of the flood would most certainly have been passed down through the lines of Shem, Ham, and Japheth to their descendants. The descendants of Shem include both the Israelites and the Mesopotamians, while the descendants of Ham include the Egyptians and Mesopotamians. Taking seriously the table of nations in Genesis 10, which I believe K.A. Kitchen (Kitchen, 2003) and Bruce Waltke (Waltke, 2001) have shown to be historical, it would not be unreasonable to conclude that the descendants of Shem and Ham who became the Mesopotamians used their oral traditions of Noah's flood to construct their own flood accounts, flexibly adapting them to their own theology and political ideas of kingship in their early city-states, and then applying them to actual local flooding events.

In other words, what am I proposing is that Noah's flood happened just like God revealed, the descendants of Shem and Ham kept the story alive through oral retelling, these descendants eventually became part of the Sumerians, local catastrophic floods happened at Kish and Shuruppak around 3000 BC, and these people, having a knowledge of the flood passed down for millennia, adapted the events of Noah's flood to local flooding events for their own religious and political purposes. Then later after the exodus from Egypt, Moses wrote down the real events of Noah's flood by revelation, oral tradition, written records, or all three to compose Genesis chapters 6-9.

The common argument leveled by most scholars is that Noah's flood is very similar to the Mesopotamian flood myths, especially with Utnapishtim in Gilgamesh, therefore later biblical writers adapted these myths and changed the theology for the purposes of their religion. Most people know that oral transmission is not the best way to maintain accuracy in reporting and relaying information. But it can be very accurate for non-literate cultures using various memory devices, like rhyming, poetry, and repeated refrains. But the easiest things for people to remember are the general theme and plot of a story, even if they forget the details. And the earliest attested flood myths from Mesopotamia demonstrate this general knowledge, in which broad themes are easiest to remember and pass on. For instance, all three flood myths from Mesopotamia teach that the flood happened because the gods were angry and decreed it, that a god informed a single man that the flood was coming, that a boat was built to save a man, his family, and the animals, that the flood appeared to cover all the land, and that a sacrifice and offering occurred when the man disembarked after the flood. While other details are much harder to keep straight by oral transmission, like why the gods were angry, how many decks and compartments the ships had, the shape of the boat, the materials used, whether precious metals and other people were loaded onboard, where the boat came to rest, and what happened to the people who survived. Yet other details are common sense and must be incorporated in order to make the story believable, like waterproofing an enormous vessel so it doesn't sink.

And this is the pattern we see in other Mesopotamian literature. Archaeologists find fragments of tablets from various times and places and sometimes they match the stories that are known and well attested, while other times they are quite distinct, yet clearly come from the same story. And then scholars often labor in vain trying to come up with a master foundational document of what the story "really" was.

So, what I am proposing is that the descendants of Shem and Ham who remained in Mesopotamia continued the oral tradition of Noah's flood and we see the general themes of that actual event recollected in the myths, while the details are foggier and more fluid, the theology is adapted to their polytheism, and the context is adapted to fit their political ideas. Moses then writes down the correct story and places it into its correct theology.

The same arguments could be made about the Sumerian king lists and the fantastically long reigns of the antediluvian kings prior to the flood and then the gradual (albeit still fantastic) reductions in the lengths of reigns after the flood. I argue that essentially what we have are the recollections and traditions of actual events that have been made more unrealistic by the Mesopotamians for their own uses. Kings that reign tens of thousands of years, giant boats that can't stably float constructed within a week, and all the water receding in an implausibly short amount of time. All corrupted stories with their details dimly remembered.

But isn't this a just-so story? It may sound good to conservative Christians who take seriously the word of God, but why believe Noah was first? Apart from the solid reason of accepting what God has revealed, in later chapters we will make a strong case for dating the garden of Eden and placing it squarely in history. And by extension of that argument, we will see that Noah and the flood predate Sumerian civilization by at least a couple millennia. But for now, we turn to the biblical text and the first chapters of Genesis.

Part 2

The Bible, science, and the claims of both

Chapter 5

Genesis 1:1-31 and Genesis 2:1-3

The first chapter of Genesis has been the most debated and controversial book in the whole Bible. Although it might be hyperbole, there seems to be as many interpretations of the text as there are dogmatic positions about rejecting anything not conforming to that particular reading. Many like me who were once young Earth creationists would have insisted that Genesis 1 consists of historical facts about how and when God created the universe using scientific language that transcended the local culture and time. I believed that it described the step-by-step divine fiat of each 24-hour day with the culmination of a universe and world that was created in a span of six days filled with all the flora and fauna on the planet prior to the flood. And I thought that a literal reading and interpretation was the most honest way to describe it.

Even when sound principles of interpretation are used, different readings on Genesis 1 are held by numerous scholars. One of the main problems I saw with some popular evangelical interpretations of Genesis 1 is that I did not think they were using the principle of interpreting Scripture in light of the type of literature that it is and not getting any light or insight from the historical context in which it was written. Many of us have a hard time putting ourselves into the cultures of the ancient near east, the intended original audience, their worldview, and trying to evaluate the type of literature we are dealing with. When people used to claim that the only honest way to interpret the text was literally, I thought this made sense. I had seen the fruits of many commentaries on this chapter, and I did not like where their conclusions ended when starting with a non-historical hermeneutic.

But was my skepticism justified? The more I read good, conservative, learned biblical scholars I began to realize that we are reading the text properly when we realize what it is and the type of communication to be received by the audience. For instance, we are not reading the text of Revelation properly if we do not realize that it is apocalyptic literature. And if we do not understand what the original audience could have understood. We would be amiss if we came away from Revelation taking it literally, in the sense of not understanding the symbolism in the book. We would be left with literal beasts, dragons, amazing creatures, and a real foreboding about what this future is going to look like and questions on why it is so disconnected with our experience.

This is not to diminish the illumination of the Holy Spirit and the fact that the Bible is a living book. It is not the same as any other book and usually it is critiquing us instead of the other way around. But God has condescended to our level to reveal Himself and His plan of redemption to us in ways that both the original audience and we can understand. Which includes various types of literature in His word.

This is not to say that the account of creation has nothing to say to us in the 21st century. God the Holy Spirit had His future church in mind, as well as the ancient Israelites, when He inspired Moses to write these verses. This is the beginning of the overarching storyline in the Bible of creation, fall, redemption, and consummation in the new creation. And it is the foundation of Chistian theology. If God is not Creator, then the Bible does not hold together.

Was Genesis 1 meant to be taken as a treatise on scientific origins like big bang cosmology, the theory of general relativity, the expansion of spacetime, or even plate tectonics? What kind of problems do we run into if we think that the purpose of Moses was to explain scientific mechanisms and theory here in Genesis 1? I think the best place to start is with the Bible itself and see what it has to say for itself. Not to start with some preconceived notions about the text. Sometimes we can assume we know what it says because we have believed something for so long. Reading the text itself has no substitute, as opposed to what someone tells us it means. We want to have a way of interpreting the text based on exegesis and not what we can read into the text.

2 Timothy 2:15

15 Be diligent to present yourself approved to God as a workman who does not need to be ashamed, accurately handling the word of truth.

Let's first lay down some basic ground rules for Biblical interpretation and then look at a few popular Bible commentary summaries of Genesis 1.

1. Interpret the verses within their context.

2. Interpret the parts in relation to the whole Bible and its theology.

3. Seek to understand what the original authors intended for their original audience.

4. Be cognizant of the type of literature it is (is it narrative, history, poetry, parables, apocalyptic).

5. Take the words in their normal, natural sense (do not make the passage say what you want it to say).

6. Does the interpretation conflict with the rest of the Bible on that subject?

If we can apply these principles of interpretation to Genesis 1, we should be able to protect ourselves from overt error. What have some popular commentaries of Genesis 1 concluded over the years?

Augustine (354-430) had four commentaries on the creation account in Genesis and numerous other works dealing with Genesis. He left open the possibility of interpreting the text as figurative instead of literal and warned against making such assertions that it produces interpretations that are too dogmatic. (Augustine & Taylor, 1982) But he posited in some of his works that the creation could have happened instantaneously and not over six literal days.

Martin Luther (1483-1546) in his own commentary on Genesis seemed to be of the opinion that none before him had adequately explained the creation account and that they had left more questions than answers. He seemed to believe that we should be content knowing that the world had a beginning, was created by God, and that it was created out of nothing. But he prefaces his work by saying that Moses taught the world did not exist before 6000 years ago, and by taking a straightforward literal view and denying that Moses was speaking figuratively.

John Calvin (1509-1564) taught that the world was formed from nothing and the idea that matter is eternal is refuted by the Bible. He says that the initial heavens and earth were a confused mass. He does not opine on the date of creation in his commentary on Genesis 1.

The Reformation Study Bible basically affirms the doctrine that God created the universe out of nothing and not from preexisting material. The emphasis being on God's ordering of a formless world. It claims that the compound phrase "heavens and the earth" in Genesis 1:1 represents the organized universe. (Sproul, Waltke, Silva, & Whitlock, 1995)

The MacArthur Study Bible teaches that God created without preexisting material, this marked the beginning of the universe in space and time, the date of creation is 10,000 years ago or less, it happened in six consecutive 24-hour days, and that the phrase "the heavens and the earth" are a summary statement which include all six days of creation. (MacArthur, 1997)

I highly respect John MacArthur. God has used him mightily in the work and advancement of His kingdom, and his ministry has been a great help to me over the years. In his study Bible from 1997 (which is the one I own) he represents the view held by many American evangelical churches and young Earth creationists. This is also undoubtably the view held by most of the Christian church since its inception. This is a good place to start, because it is often asserted that the opening verse in Genesis is a title or a summary statement about the six days of creation.

Genesis 1:1

1 In the beginning God created the heavens and the earth.

God is assumed in the first verse of the Bible. There is no justification or explanation for Him, and the rest of scripture makes clear that He is eternal and uncreated. God is referred to here by the Hebrew Elohim which is plural. He is the transcendent creator God. The plural is interesting, and I believe already points to the trinity in the Godhead. In the next verse we are already introduced to the Spirit of God. And the rest of the Bible describes God's triune nature, even at creation (John 1). In 1:26 the text says, "let Us make man in Our image", and then in the very next verse it says, "God created man in His own image", so there is already a plurality of Persons in one God in the first chapter of the Bible.

If this indeed is the title or summary for the rest of Genesis 1 then we are left with the uncomfortable idea that beginning with verse 2 the earth is already created, and the Spirit of God is hovering over the surface of the waters. What are we to make of this? Why do we not have the common, "Then God said" and, "it was so"? Should not 1:2 start with God calling into existence the planet Earth and the deep on which He was about to begin His work? If not, who made it? Was it eternally there? I don't think we can view 1:1 as a title or a summary statement because then we are left with the theologically troublesome position that God shows up to prepare the world as a habitable place for plants, animals, and humans, while the mass of the earth and waters are already waiting. C. John Collins argues that the grammatical structure and verb tenses in Genesis 1:1-2 point to this not being a summary statement of the creation week, but as background material to the creation narrative and giving us information preceding the beginning of God's workweek, which commences in verse 3. (Collins C. J., 2006)

The late John Sailhamer points out that the Hebrew word used for beginning here, Re'shiyth, is often used in Scripture to denote an unspecified or indeterminate duration of time, like in Job 8:7 talking about the early part of Job's life, the early part of Nimrod's kingdom in Genesis 10:10, and the initial part of the reign of a king of Israel before his official office was counted, like in Jeremiah 28:1. And that Moses could have used another Hebrew word to imply a start or initial point. (Sailhamer, 1996) The Hebrew word for created is bara' which is always used with God as the subject and never with Him using preexisting matter to form things. This Hebrew word is dedicated for God alone in His creative work.

The term "heavens and the earth" is verse 1 is considered a merism, which is a way of bringing together opposites as a way to signify everything. It is a figure of speech which conveys the meaning of totality, because the ancient Hebrews did not have a word for universe. The two words used by themselves do not mean the same thing as they do used together in this context. The word heavens (shamayim) is usually translated sky, heaven, or heavens, while the word earth ('erets) is usually translated earth, ground, or land. But both are dictated by the context in which they are found. This distinction between the uses of these words in very important throughout the Bible and especially in these early chapters of Genesis to keep our minds from wandering to a mental picture of the planet Earth set in outer space, when sometimes the context dictates the land and the sky. Which is often what the author is envisioning from his point of view.

I then conclude that Genesis 1:1 teaches that the eternal creator God created all things in the universe out of nothing prior to the ordering of the Earth for his creatures and His glory.

Some will argue that creation from nothing is not implied in the text. Many theologians would disagree as we have just seen, as the context and word usage imply as much. But we can look to other verses to inform our interpretation.

John 1:1-3

1 In the beginning was the Word, and the Word was with God, and the Word was God. 2 He was in the beginning with God. 3 All things came into being through Him, and apart from Him nothing came into being that has come into being.

Hebrews 11:3

3 By faith we understand that the worlds were prepared by the word of God, so that what is seen was not made out of things which are visible.

Nehemiah 9:6

"You alone are the Lord.

You have made the heavens,

The heaven of heavens with all their host,

The earth and all that is on it,

The seas and all that is in them.

You give life to all of them

And the heavenly host bows down before You.

Hebrews 1:10

10 And, "You, Lord, in the beginning laid the foundation of the earth,

And the heavens are the works of Your hands;

Psalm 89:11

11 The heavens are Yours, the earth also is Yours; The world and all it contains, You have founded them.

Acts 17:24

24 The God who made the world and all things in it, since He is Lord of heaven and earth, does not dwell in temples made with hands;

Job 38:4

4 "Where were you when I laid the foundation of the earth?

Tell Me, if you have understanding,

The Bible tells us that God is the One responsible for all that we see. And that it was formed only by His Word out of nothing. When was God laying the foundations of the earth? The most obvious way to read the text is that He was doing it "In the beginning". When He created all space, time, matter, and energy. Otherwise, we have a hard time maintaining the doctrine of creation out of nothing from this verse alone if we use 1:1 as a summary statement. As we move from verse 1 to the rest of the chapter, we will see the theological significance shift from the creation of all things to the ordering of those things to be a habitable place for man and beast to dwell. This idea of creation out of nothing is a novel concept from Genesis alone and from no other ancient religious text.

Genesis 1:2

2 The earth was formless and void, and darkness was over the surface of the deep, and the Spirit of God was moving over the surface of the waters.

The translation "formless and void" (Hebrew tohu wabohu) is a good rendering and coveys the idea that in its pre-ordered state, the earth is uninhabitable and uninhabited. What follows is making the various realms of creation habitable and then filling those realms of creation with plants, animals, and mankind so they are no longer void. It could also be said to be unformed and uninhabited. We have seen this concept in the other ancient near eastern creation accounts of starting with waters, sometimes with negative attributes, and then proceeding to order the cosmos by separating and distinguishing. I think the same overall theme is present here. But there is a huge difference in a single all-powerful God with His Spirit hovering over His creation, even if it is dark, formless, empty, foreboding, and not yet fit for life. A similar concept is found in Isaiah.

Isaiah 45:18

18 For thus says the Lord, who created the heavens (He is the God who formed the earth and made it, He established it and did not create it a waste place, but formed it to be inhabited), "I am the Lord, and there is none else.

Bruce Waltke makes some insightful comments in his commentary on Genesis here. He points out that there is darkness and the deep from the beginning of creation. He also points out that even after the world is separated and filled, there is still night and the sea. Even though forces hostile to life are present in the pre-created state and after the ordered creation, God has set their limits and bounded them. He also argues that we can infer from this that not everything that is hostile to life and causes pain and suffering can be a result of sin. (Waltke, 2001) This will be important for our thesis in this book. But why should we think that the night and sea imply forces hostile to life? Firstly, because there is only darkness and the deep making up the uninhabitable world. And secondly, because of how Revelation describes the new heavens and earth, and the new Jerusalem.

Revelation 21:1

Then I saw a new heaven and a new earth; for the first heaven and the first earth passed away, and there is no longer any sea.

Revelation 21:22-26

22 I saw no temple in it, for the Lord God the Almighty and the Lamb are its temple. 23 And the city has no need of the sun or of the moon to shine on it, for the glory of God has illumined it, and its lamp is the Lamb. 24 The nations will walk by its light, and the kings of the earth will bring their glory into it. 25 In the daytime (for there will be no night there) its gates will never be closed; 26 and they will bring the glory and the honor of the nations into it;

Revelation 22:5

5 And there will no longer be any night; and they will not have need of the light of a lamp nor the light of the sun, because the Lord God will illumine them; and they will reign forever and ever.

We can already see some important theology taking place for our overall thesis. God has no equal and opposite evil power like in some other religions. Even the deep and darkness are part of His plan. And what some may call "natural evil", like natural phenomena hostile to life and flourishing, are not the result of sin and moral evil. We can also infer that the new heavens and earth and new Jerusalem are qualitatively different than the original creation.

Another thing that skeptics sometimes try to do is tie this verse in with the ideas from other ancient near eastern creation myths that describe a cosmic battle against disorder and warring gods by claiming that Tiamat in the Enuma Elish is what the Hebrews are referring to here by deep. But the etymology of the Hebrew word for deep (tehom) is simply a derivative from many Semitic languages for ocean. There is nothing in this text to suggest that God is somehow subduing chaos, battling with disorder, or locked in battle with a water deity. It is His creation and the earth, ocean, and darkness do His bidding.

So as God is about to begin His work of ordering the Earth for the plants, animals, and the creation of mankind made in His image, we need to switch gears and look at the literary structure of the rest of Genesis 1.

One of the things that theologians have noticed about the creation account in Genesis 1 are the obvious literary devices. One involves parallelism to compare days 1 and 4, days 2 and 5, and days 3 and 6. These parallels were noted as early as Augustine in his *City of God* and have been fleshed out throughout the centuries by scholars like J.H. Kurtz, Gottfried Herder, Hugh Miller, Samuel Driver, Henri Blocher, and more recently by scholars Meredith Kline, Mark Futato, and Lee Irons.

The second thing that we notice about Genesis 1 is the repeating phrases "then God said" or "let there be" coupled with "and it was so" or "and there was". It is important to note here that the use of literary devices in and of themselves does not demand that the subject matter be taken as figurative, metaphorical, or symbolic, as opposed to historical narrative. Or to impose some literary style or genre on the text. Literary devices and different literary genres are employed extensively throughout the Bible.

We see that on day 1 the light and darkness are created while on day 4 the sun, moon, and stars are appointed to govern the day and the night. On day 2 the sky and the seas are created to be filled with birds and the sea creatures on day 5. Whereas the parallels of days 1,4 and 2,5 have single creation events, days 3 and 6 both contain two separate creation events. On day 3 God creates the dry land and the plants and on day 6 He creates the land animals and mankind. And then He gives the seed-bearing plants and fruit to man and green plants for animals.

What are we to make of this parallelism and repeating refrains? Do these uses of literary devices mean that the narrative was not meant to be taken as sequential periods of creative events? Many who hold to this type of framework interpretation see the parallel days as the same event for the triad of days described from different perspectives, or at the very least, that days 1 and 4 are describing the same day. In the end, many of them have 3-5 days total in which God is doing His work. But as the 4th commandment states the reason for observing the Sabbath is that in six days God made the heavens, earth, sea, and all that is in them and rested on the 7th day, this would seem to pose a problem for condensing the days into fewer than six. According to many scholars, the Hebrew grammar used in Genesis 1 is indicative of a sequential ordering of the days that would limit our flexibility of interpretation otherwise. Then what is Moses doing with these verses about God working for six days and resting on the seventh? Why would God need to rest from His work? And can God only work during the daytime? Agreeing with Dr. Collins, I believe what Moses is doing is making an analogy about the work week as it applies to God's image bearers. (Collins C. J., 2006) And surely the Israelites would have understood it in this sense after they had received God's covenant on Mt. Sinai and were waiting to enter the promised land across the Jordan river.

Days 1 and 4

Day 1

Genesis 1:3-5

3 Then God said, "Let there be light"; and there was light. 4 God saw that the light was good; and God separated the light from the darkness. 5 God called the light day, and the darkness He called night. And there was evening and there was morning, one day.

In Genesis 1:3 we get the first proclamation of God speaking to bring about light, which He calls good, and separates it from the darkness. This is also the first time we are told that there was evening and there was morning. Another important point from the text is that the Hebrew says, "one day". Some translations make this verse say "the first day" but that is not in the Hebrew. Why would Moses use "one day" instead of "the first day" if the earth itself and the deep ocean covering it was formed on day one in verse 1:2? Bruce Waltke points out that the first five days of creation lack the definite article, so don't necessarily have to indicate chronology. (Waltke, 2001)

Why does God call the light day instead of just light? John Walton argues that the ancient near eastern people, including the Israelites, did not think of creation and existence solely in terms of material existence like we modern people do. Instead, they thought about ontology, existence, and creation as functional definitions instead. So, he argues that God calling light day instead of simply light reflects this different ontology. (Walton, 2010) I think this is a keen observation and fits in with the general literary purpose of Genesis 1. But I don't necessarily agree that ancient people had only this view of the world. I think they were cognitively just like we are and I'm not sure the evidence is great that they did not take their sense perception and reason back to materiality. Put differently, I find it hard to believe they didn't understand cause and effect. The ancient pagans may have believed that the gods influenced how productive their crops and harvest were, but they still knew that if they failed to plant seeds and water the ground, they would have no crops. But to be fair to Dr. Walton, he does point out that the ancients could know something was physically there and perceived by the senses, but it still did not "exist" in the categories of their minds apart from having a function, primarily relating to human society. (Walton, 2010) And I acknowledge that Dr. Walton has far more knowledge about the culture and beliefs of the ancient near east than I do. But I think more to the point, they don't have to be mutually exclusive options. I believe that the text is telling us that God created the material light for this world and then He named it after its function, to separate light from darkness, and then He called them day and night. Thus, He begins the process of making the earth habitable and fruitful.

Again, we make the quick point that darkness is still present after one day. God alone calls the light good, but the darkness is still part of the created order.

Day 4

Genesis 1:14-19

14 Then God said, "Let there be lights in the expanse of the heavens to separate the day from the night, and let them be for signs and for seasons and for days and years; 15 and let them be for lights in the expanse of the heavens to give light on the earth"; and it was so. 16 God made the two great lights, the greater light to govern the day, and the lesser light to govern the night; He made the stars also. 17 God placed them in the expanse of the heavens to give light on the earth, 18 and to govern the day and the night, and to separate the light from the darkness; and God saw that it was good. 19 There was evening and there was morning, a fourth day.

Then applying the parallelism, we see that on a fourth day God appoints the sun, moon, and stars to separate the day from the night, to separate the light from the darkness, and to be for seasons, signs, days, and years. The light created on Day 1 is governed by the light bearers on day 4. On day 4 God is appointing the celestial bodies to give light unto the earth and serve as markers to give humans a sense of time and season, which gives us further confirmation from the context of appointing the heavenly bodies for their functions. I say that He appoints the celestial bodies instead of creates them because in verse 16 the Hebrew word is 'asah and the context does not dictate its meaning as create opposed to appoint. 'Bara would have been the appropriate word to use if God was creating them from nothing for the first time, but this is probably stretching the interpretation. Surely the concept of God creating and appointing them is implied, but the context is not for the universe but as it applies to this world to make it formed and filled. And God saw that it was good.

Psalm 104:19

19 He made the moon for the seasons; The sun knows the place of its setting.

Jeremiah 31:35

35 Thus says the LORD, Who gives the sun for light by day And the fixed order of the moon and the stars for light by night, Who stirs up the sea so that its waves roar; The LORD of hosts is His name:

As we look at a 4th day of creation, we notice several interesting things about the way Moses writes about the celestial bodies. The first thing to notice is that he designates the purpose of the sun, moon, and stars as being light for the earth and to mark off time. The next thing to notice is that Moses does not even name the sun or the moon. These are very unusual Hebrew phrases for referring to the sun and moon. He calls them the greater light and the lesser light. Finally, the appointment of the stars is almost an afterthought in the text.

Why does Moses seem to downplay the sun, moon, and stars? It is clear to many interpreters, including this one, that Moses is directly denouncing the pagan worship of the heavenly bodies in the ancient near east and creating a polemic against their creation myths and religions. The cultures of Mesopotamia and Egypt were both polytheistic religions as we have seen. They had deities that represented the sun, moon, and stars, among many, many other gods. And these sun and moon gods were often supreme and powerful in their polytheistic religions.

In Egypt some of the sun gods were Atum, Ra or Re, Atum-Re, Amun-Re, Khepri, and Aten (they were usually one of the chief deities associated with creation, with Aten being **the** chief deity during the reign of the so-called heretic Pharaoh Akhenaten), the moon god Khonsu, and different gods for the stars like Sah and Sopdet.

In Mesopotamia, the sun god was Utu/Shamash (Sumerian/Akkadian), the moon god Nanna/Sin, and the god of constellations An/Anu. When Abram left Ur with his family, the patron deity of that city was the moon god Sin. And there were many cultic centers to this moon god throughout Mesopotamia and Syria, including in Haran, into which Abram and his family traveled and then settled. It is interesting that they settled in Haran, another major cult center for the moon god Sin. And from Haran, God calls Abram, gives him promises, and Abram sets out for the land of Canaan. In most Mesopotamian theologies, Sin was the father of the sun god Shamash. And what I really think is telling is that the temple of the moon god Sin was called “the house of the great light” by the Mesopotamians.

So, Moses is denouncing some of the most revered deities in Egypt and Mesopotamia by saying that the sun, moon, and stars are nothing but material objects created by God. And not even worthy of their proper Hebrew names. He is calling the moon the lesser light, in one sense because it is less bright than the sun, which he calls the greater light, but also denouncing the Mesopotamians who are calling their moon god the great light and the father of their sun god. Abraham and his descendants surely would have recalled the family idolatry with the moon god Sin before the true God called them. On top of this, many ancient near eastern people believed that the stars were associated with directing men’s destinies. (Waltke, 2001) In not even naming the sun and moon and making passing mention of the stars, this can be seen as a strong polemic against the false religion of the Babylonians, Assyrians, Egyptians, and inhabitants of Canaan. God is communicating through Moses that there is only one true God, nature is His creation, and it is not to be worshipped.

Deuteronomy 4:19

19 And beware not to lift up your eyes to heaven and see the sun and the moon and the stars, all the host of heaven, and be drawn away and worship them and serve them, those which the Lord your God has allotted to all the peoples under the whole heaven.

Deuteronomy 17:3

3 and has gone and served other gods and worshiped them, or the sun or the moon or any of the heavenly host, which I have not commanded,

Isaiah 47:13

13 “You are wearied with your many counsels;

Let now the astrologers,

Those who prophesy by the stars,

Those who predict by the new moons,

Stand up and save you from what will come upon you.

Jeremiah 8:2

2 They will spread them out to the sun, the moon and to all the host of heaven, which they have loved and which they have served, and which they have gone after and which they have sought, and which they have worshiped. They will not be gathered or buried; they will be as dung on the face of the ground.

Psalm 148:3

3 Praise Him, sun and moon;

Praise Him, all stars of light!

Days 2 and 5

Day 2

Genesis 1:6-8

6 Then God said, "Let there be an expanse in the midst of the waters, and let it separate the waters from the waters." 7 God made the expanse, and separated the waters which were below the expanse from the waters which were above the expanse; and it was so. 8 God called the expanse heaven. And there was evening and there was morning, a second day.

On a second day, God creates the expanse or firmament to separate the waters below the expanse from the waters above the expanse. What are we to make of this expanse or firmament? Many Bible commentaries say that it is the sky or atmosphere and the reference to the waters above the expanse is simply a reference to the clouds from which rain falls. But is that letting the text speak for itself or our modern idea of meteorology and atmospheric science reading something into the text? On the other hand, are we justified in assuming that God is relating an idea of a solid dome structure, and this is what the Israelites believed about how the heavens were constructed?

The Hebrew word for expanse or firmament is Raqiya`. This word is derived from a Hebrew verb meaning to beat or stamp out. Some scholars think that this is describing a solid dome-like structure (which is why it is sometimes translated

firmament in older translations) that has waters above it. They prop up their argument by saying that not only does the etymology of the Hebrew word point to this meaning, but that the words used in the Greek Septuagint and Latin Vulgate translations of the Old Testament make use of words in those languages to mean make firm or solid.

Both Augustine and Aquinas appear to have understood it to be a solid structure, but Calvin warns that Moses was talking about how things appeared, and that we were not to learn astronomy from these verses on the firmament.

It could be argued that this same idea of waters above the firmament is present in the flood account when describing the floodgates of heaven being opened at the start of the flood and the floodgates closed at the end of 40 days/nights. The Israelites may not have understood many things about the atmosphere, including the composition of molecules and particles in the atmosphere and how shorter wavelength blue light get scattered more than the other wavelengths of visible light, making the sky appear blue, but they were not stupid. There are numerous passages in the Old Testament about God bringing rain from the clouds and I am positive that they knew this from experience. But that doesn't mean they did not believe in waters above the expanse of the sky. I think both of those things could be true.

But I maintain that it could just as easily be argued that this description is based on how things looked. God names the Raqiya` (expanse) Shamayim (heaven or sky), which all the Israelites were familiar with. If we look at the text itself, God creates the expanse to separate the waters and then names it heavens or sky. On a fourth day God places the sun, moon, and stars in the expanse of the heavens. And in verse 20, the birds are commanded to fly in the open Raqiya` (expanse) of the Shamayim (heaven or sky). So, the atmosphere and sky are definitely part of what was created and separated on a second day. It is the sphere in which the birds will inhabit.

Many scholars claim that the idea of a solid dome with an ocean of water above to be the common cosmology of the day at the time of Moses and for millennia afterwards. Other Christians have developed models that a thick canopy of water vapor was over the earth from creation to the flood, and this canopy then collapsed contributing to the water needed to flood the whole Earth. This is far more information than the text would warrant. But the idea of waters above the expanse of the heavens was still around when the Psalms were composed.

Psalm 148:1-6

1 Praise the LORD! Praise the LORD from the heavens; Praise Him in the heights! 2 Praise Him, all His angels; Praise Him, all His hosts! 3 Praise Him, sun and moon; Praise Him, all stars of light! 4 Praise Him, highest heavens, And the waters that are above the heavens! 5 Let them praise the name of the LORD, For He commanded and they were created. 6 He has also established them forever and ever; He has made a decree which will not pass away.

But we will see that whether the text is implying a solid dome holding back waters above, giving a phenomenological description of the sky, or what the Israelites and their contemporaries believed about how the sky, waters, sun, moon, and stars were arraigned, it is immaterial to our thesis when we examine Genesis chapter 1 as a whole.

After God creates the expanse on a second day, the expression "God saw that it was good" is missing. What are we to make of this admission? Many Bible commentators have realized that the waters below had still yet to be gathered into one place, as on a third day, and I think they are partially right. I think the overarching message God is giving us in Genesis 1 is that what He is doing is ordering His creation for the benefit of His creatures and specifically mankind. And this ordering of the heavens and earth is what God is seeing as good at each step in the process. Not that the physical elements are in themselves good or morally pure, but that the materials are being constructed in such a way as to confer benefits to the creatures and man. On a second day we see that the waters are not yet in their proper abode for the benefit of the sea creatures, and therefore "it was good" is omitted.

Day 5

Genesis 1:20-23

20 Then God said, "Let the waters teem with swarms of living creatures, and let birds fly above the earth in the open expanse of the heavens." 21 God created the great sea monsters and every living creature that moves, with which the waters swarmed after their kind, and every winged bird after its kind; and

God saw that it was good. 22 God blessed them, saying, "Be fruitful and multiply, and fill the waters in the seas, and let birds multiply on the earth." 23 There was evening and there was morning, a fifth day.

As we move on to the parallel fifth day, God creates the birds to fly on the "face of the expanse" or "in the open expanse" of the heavens/sky and the sea creatures to fill the waters. God blesses the birds and sea creatures and tells them to be fruitful and multiply. What is the purpose of audibly blessing creatures devoid of understanding we may ask? We begin to see here for the first time in the Bible that God cares for all the creatures on His Earth. And it is the essence of life according to the Bible to have God's blessings and have His face shine upon us. Filling their realms of creation by being fruitful and multiplying is a blessing from God and shows His favor. Just like when God makes the covenant with Noah, his family, and his descendants after the flood, His promises and blessings also extend to the creatures who were with Noah on the Ark. And we will see that Noah is a type of Adam, a second Adam if you will. And a type of re-creation takes place again after de-creation during the flood. To have population numbers decreased due to cursing and to be fruitful and multiply as blessing is a common theme throughout the Bible.

Days 3 and 6

Day 3

Genesis 1:9-13

9 Then God said, "Let the waters below the heavens be gathered into one place, and let the dry land appear"; and it was so. 10 God called the dry land earth, and the gathering of the waters He called seas; and God saw that it was good. 11 Then God said, "Let the earth sprout vegetation, plants yielding seed, and fruit trees on the earth bearing fruit after their kind with seed in them"; and it was so. 12 The earth brought forth vegetation, plants yielding seed after their kind, and trees bearing fruit with seed in them, after their kind; and God saw that it was good. 13 There was evening and there was morning, a third day.

A third day begins with the triad of days in which there are two separate creation events. God orders the water below the heavens to be gathered into one place, allowing the dry land to appear. He names the dry land earth, and the waters He names seas. And God saw that it was good. We note again that the seas exist and are a part of God's original creation, in contrast to the new heaven and new earth.

Revelation 21:1

Then I saw a new heaven and a new earth; for the first heaven and the first earth passed away, and there is no longer any sea.

The second creation event on a third day is God calling on the dry land just created to bring forth vegetation, plants yielded seed, and fruit trees yielding seed. The land itself brings forth the vegetation under the command of God and He saw that it was good. With the dry land in place and the earth bringing forth vegetation and fruit, we jump to the parallel day six.

Day 6

Genesis 1:24-31

24 Then God said, "Let the earth bring forth living creatures after their kind: cattle and creeping things and beasts of the earth after their kind"; and it was so. 25 God made the beasts of the earth after their kind, and the cattle after their kind, and everything that creeps on the ground after its kind; and God saw that it was good. 26 Then God said, "Let Us make man in Our image, according to Our likeness; and let them rule over the fish of the sea and over the birds of the sky and over the cattle and over all the earth, and over every creeping thing that creeps on the earth." 27 God created man in His own image, in the image of God He created him; male and female He created them. 28 God blessed them; and God said to them, "Be fruitful and multiply, and fill the earth, and subdue it; and rule over the fish of the sea and over the birds of the sky and over every living thing that moves on the earth." 29 Then God said, "Behold, I have given you every plant yielding seed that is on the surface of all the earth, and every tree which has fruit yielding seed; it shall be food for you; 30 and to every beast of the earth and to every bird of the sky and to every thing that moves on the earth which has life, I have given every green plant for food"; and it was so. 31 God saw all that He had made, and behold, it was very good. And there was evening and there was morning, the sixth day.

On the sixth day God commands the earth to bring forth cattle, creeping things, and the beast of the earth after their kind and it was so. And He saw that it was good. He makes the land animals that can be domesticated, animals that creep and slither, and wild beasts of the land. The land realm separated on a third day is filled with land animals on the sixth day and is no longer void or empty.

The second act of creation on the sixth day is the culmination of creation, that of mankind. Here we switch from God directly commanding the creation of realms and the creatures or celestial bodies to fill those realms, to one of consultation within the Godhead. Some Bible commentaries believe the plurality of Us and Our in verse 26 refer to the three Persons of the Trinity, while others see it as a reference to the angels surrounding God's throne, while yet others interpret it as a royal "we". I take the plural to refer to the plurality of Persons within the Godhead, as this seems like the most solid interpretation based on immediate context and what the rest of the Bible teaches. Especially in John's gospel chapter 1, speaking of Jesus being in the beginning with God, and all things coming into being through Him. And in the immediate context of the chapter, we have Genesis 1:2 speaking about the Spirit of God hovering over the waters, verse 1:26 uses the plural pronouns Us and Our when describing creating man in the image of God, while verse 27 uses the singular male pronoun for God to describe the same creation of man in the image. We therefore have plurality of Persons and one God right in the text.

Some commentaries have suggested that God was consulting with either angels or the earth itself before He began His creation of mankind, but this seems to be very much at odds with orthodox Christian doctrine. Mankind is never said to be created in the image of angels and the earth itself would produce a strange meaning. Some secular scholars think that the Israelites forgot to edit out the polytheism in which this creation account supposedly borrowed from and that is why there are a plurality of gods in this chapter. Or that tradition was so strong that they didn't want to rock the boat by changing it. Apart from the problem that the editors would indeed have to be rather daft to leave that in the final form of the Old Testament, it is in contradiction with the theology of the rest of scripture. I say that we can confidently interpret the plural as a hidden beginning to the understanding of the nature of God Himself.

The word used for man on the sixth day varies in grammar and references in this passage. In verse 26 it refers to mankind in general, "man, them", in verse 27 it refers to a male and then both male and female, "man, him, male and female, them", and in verse 28 God blesses them. So, God creates man in His image and likeness, and this man represents a particular man, and mankind in general, male and female. God gives humanity, created in His image and likeness, dominion over the creation and the creatures He has created and made. God reiterates that they are made in His image, and tells them to be fruitful and multiply, fill the earth, subdue it, and rule over every living thing that moves. They are given plants yielding seed and fruit trees yielding seed to be food for them and the green plants for the animals. We will address the concepts of the image of God and original sin much more in depth in later chapters. Finally, God saw all that He had made, and it was very good.

Day 7

Genesis 2:1-3

Thus the heavens and the earth were completed, and all their hosts. 2 By the seventh day God completed His work which He had done, and He rested on the seventh day from all His work which He had done. 3 Then God blessed the seventh day and sanctified it, because in it He rested from all His work which God had created and made.

Some would say that we were hasty in calling the sixth day the pinnacle of the creation account, as they see the seventh day as the pinnacle. Many recognize the chiastic structure of these verses and see them as the summary statement for the whole creation narrative. I think this interpretation is best and makes the most sense because verse 4 starts a toledot section that is an introduction to the rest of chapter 2. And the toledot sections are always introducing new material, not looking back. (Kidner, 1967)

God completes His work of creation, and He rests. God sets the seventh day apart as holy and blesses it. Many holding to the framework hypothesis bring up the fact that the seventh day is missing the phrase "and there was evening and there was morning" and that if this day does not refer to a literal 24-hour day then this leaves open the possibility that the other six days should possibly not be taken as 24-hour solar days either. But I think this advances very little discussion and understanding of the text, which seems to clearly indicate six 24-hour sequential days.

Exodus 20:8-11

8 "Remember the sabbath day, to keep it holy. 9 Six days you shall labor and do all your work, 10 but the seventh day is a sabbath of the Lord your God; in it you shall not do any work, you or your son or your daughter, your male or your female servant or your cattle or your sojourner who stays with you. 11 For in six days the Lord made the heavens and the earth, the sea and all that is in them, and rested on the seventh day; therefore the Lord blessed the sabbath day and made it holy.

Hebrews 4:3-11

3 For we who have believed enter that rest, just as He has said,

"As I swore in My wrath,

They shall not enter My rest,"

although His works were finished from the foundation of the world. 4 For He has said somewhere concerning the seventh day: "And God rested on the seventh day from all His works"; 5 and again in this passage, "They shall not enter My rest." 6 Therefore, since it remains for some to enter it, and those who formerly had good news preached to them failed to enter because of disobedience, 7 He again fixes a certain day, "Today," saying through David after so long a time just as has been said before,

"Today if you hear His voice,

Do not harden your hearts."

8 For if Joshua had given them rest, He would not have spoken of another day after that. 9 So there remains a Sabbath rest for the people of God. 10 For the one who has entered His rest has himself also rested from his works, as God did from His. 11 Therefore let us be diligent to enter that rest, so that no one will fall, through following the same example of disobedience.

Literary type and audience

Now that we have gone through the creation account, let us try to draw some conclusions about the type of literature we are dealing with. Most Old Testament scholars agree that the Hebrew grammar and verb usage starting at verse 3 through the end of the creation account is what we find in other narrative in the Old Testament. They also generally agree that the way that the Hebrew yom (day) is used throughout the passage in consistent with the normal use of the word day. Meaning that it reads like six 24-hour days. Further confidence is attained by noticing that each day is an evening and a morning. So, we can reasonably conclude that we are dealing with normal 24-hour days and a consecutive series of events like in normal biblical narrative. I therefore think the young Earth creationists are right when they say that the Hebrew of the creation account points to narrative literature. But is it a historical narrative?

Some scholars say that the other ancient near eastern creation stories are written in epic poetry and this is not epic poetry. And I agree that it is neither Hebrew poetry nor epic poetry. But it is certainly highly stylized and magisterial in its presentation. We have seen that this chapter is highly stylized with the parallelism and repeated formulaic refrains. We have also seen that the word choices for describing creation are often not the typical Hebrew words for such things. We have anthropomorphic images of God working during the day and resting at night. And we see Him resting from all His work on the seventh day and sanctifying and blessing it.

Therefore, because of the unique literary characteristics of the account, I believe we can safely call it something besides normal historical prose or narrative. We have here a unique genre of literature in the Bible. One that is similar in certain respects to other ancient near eastern creation stories in some respects, while quite different in its theology.

But an important caveat. Just because some biblical literature is not historical narrative, it does not follow that it does not contain theological truths. Just like we would be amiss if we interpreted **Psalm 96:11-12 "Let the heavens be glad, and let the earth rejoice; Let the sea roar, and all it contains; 12 Let the field exult, and all that is in it. Then all the trees of the forest will sing for joy"** as teaching that the heavens, earth, sea, field, and trees have emotions and can make audible utterances. Instead of seeing this poetry as a way to exalt the Creator, His works, and His coming in judgement.

What then would the original audience of the Israelites led out of Egypt and more than likely having already received the covenant on Sinai thought of this revelation from God? I am assuming more than just the creation account in this exercise but trying to stay as focused on this portion of scripture as possible. They would already have had the core writings of the first five books of the Bible, including the Fall, flood, tower of Babel, table of nations, choosing of Abraham, Isaac, and Jacob, the promises and covenants, the slavery in Egypt, the covenant with Israel given to Moses, the tabernacle and priests, and other revelation given to Moses and written down prior to crossing the Jordan into Canaan and the death of Moses. Not to mention their own experiences in the desert and tabernacle worship.

I propose they would have immediately made the connection between this creation account and the ones they had heard from the surrounding cultures. Including Egypt, where they were just redeemed from by Yahweh. They could hear the similarities about how the world was separated and differentiated and how disorder was turned into order. And they would have immediately recognized the polemic nature of this creation account critiquing those of Egypt and Mesopotamia. They could see that their God, Yahweh Elohim, did not have to battle amongst the gods or subdue chaos in the form of deities. Their God was a single all-powerful God who created the world from nothing and then ordered it for the blessings of His creatures by the power of His will alone. And there was no pushback from the creation or evil forces. They could tell that even if hostile forces like the darkness and sea remained, their God had bounded them, and they did His bidding. (Waltke, 2001)

They could recognize that there was not a pantheon of gods each generating the next, usually by sexual reproduction, but a single eternal God. They would have recognized that the different parts of nature were not deities themselves, but material creations that God named for a fruitful and productive earth. They would have seen that the sun and moon are not even given proper names, much less to be worshipped. And that the power of God produces the stars almost incidentally. They would see that this God was a good, kind, and benevolent God. And that He ordered the cosmos for the benefit of His creatures, with man given a special place on the top of the hierarchy.

They would have seen the stark contrast in theology with the pagan nations, where man was created to do the work the gods refused to do and to produce food for them. That the gods of the pagans were constantly annoyed with mankind and trying to limit their population, as opposed to the theology of their benevolent God who blessed them and told them to be fruitful and multiply. They would have immediately recognized that although the pagan religions made idols in the image and likenesses of their gods, the Israelite God made them in the image and likeness of Himself. The Israelites coming out of Egypt already knew that Pharaoh claimed to be a god on earth, the son of Re, the manifestation of Horus. And by the time of the Exodus in the New Kingdom of Egypt, the living royal ka (part of the Egyptian understanding of the soul) of Amun-Re. And they likely knew of the Mesopotamian culture and others around them where the kings claimed to rule by divine appointment and authority. This democratization of the image of God was revolutionary. They now knew that because they were made in the image of God, they were called to have dominion over the earth and its other creatures as wise and righteous vicegerents. To sum it up, they were familiar with literature very close to this, but could see that they alone were chosen to worship and serve the one, true, living God.

But did they take it literally, in the sense of hearing it as a historical account? While it is probably impossible to answer such a question with absolute certainty, I believe there are good reasons to think that they would not have viewed it as something that happened, say 2000, 5000, or 10000 years ago in the historical sense. They knew that God is spirit, and He does not tire like a man. So, He would not need to rest. They could reason that He did not need six days to create but could do it in a moment of time. C. John Collins argues that the author of Genesis 1 is presenting the week as God's workweek, working through the day and resting at night, to culminate with the Sabbath rest after God's workweek was completed. He also argues that "kinds" in this chapter are more like categories that are based on appearances. Not some scientific taxonomy, but what was familiar to a normal person used to working with domesticated plants and animals. (Collins C. J., 2006)

And I think Dr. Collins's insights are correct. They would have seen the connection between the kind of workweek they were commanded to follow and the analogy of God having a six-day workweek and resting on the Sabbath, according to the 10 commandments they had just received on Mt. Sinai. They could see that the type of living they were used to with planting crops from seed and harvesting them in Egypt made sense within the context of the creation narrative. They could see that the categories they were familiar with for plants and animals corresponded well with the "kinds" in the account. And that like kinds did indeed beget like kinds.

But to suggest that it wasn't revealed by God to be interpreted in a historically literal sense is not the same thing as saying that it was not to be believed. God condescended to communicate to His chosen people in a form and type of literature they were familiar with and could understand. A type of literature that gave them true knowledge and right theology of who God was and what He had done based on how the literature was meant to be understood. Just like the song of Deborah in Judges 5 gives us true information in a different type of literature about historical events that happened in Judges 4. Or how 2 Samuel 22 gives poetic utterance to 2 Samuel 21. But stick with me, because we will see a different type of literature beginning in Genesis 2.

In summary then, what are we to make of our reading through Genesis 1? The purpose of looking at Genesis 1 from the text itself is to see if it is intended to be read as historical narrative or to teach science that transcends time. With the literary structure of the chapter, the parallelism, the repeated refrains, the Hebrew word choices, the enigmatic nature of the chronology, and the clear polemical nature of the creation account, we can safely say it is not historical narrative. But I hope to show that this chapter stands in contrast to the type of literature of Genesis 2, which I believe is to be taken as historical narrative. What is happening here in Genesis 1 is a phenomenological (the way it looked from human eyes) view of the world from the perspective of the Israelites and their contemporaries. God is speaking to the Israelites in a way that they can understand. If God was to compose the creation account describing the expansion of spacetime, the formation of galaxies, the nuclear fusion of the sun, the tidal forces of gravity, and the fabric of spacetime being influenced by the mass of planets and stars, the Israelites would have been profoundly lost. God is accommodating to the Israelites what could be understood and believed in order to teach the theological facts that all things exist because of His Word, He is in control and sovereign over His creation, nature is not to be worshipped, they are to follow God's example of working six days and resting on the seventh, and that man is the penultimate end (with God's glory being the ultimate) of the creation. That they are to be God's image bearers, to have dominion over the creation as righteous vicegerents, and ultimately to expand His kingdom. All of this while teaching the core historicity of the account.

What do I mean by core historicity? By this I mean that there is only one, true, eternal God who is distinct from His creation, creation out of nothing, that God alone is in control, that worshipping the creation and creatures is idolatry, that this God is good, and that the goal of humankind is live in communion with Him and to reflect God to the world. All of this communicated to any unlearned hearer with the use of a genre of literature known by these people in the second millennium BC. God not only had the ancient Israelites in mind when Moses composed the text but His church throughout all time. It is an account that anyone can understand and know that God is above all other so-called gods. And it is sublime compared to the other pagan creation accounts.

Many may be objecting and saying at this point, "This kind of interpretation of the text has only come about because science has become the hermeneutic by which the text is interpreted, not the Bible itself". But this is the reason we needed to look at the text first to see what kind of literature it is, what the original audience would have thought, and what it actually says for itself. Is this just a modern way of looking at the text to resolve perceived conflicts with geology, astronomy, and biology?

The reading of the creation text as a historical narrative giving us an age of the universe of 6000-12000 years is seen by many faithful Christians as the only way to interpret the text. But I think we will see that this view is not as simple as it sounds. I use the word literally in this section to mean reading the text as literal history, or as historical narrative. But we have seen that reading a text, including scripture, literally simply means reading the text for the type of literature that it is and the communication it is meant to convey.

Many holding this literal view see 1:1 as a summary statement for the six-day week. That means that the earth and deep were already present when the narrative starts in Genesis 1:2. So, either the formless Earth was eternal, created in the past during an undetermined period, or it was created on the first day along with the ocean. But we are not told in the text that God said let the earth and the deep be created. God calls light into existence and separates it from the darkness and names them. This is all the text says about God's creating activities on one day. Many say that God was the source of the light and yet He speaks of it as something separate from Himself. We must always be on guard against making God and the creation of the same substance. This leads to pantheism. The false idea that God is nature and nature is God. And this contrasts with the new Jerusalem in Revelation where God is the source of the light. We are then led to conclude that it was an independent light source not coming from God or the sun. This light source was capable of producing evening and morning on the planet. Many speculate that when God started working the light began to shine, and when He completed His work for the day the light went out. We are not sure if the light source itself goes out or if the rotation of the Earth just brings darkness to one side. But the text seems to imply that the entire planet is being worked on at once. Then the day ends with one day. The Hebrew does not say the first day, so we must conclude that we don't know if it was the first day of the "universe" or not.

On a second day, God makes the firmament (or expanse) and separates the water below it from the water that are above it. As we have seen, many scholars believe the firmament was understood by the Israelites as a solid dome structure holding back a body of water above it. If we are then to read the second day literally, we might have to say that there was a body of water above a solid dome structure. If we claim, like some do, that there was a dense water vapor canopy in the sky that is being referred to on day two that came down to the surface of the earth during the flood and after the flood it no longer existed, we first must admit that this is not reading the text literally. And then we must deal with Psalm 148 and decide why the Psalmist apparently still talked about the waters above the heavens. It is not reading the text of the second day literally, it is reading it phenomenologically, when we make the Hebrew say something it doesn't and by introducing a novel idea about the atmosphere that is not in the text. But to be charitable, some scholars think that it is unnecessary to interpret the text to mean a solid dome with waters above it. This day ends with the independent light source going out.

On a third day, the waters are gathered together into the seas and the dry land appears. The dry land produces vegetation, and the light source once again goes out.

On a fourth day, God creates the sun, moon, and stars. Many biblical scholars believe that Moses is describing God creating the celestial objects and placing them in the solid firmament itself, which would be below an ocean above it. And that is how the ancient Israelites, along with their contemporaries, thought about the sun, moon, and stars. But for the sake of argument and to be charitable to other interpretations which I think are tenable, let's assume that the text teaches that the stars are placed in outer space and can be somehow seen through the waters above to give light to the Earth. God makes the greater light to govern the day and the lesser light to govern the night. Here to be literal we might read the text as teaching that the moon itself was a source of light but not as bright as the sun. Given this literal interpretation, where can we look to the text that the moon was simply a reflective body for the sun's light? This would be reading a modern understanding of astronomy into the text and not out of it. A literal reading of the text suggests that the moon itself was an independent light source. At the end of the fourth day the sun and moon can take up the duties of separating the light from the darkness and the independent light source from days 1-3 is no longer needed.

On a fifth day, God creates the birds to fly on the face of the expanse of the heavens. He also creates the sea creatures and blesses them both.

On the sixth day, God creates the land animals and then man in His own image, blesses them, and commands them to exercise dominion. Man is given plants, seeds, and fruit as their diet. The green plants are given to birds and land animals, presumably including lions, tigers, and carnivorous dinosaurs.

My point is not to mock this view, but to show that reading it as a historical narrative (literally) is not as straightforward as we may think. I held this position myself for a very long time and have defended it in the past. But I hope that we can see that a literal reading of Genesis 1 as pure historical narrative, like a reporter watching the building of a great monument, is not as simple as we may think. When many people say they are interpreting it literally, they are actually interpreting many things in the passage phenomenologically, that is, based on how things appear to us from the ground. Or they use a modern understanding of things to put meaning into the text. Or introduce novel concepts and notions not in the text. That's why I think it's vital to understand the type of literature that we are dealing with, what God was trying to communicate to the Israelites with the creation account, and then try to ascertain what they might have thought hearing it.

The bottom line is not to cast doubt on God's word, but instead to interpret Genesis 1 in light of the type of literature that it is and free the Christian from endless controversies stemming from arguments that the text itself does not warrant. There are tons of passages in the Bible that are not meant to be interpreted literally, this is just one of them. Because I have been there myself, I know that some readers will still be thinking at this point that I am saying Genesis 1 is telling us lies. If the sun, moon, and stars were not created on the fourth day of the universe, then isn't Genesis 1:14-19 false? Absolutely not. Let's look at just a couple examples out of many to choose from to make the point.

Job 38:31-32

31 "Can you bind the chains of the Pleiades,

Or loose the cords of Orion?

32 "Can you lead forth a constellation in its season,

And guide the Bear with her satellites?

This is part of the response of the Lord answering Job. These verses imply that God binds and loosens the constellations of stars with chains and cords. It also implies that He leads them around the heavens in their seasons. Do we think that God literally pulls out chains and cords to bind the constellations and then physically leads them around the sky? If not, then do we claim that Job 38 is false? No, because we understand this is poetic language.

Exodus 15:8

8 "At the blast of Your nostrils the waters were piled up,

The flowing waters stood up like a heap;

The deeps were congealed in the heart of the sea.

This is a verse from the song that Moses and the sons of Israel sang to the Lord after the Israelites walked through the Red Sea on dry land and the waters returned and destroyed Pharaoh's army. Do we think that God has a body and nostrils? And that the air from God's nostrils caused the Red Sea to part? If not, do we claim that Exodus 15 is false? No, because we understand this is part of a song and songs in the Bible often use poetic verse.

Revelation 17:3

3 And he carried me away in the Spirit into a wilderness; and I saw a woman sitting on a scarlet beast, full of blasphemous names, having seven heads and ten horns.

Do we think this is literal? Do we think a woman is somewhere in the wilderness sitting on a red beast with multiple head and horns? When the angel interprets the meaning for the apostle John in verses 7-13, do we then think that verse 3 is false? No, because we understand that Revelation is apocalyptic literature and is not to be understood literally. It is to be interpreted symbolically. And I will give other examples as we work our way through the book.

Hopefully we can begin to see that to get the correct meaning from a text in the Bible, we need to know what type of literature it is and how it is meant to be understood. And that's why it was vital to show that Genesis 1 was not written by Moses as a historical narrative. God really did create the universe and all things in it by a free act of His will. God really did say "let there be", and it was so. By His eternal decrees and divine power to bring to pass whatsoever He planned, including a habitable planet Earth. He really did order the planet Earth so that it was formed and filled. He really did create man and beast to live on this planet. And He really intends for us to understand that this creation is a material creation made up of matter and energy, not gods, and that it is idolatry to think otherwise. But God teaches us these things through the inspired writing of Moses using a unique genre of literature employing an analogy of God's workweek and Sabbath rest.

I don't believe that Genesis 1 was meant to teach unassailable science, nor to be a literal historical account of exactly how the Earth and its life came to be. If you are like I was, it is sometimes hard to grasp how God could reveal truth to us through an account that is not literal history. And then square that with the way Genesis chapter 1 is referred to in other places in the Bible. But once we see that these other inspired authors reiterate that God is eternal, that all things exist because of His will, whether in the spiritual or material world, that the material creation is not to be deified nor worshipped, that He ordered the universe and world for His glory and our joy and flourishing, that we are made in His image, that He gives all living beings life and breath, that we are to have dominion, and that we are to work for six days and rest and hallow the Sabbath, then it is easier to be comfortable with other biblical authors employing the language of Genesis chapter 1.

Others have gone the other way and tried to reconcile the text with science by maintaining a historical narrative genre for chapter 1 and then tried to force it to say things foreign to the text. But is this wise?

In the Middle Ages people tried to force Genesis 1 to conform to the astronomy of Ptolemy, after that to the astronomy of Copernicus, and they are still attempting to do it today with big bang cosmology, modern astronomy, geology, atmospheric science, and biology. Scientific theories can and do change when new observations contradict the models. We can keep adding ad hoc additions to our scientific models to account for the data, but eventually the model must (or should) change. I believe we are justified in not interpreting chapter 1 as historical narrative and then we can let science go its own way to try to understand God's creation.

Let us allow the Bible to speak for itself and realize that the God of the universe always speaks in a way that can be understood to lead us to faith in Christ and to keep us from worshipping the creation instead of the Creator who is blessed forever. I hope that we will see that the transition to Genesis 2 and the all-important topic of Adam and Eve allows us to see how this unique literary type of Genesis 1 back into normal historical narrative gives us every reason to believe Adam and Eve were meant to be understood as real, historical people created by God.

Chapter 6

Genesis 2:4-25

Many people recognize that the chapter break should probably start in verse 4 of Genesis 2, as verses 1-3 naturally conclude the creation account. Verse 4 is the first of ten toledot sections in Genesis. They are usually translated "this is the account of" or "these are the generations of". But they all function to start a new section of Genesis as the introductory clause of that section. And in general, the Hebrew toledot means offspring, descendants, or that which is brought forth from.

Genesis 2:4

4 This is the account of the heavens and the earth when they were created, in the day that the Lord God made earth and heaven.

We here begin the account of what the heavens and the earth brought forth. As opposed to some interpretations, this is best not to be viewed as a conclusion to the creation account, like in verses 1-3, but the introduction of a new section. Day here is not used in the definite sense of a 24-hour day, but instead more along the lines of indeterminate time.

Here we go from the transcendent Creator Elohim to the conjunction Yahweh Elohim. Here we have the first mention of the covenant keeping God of Israel, Yahweh. And this most sacred of divine names is tied back to the name of God in chapter 1. Contrary to critical scholars who see this change in divine names as evidence for different authors and sources who only know and use one divine name, Moses is telling us that the same God who is the all-powerful, eternal, self-existent Creator is also the personal, immanent, covenant-keeping God of the Hebrews.

Genesis 2:5-6

5 Now no shrub of the field was yet in the earth, and no plant of the field had yet sprouted, for the Lord God had not sent rain upon the earth, and there

was no man to cultivate the ground. 6 But a mist used to rise from the earth and water the whole surface of the ground.

Here we are transported to an environment that is not like the high, lofty, magisterial language of Genesis 1, but to a more familiar setting where the rains have not yet fallen. Where the wild and cultivated vegetation has not yet germinated and grown in the land, which is a better rendering than earth here, so we don't picture the whole planet in our minds. The land that can be cultivated and planted was unproductive because the Lord God had not yet brought the rains, and there was no man to work it. A mist (or flow) used to rise from the earth to water the ground. This curious verse will become important when we look at the evidence for the garden.

Some argue that these first three verses show us that the days in chapter 1 are not to be taken as literal 24-hour days because if this is a zooming in on day six and the creation of mankind, then the length of climate cycle described in these verses (not to mention the time to create man, woman, the garden, and all their activities) would be inconsistent with ordinary days. Others do think that this is an expansion of the sixth day, but that we can still assume the 24-hour days in chapter 1. We have already concluded that the evidence points to six consecutive 24-hour days in the first chapter, but that the type of literature is not to be interpreted literally. While I agree that Adam and Eve are in view on day six in Genesis 1 on account of Adam being associated with the rest of humanity and the image of God from that chapter, I don't feel the need to make this an explicit elaboration of day six because of the change of genre. Neither do I think it is warranted to try to synthesize the chronology, because Genesis 1 was not meant to give us historical chronology. It tells us God created all things, formed the world to be habitable, made all the creatures, and then made man in His image and gave him certain roles and blessings. While Genesis 2 changes genre and gives us the historical narrative of the events that transpired with Adam and Eve made in His image.

Genesis 2:7

7 Then the Lord God formed man of dust from the ground, and breathed into his nostrils the breath of life; and man became a living being.

God forms man from the dust of the same ground that has no one to cultivate it and make it productive. And God animates him by breathing the breath of life into him. There is a word play here between the man (adam) and the ground (adama). The man is from the earth, earthy. (1 Corinthians 15:47-58)

Some use this verse to support the idea of God giving us souls. It depends on what we mean by soul. If we mean the immortal, immaterial part of man, then I think it is hard to read that out of this verse alone. The idea that man is both material and immaterial is clear in the Bible, but even more so in the New Testament. But if we mean the idea that God gives the spirit of life and takes it away, then I think yes. It is common in the Old Testament to equate the soul (nephesh) with the animating life of a creature, not only for man but other creatures as well. But even the Old Testament equates the soul with the person. It is not only an animating principle, but the very person themself, including the inner life. And I will argue later that it is the same as the soul that is articulated clearly in the New Testament.

I will also argue further when we look at the image of God, that the Bible's view of man is more holistic. Man is both body and soul, but the two are so interconnected and interdependent that it is more accurate to say that man *is* body and soul, versus saying he *has* a body and a soul. One without the other is an unnatural state for man. But it is important to see that God is personally animating this man with His breath. This will be important for the image of God.

But I think the main point being communicated is lost on most of us modern readers (including myself) until I came across a lecture by Sandra Richter. In a lecture she gave at the Carl F.H. Henry Center for Theological Understanding at Trinity Evangelical Divinity School, Dr. Richter contrasts religious rituals from ancient Mesopotamia that were used for animating and preparing the statue of a god with the garden of Eden and the theology of Isaiah that denounced this pagan idolatry. These Mesopotamian rituals have been elucidated by ancient near eastern scholars such as Michael B. Dick, Christopher Walker, and others. (Dick, 1999)

Dr. Richter explains that the Mesopotamians had religious rituals (that appear to be attested from at least 2150 BC to 586 BC) that functioned as a way to make a god from an image or idol. Among the various rituals, they would practice divination to ask the god if they can make the image (statue), if approval is given by the deity, craftsmen and artisans would make the image (statue), and the image would then be moved to a sacred garden next to the temple facing a river and the sunrise to the east (Mesopotamian temples were usually large complexes with green spaces or gardens outside of the temple situated adjacent to a river). The priests would leave it there overnight and come back the next morning to declare that the god had birthed an image through varied rituals, including ones meant to symbolize the cleaning of a newborn after birth. They would wash the statue, and then install it in the temple to be cared for daily. Therefore, one of the main theological points being made in this verse and the next is that God makes an image of Himself and He then animates him, versus the pagans making images of the gods who are then animated by the gods and incarnated to be placed in their temples. God takes this image (man) and places him in the sacred garden, which is also a temple. (Richter, 2018)

Ancient Egypt also had similar rituals for animating and enlivening the statues of the gods. This would take place by their opening of the mouth rituals, which was performed in a similar manner after the mummification of a dead body. The Egyptians would make statues that were meant to represent the exact form and likeness of the god or goddess. They would use the most precious materials to construct the statue and then "animate" the deity by this opening of the mouth ritual. At which point it would then be a place of manifestation for the god, housed in the sanctuary of the temple, and a point of interaction with the priests and human world. Assuming the Egyptians continued to care for it, honor it, and offer sacrifices and prayers. Otherwise, the god or goddess had no obligation to continue to dwell in the statue. (Walls, 2005)

We can then see that this historical event was also a polemic against the Egyptians and Mesopotamians.

Genesis 2:8-9

8 The Lord God planted a garden toward the east, in Eden; and there He placed the man whom He had formed. 9 Out of the ground the Lord God caused to grow every tree that is pleasing to the sight and good for food; the tree of life also in the midst of the garden, and the tree of the knowledge of good and evil.

God plants a garden in the east. East from where? I purpose that the Israelites hearing this story would envision a land east of Canaan, the promised land. And this is the geographical setting of the first 12 chapters of Genesis, namely in Mesopotamia. Eden is generally thought to mean delight or pleasure. Some linguists claim that the word Eden was found in the Sumerian language and was probably borrowed from the earlier Ubaid people. Which is interesting considering the antiquity of the garden. One thing to take note of is that God plants a garden in a land to the east which He calls Eden. In other words, in the large lands to the east there is a smaller territory of land called Eden, and within Eden God creates a garden. He places the man He had created into this garden delight. The good and benevolent Lord God causes all kinds of good and pleasing trees to grow in the garden to be food for the man. We also find two other curious trees. The tree of life and tree of the knowledge of good and evil.

Genesis 2:10-14

10 Now a river flowed out of Eden to water the garden; and from there it divided and became four rivers. 11 The name of the first is Pishon; it flows around the whole land of Havilah, where there is gold. 12 The gold of that land is good; the bdellium and the onyx stone are there. 13 The name of the second river is Gihon; it flows around the whole land of Cush. 14 The name of the third river is Tigris; it flows east of Assyria. And the fourth river is the Euphrates.

Here we begin to really see the differences of the types of literature we have between chapters 1 and 2. We have the garden of Eden situated among two well-known rivers and two lesser-known rivers. This squarely puts us in the modern-day nation of Iraq and the ancient area of Mesopotamia. What Genesis calls Shinar. Some will claim that we know of no large river that becomes four rivers as it flows downstream. But the verses don't give us perspective on whether we are looking up or down river. The text does say that the rivers flowed out of Eden. And we must keep in mind that the garden is a smaller place within the larger territory of Eden. (Kitchen, 2003) And rivers in general go from smaller ones that converge together to make a larger river. Later when we look at evidence for the garden of Eden, we will see this makes perfect sense.

Genesis 2:15-17

15 Then the Lord God took the man and put him into the garden of Eden to cultivate it and keep it. 16 The Lord God commanded the man, saying, "From any tree of the garden you may eat freely; 17 but from the tree of the knowledge of good and evil you shall not eat, for in the day that you eat from it you will surely die."

God takes the man and places him in the luscious garden He prepared for him. He is given duties to cultivate and keep it. Many biblical scholars, most notably G. K. Beale, argue convincingly that the garden represents a temple and Adam as a priest in that temple. Dr. Beale argues that the two words translated here "cultivate and keep" are always translated "serve and guard" later in the Old Testament when they occur together in the text. And most often the references are to priests serving in the temple and guarding it from the profane and unholy. (Beale, 2018) Priests in the tabernacle and temple were required to maintain and guard the temple. To keep it holy and undefiled. The word garden in Hebrew (gan) has the meaning of a separate, distinct space that in enclosed and protected. (Waltke, 2001) Many also point out that the decorations in the later tabernacle and temple pointed back to the garden, including decorations of palm trees and open flowers, that the cherubim guarded the holy of holies on the veil separating the holy place from the most holy place, and were covering the mercy seat on the ark of the covenant.

God warns the man with certain death upon eating the tree of the knowledge of good and evil, amidst the bounty of pleasure and food the good Lord God had provided.

Genesis 2:18-20

18 Then the Lord God said, "It is not good for the man to be alone; I will make him a helper suitable for him." 19 Out of the ground the Lord God formed every beast of the field and every bird of the sky, and brought them to the man to see what he would call them; and whatever the man called a living creature, that was its name. 20 The man gave names to all the cattle, and to the birds of the sky, and to every beast of the field, but for Adam there was not found a helper suitable for him.

This is the first instance of something not being "good" since the start of the Bible. We had the continual refrain of God evaluating His work and pronouncing it good. But here we find something that is not good. But it is important to note that it does not refer to moral evil or sin. The man is incomplete, and he has no helper suitable for him. The function of man within creation is lacking.

God forms animals out of the ground, brings them to the man, and he names them. Part of naming in the ancient near east was a sign of authority and dominion. We should not look here to find a treatise on taxonomy or try to imagine how much time it would take to name all the cattle, birds, and wild land animals. That is not what the narrative is communicating. The man has dominion over the other creatures, but none are like him.

Neither should we try to harmonize these verses with chapter 1 and give in to the skeptics that say the order of creation here is all wrong compared to the chronology of chapter 1. We have already concluded that harmonizing the two accounts is unwarranted and unnecessary. But even those who think they should be harmonized are still on solid ground because the perspective in these verses is the garden of Eden, not the entire world.

Genesis 2:21-25

21 So the Lord God caused a deep sleep to fall upon the man, and he slept; then He took one of his ribs and closed up the flesh at that place. 22 The Lord God fashioned into a woman the rib which He had taken from the man, and brought her to the man. 23 The man said,

"This is now bone of my bones,

And flesh of my flesh;

She shall be called Woman,

Because she was taken out of Man."

24 For this reason a man shall leave his father and his mother, and be joined to his wife; and they shall become one flesh. 25 And the man and his wife were both naked and were not ashamed.

The Lord God puts the man into a deep sleep, takes a rib from his side, closes the wound, and builds a woman from the rib (or side) of the man. God brings the woman to the man, and he breaks out in a poem calling her woman because she was taken from man and is bone of his bones and flesh of his flesh. We are told that this is the reason a man should leave his parents and be joined to his wife, and they shall be considered one flesh. And we see that they had no shame in being naked. There is no inkling of sin, shame, guilt, and hardship here. The two are in harmony and living in a paradise supplied by their good and kind Creator and Lord.

As we noted before, many biblical scholars think that Genesis 2 is a zooming in or expansion of day six from the creation account. There are different reasons they make this argument, but a few reasons are that they don't want to agree with the skeptics that there are two different creation accounts, Genesis 2 applies to the creation of man and woman, and Jesus Himself seems to tie the two together.

Mark 10:6-8 (see also Matthew 19:4-5)

6 But from the beginning of creation, God made them male and female. 7 For this reason a man shall leave his father and mother, 8 and the two shall become one flesh; so they are no longer two, but one flesh.

Here Jesus quotes either Genesis 1:27 or Genesis 5:2 and ties it in with Genesis 2:24. I have already argued that our interpretation of Genesis 1 does not constrain us to see chapter 2 as an expansion of day six in chapter 1. So, what do we do with the words of our Lord here? We have seen that Genesis 1 teaches truth, even if it is not historical narrative. However Jesus may have viewed the literary type of Genesis 1 (He never tells us), He upholds the authority of the word of God (which He knew the Pharisees whom He was addressing believed as well) that teaches God created man and woman in His image and that the covenant of marriage is based in the historical narrative of Adam and Eve. This was to silence their questions about the legality of divorcing their wives, for they appealed to Deuteronomy 24:1-4 where Moses permitted a husband to write his wife a certificate of divorce if he found some indecency in her. The different rabbinical schools of Jesus's day varied in their teaching on divorce. From it being allowable in any situation (for any reason, Matthew 19:3), to its being permissible only for adultery. Jesus says Moses allowed it in the old covenant because of their hardness of heart, but then appeals to prior scripture to show that it was not the original intent of God for men and women.

That is a long-winded way of saying that we can't use these verses from our Lord as proof texts to show that Genesis 2 is an expansion of day six from chapter 1, or that He viewed Genesis 1 as historical narrative. But it does mean that Adam and Eve are in view in chapter 1, and they are imperative for understanding the image of God, the federal headship of humanity, the institution of marriage, and the blessings and imperatives given to them. But in my view, it does not follow that we must take chapter 2 as an explanation of what occurred on day six in chapter 1. They are different genres and should be interpreted differently.

Literary type and audience

Some claim that the literature is stylized and figurative, but more agree that the Hebrew grammar would lead any competent Hebrew listener into viewing it as historical narrative. We then are not going to try to interpret this chapter, or the subsequent ones, in any different way than other historical narratives in the Old Testament.

The audience of the Israelites led out of Egypt, especially the educated ones, would recognize that the account was of a historical nature, and that the toledot in verse 4 was an introductory clause to the following account. Those familiar with the lands east of the Mediterranean would have caught the way that the crops needed to wait for the rains from God, versus growing crops on the predictable flood plains of the Nile in Egypt. What they understood about a mist (or flow) coming up to water the ground is less sure. They would have believed, just like the modern reader, that God creating the man from the dust of the ground was a miracle. They probably would have been familiar with myths from Egypt and Mesopotamia about the creation of mankind but had never encountered a story of just a single person made by the gods. They were almost certainly familiar with the opening of the mouth ceremonies in Egypt used to enliven the cult statues of the gods/goddesses and in preparing the dead by giving the mummified bodies access to their senses again so they could navigate to the afterlife. But the more educated, and Moses himself, probably also knew about similar practices from Mesopotamia, which included the rituals for animating the statue of the gods before being placed in their temples. This narrative in contrast is the account of God animating man before placing him in his temple garden. (Richter, 2018)

They could see the love that the Lord God had for this man and how He graciously and generously put him in a paradise with all he needed to flourish. They would immediately have recognized the geographical location of the garden in Shinar (Mesopotamia) and known that it was a real place. They would have recognized the priestly duties given to the man in his temple garden and knew that this garden was holy and set apart from the surrounding land. They would have been aware that their God, Yahweh Elohim, was a God making covenant with the man and his wife. And that this covenant had conditions, with blessings for obedience and sanctions and penalties for disobedience. They would have seen the dominion that this man exercised in naming the animals that were brought to him. And that the reason for marriage and sexual fidelity was rooted in God's benevolence for man's wellbeing, pleasure, and to fulfill the commission given to him as a wise and righteous vicegerent. They could see that part of the image of God consisted in relationships and social organization. And probably most of all, they would wonder why things are not like this anymore. Why the couple who were naked and unashamed and stood in close communion with their God was not their own experience? In the following chapter we will take a brief excursus from the text to look at the evidence for the garden of Eden.

Chapter 7

The Garden of Eden

We may have at some point in our lives seen paintings by Renaissance artists or others depicting the Garden of Eden. It is often a scene of wild animals like lions, cheetahs, deer, and cattle lounging around eating grass and generally enjoying each other's company. Adam and Eve are naked somewhere in the picture and the whole scene is a place of idyllic beauty and peace. This is often the place people's minds envision when someone brings up the Garden of Eden or they read the second chapter of Genesis. But the suggestion that it is an actual historical place that Adam and Eve dwelt in is usually met with eye-rolling or a condescending smile. Let's look at the text again describing the four rivers.

Genesis 2:10-14

10 Now a river flowed out of Eden to water the garden; and from there it divided and became four rivers. 11 The name of the first is Pishon; it flows around the whole land of Havilah, where there is gold. 12 The gold of that land is good; the bdellium and the onyx stone are there. 13 The name of the second river is Gihon; it flows around the whole land of Cush. 14 The name of the third river is Tigris; it flows east of Assyria. And the fourth river is the Euphrates.

We are given specific information about the four rivers. We are told that it is one river that branches off into four other rivers. We are not told whether the four rivers were upstream or downstream of the garden, but very few rivers branch off downstream. They almost universally come together upstream to form a larger main river. And the text says that the river flowed out of Eden (larger territory) to water the garden (smaller area). (Kitchen, 2003) Therefore, I think we can confidently say that the meaning of the text is that four smaller rivers come together upstream of the garden in the land of Eden and then flow through the garden.

The easiest river to locate, and for which the text gives no other information, is the river Euphrates. This river today starts in the mountains of eastern Turkey, runs through Syria, and then Iraq before entering the Persian Gulf on the border of Iraq and Iran. Where this river runs through north central Syria was the northern border of the land promised to Abraham. (Gen. 15:18, Ex. 23:31) And the land up to the Euphrates was finally subdued and administered under David and Solomon (2 Samuel 8:3-5, 1 Chronicles 18:3, 1 Kings 4:21-25).

The second river that is easily identified is the river Tigris. The Tigris also starts in the mountains of eastern Turkey, clips a little part of Syria, runs through Iraq, and in the present day meets up with the Euphrates to form the Shatt al-Arab river about 120 miles northwest of the head of the current Persian Gulf. We are told that "it flows east of Assyria". The Assyrians were a Semitic people who spoke an eastern Semitic language that modern scholars generally classify as Akkadian, grouping the dialects of Assyrian and Babylonian together. It is believed that the capital city of Ashur (or Assur) was founded around 1900 BC. According to the Bible, it is said to be named for a descendant or nation (son) of Shem (one of Noah's sons) named Asshur (Genesis 10:22), and Genesis 10:8-12 describes Nimrod (Sargon the Great) conquering the land and future locations of the Assyrian empire. Ashur (the city) sat on the west bank of the Tigris River about 50 miles south of Nineveh. Ashur was also a local (and then national) deity that the Assyrians worshipped. The Assyrians would later dominate the region from present day Iran, to Turkey, through northern Saudi Arabia, and all the way to Egypt beginning around the end of the 10th century BC. We can then be confident that the text is describing the Tigris River.

The first river referred to in our text is called the Pishon. This has long been an enigmatic river with various interpretations as to its location. Josephus thought it was the Ganges River in India. (Josephus, 1999) Others have suggested that it was the Nile, mountain streams, the Golden River in Iran, the Blue Nile in Africa, Mongolia, and even in the United States! Not exactly narrowing it down.

We are given some information from the text, so let's start there. We are told **"it flows around the whole land of Havilah, where there is gold. 12 The gold of that land is good; the bdellium and the onyx stone are there."** So where is Havilah? Let's start with the clearer indications from the Bible about the land of Havilah. Genesis 25:18 says, **"They settled from Havilah to Shur which is east of Egypt as one goes toward Assyria; he settled in defiance of all his relatives."** 1

Samuel 15:7 states, **"So Saul defeated the Amalekites, from Havilah as you go to Shur, which is east of Egypt."** I am confident that Shur was on the eastern edge of the kingdom of Egypt (which included the Sinai from the time of Ismael to king Saul), i.e. northwestern Saudia Arabia and Jordan. From Havilah to Shur (i.e. the Arabian Peninsula to Jordan and southeast of Canaan) is where the Ishmaelites, Midianites, Amalekites, and all the sons of the east (Judges 6:3, Judges 6:33, Judges 7:12, 1 Kings 4:30) are from in the Old Testament. It is the dwelling place of Joktan and his descendants (Genesis 10:25-29). And Havilah is still further east in the Arabian Peninsula than Shur, as you journey towards Assyria (a long way from Egypt). So, we can confidently conclude that Havilah would correspond to modern day Saudi Arabia, Kuwait, and western Iraq. We also know that this river flowed around the "whole" land of Havilah.

The late Juris Zarins, a Professor Emeritus of Anthropology at Missouri State University, had a Ph.D. in ancient near eastern languages and archaeology. In an article in the issue of *Smithsonian Magazine* for May 1987 written by the author Dora Jane Hamblin, Hamblin interviews and lays out Dr. Zarins' theory and discusses his findings from a multi-year interdisciplinary approach using archaeology, ancient near east literature, geology, and LANDSAT images from NASA. Although Dr. Zarins' and Dora Jane Hamblin's conclusions about the origins and purpose of the garden of Eden account differ widely from the ones I am presenting here, Hamblin's article and Dr. Zarins' work on locating the rivers is both helpful and pertinent to our discussion.

He proposed that the ancient river (long dried up) known by the modern Saudis as Wadi Riniah and as Wadi Batin by the Kuwaitis (a Wadi is a dry riverbed that only fills when it rains) is the actual Pishon river based on geological evidence and satellite images from space. They add to this evidence the facts that bdellium and gold were abundant in this area. Bdellium is still found there and gold had been mined there up until the 20th century. (Hamblin, 1987) The Wadi al Batin river last flowed consistently between 3500-2000 BC. (Sauer, 1996)

Chalcedony, of which onyx is one variety, was prized by the ancient Egyptians and Mesopotamians. It is also found in regions of the Arabian Peninsula. We can then conclude that the Pishon river is the Wadi al Batin River which starts in central Saudi Arabia, flows northeast through Saudi Arabia,

Kuwait, and into extreme southwestern Iraq. It flows around the "whole" land of Havilah and was a consistently flowing river prior to 3500 BC.

We can therefore ask, "How would Moses have known about the river Pishon, which may have run dry as early as 3500 BC, when these things are not mentioned in any of the texts of Mesopotamia found to date?" We'll return to those questions.

What does the Bible tell us about the last river we are to examine, the Gihon? **"13 The name of the second river is Gihon; it flows around the whole land of Cush."** This has been a difficult geography to reconcile, primarily because of the way the Bible has dealt with the Hebrew words (Kuwsh, Kuwshiy, Kuwshiyth), usually translated (Cush or Ethiopia, Cushite or Ethiopian, and Cushite). Although many Bible translations have interpreted Cush as Ethiopia, in most verses it actually refers to the ancient kingdom of Kush (later called Nubia) which is modern day northern Sudan and southern Egypt. And indeed, that seems to be the correct geography for Cush in many passages. (2 Kings 19:9, 2 Chronicles 14:9-14, Isaiah 18:1, 20:3-5, Jeremiah 13:23, Ezekiel 29:10, Daniel 11:43, etc.) The kingdom of Kush was a long-lived kingdom along the middle part of the Nile and had a long history of interaction with the kingdom of Egypt. The Pharaohs of Dynasty 25 of Egypt during the third intermediate period came from this region. So, Cush in most passages refers to the kingdom of Kush (later called Nubia), but it does not seem to apply to others, including this one in the garden account.

The Genesis account is clear that all four rivers had confluences to become one river. This particular land of Cush would have been known to the Israelite audience that Moses was writing for. One of Noah's sons Ham was the "father" of Cush, (Gen. 10:6) who was the "father" of Nimrod (Gen. 10:8). Genesis 10:10-12 tells us where Nimrod established his cities: **"10 The beginning of his kingdom was Babel and Erech and Accad and Calneh, in the land of Shinar. 11 From that land he went forth into Assyria, and built Nineveh and Rehoboth-Ir and Calah, 12 and Resen between Nineveh and Calah; that is the great city."** As we have seen, Cush is generally believed to refer to the land and peoples of the ancient kingdom of Kush (modern northern Sudan and southern Egypt). As the other "sons" of Ham refer to Egypt, Libya, and Canaan, that makes sense for the geographic location. But all these cities of Nimrod point to locations within Mesopotamia. The land of Shinar is a reference to Mesopotamia.

Citing the late renowned archaeologist, Assyriologist, and biblical scholar Ephraim Speiser, Hamblin is her article says that Speiser had suggested the connection between Gush or Kush (Cush) and the Kashshites (Kassites). (Hamblin, 1987) Also citing Speiser in his essay, John C. Munday Jr. points to a possible location east of Babylonia for the Kassite "Cush" peoples. He argues that the word Kussu for the Kassite people is found among various Nuzi tablets who lived in the lands east of Babylonia during the 2nd millennium BC. And that Dr. Speiser identified the Kassites with Cush. (Munday Jr., 2013) Most historians believe that the Kassites originated in the Zagros Mountains east of Babylon, in what is now modern-day Iran. Derek Kidner in his commentary on Genesis also identifies this Cush in Genesis 2:13 with the Kassite lands, not with Ethiopia. (Kidner, 1967)

I think another possible interpretation that at first seems plausible is that Cush could refer to the city of Kish, which is said to be where kingship descended after the flood in one the Sumerian king lists. Kish is thought to have a much earlier occupation though. It appears to have been occupied going back into the Ubaid period, before any city-states of Sumer. Many think that Nimrod should be identified with Sargon of Akkad, and I think they are right. In our passage from Genesis 10:10, most English Bibles translate the city Accad or Akkad. This was the capital city of Sargon the Great's empire, the city of Akkad or Agade. Another circumstantial clue for the Cush/Kish hypothesis is that one of the Sumerian King lists names Sargon as king of Kish, who became king after being a cupbearer to Ur-Zababa, a king of Kish. And in one of his royal inscriptions Sargon calls himself, among other titles, king of Kish. We could therefore offer an alternative hypothesis that the Cush of Genesis 2:13 and 10:8 is the city of Kish. Although I think the argument made by Ephraim Speiser to associate Cush with the Kassites is the correct one.

This makes sense as Sargon's empire was the first true empire in the world and encompassed all the cities and city-states mentioned in our text. Dr. Douglas Petrovich believes that the tower of Babel from Genesis 11 refers to the first city in Sumer, Eridu. He notes that several places in Mesopotamia were referred to as Babel, including Eridu, but that only Eridu meets all the criteria for the tower of

Babel, including the beginning levels of a giant ziggurat that was never completed, making Dr. Petrovich sure this is the location of the tower of Babel. (Petrovich, 2020) I then conclude that this Babel in Genesis 10:10 is the same Babel which is famous for the tower in Genesis chapter 11. Which seems correct because Eridu was considered as the first city by the Mesopotamians themselves.

If we put it all together, we have Cush referring to either the Kassites coming from western Iran (most probable) or referring to the ancient city of Kish. We also have the first city in Sumer, Eridu, identified as Babel (not Babylon), which is also the first city on our list from Genesis 10:10. And we have Nimrod referring to Sargon the Great as the creator of the world's first empire, starting with the ancient cities of Eridu and Uruk in Sumer, which Sargon first conquered by defeating Lugalzagesi, the king of Sumer. He then conquered the rest of central Mesopotamia (where Kish was and where he built his capital Akkad or Agade), and then proceeded to conquer and bring into submission the land to the north, which would be the lands of the later Assyrian empire. So, I think we can affirm that Genesis 10 is correct.

Using the arguments made by Ephraim Speiser that Cush refers to the Kassites, and by employing his multidisciplinary approach, Juris Zarins proposed the Karun River was the Gihon River, which originates in western Iran and flows southwest to the Persian Gulf. This river was dammed in the 1970s, but the satellite photos show its original natural riverbed. (Hamblin, 1987)

We now then have strong evidence from satellite images, geology, linguistics, the precious elements in the land, and the Bible's own testimony of the lands and people to have good reasons to conclude that the garden of Eden was a historical place at the head of the then lower and drier Persian Gulf. Before we start to talk about possible dating for this biblical event and its implications, let's look at more evidence from science and geology of another key detail in chapter 2.

Genesis 2:5-6

5 Now no shrub of the field was yet in the earth, and no plant of the field had yet sprouted, for the Lord God had not sent rain upon the earth, and there was no man to cultivate the ground. 6 But a mist used to rise from the earth and water the whole surface of the ground.

Dr. Jeffrey I. Rose is an archaeologist and anthropologist. In his paper for *Current Anthropology* in December 2010, Dr. Rose relates some interesting points for our discussion. While his paper is mainly targeted to questions about the migration, settlements, and activities of prehistoric mankind, his model requires arguments based on local environment and climatology, and he therefore goes into depth about the ancient environment of the region around the Persian Gulf. And this is germane for our discussions about the garden of Eden.

He presents evidence that the Persian Gulf was a relatively dry flood basin between 72,000 and 10,000 BC because sea levels were much lower than today, with evidence for the lowest sea levels during the last glacial maximum around 16,000 BC. Prior to approximately 10,000 BC, this flood basin, which is now under the waters of the Persian Gulf, was fed by four major river drainages and by subterranean aquifers. And he names the same four rivers that we identified above. The Euphrates, Tigris, Karun, and Wadi Batin as the main river sources of water. The confluence of these 4 rivers became one major river that eroded the main channel in this ancient flood basin. This ancient river channel can still be distinguished in the Gulf today. Further downstream would have been fed by additional runoff from the Zagros mountains in Iran to the east and from rains falling in Arabia to the west. He notes that upwelling springs linked to two regional aquifers would have flowed up into the dry floodplain river basin and that they continue to deliver freshwater today from underneath the Gulf.

Rose says that from 7000-6500 BC geological evidence shows that the area was becoming a freshwater marsh, and the shoreline was advancing. By approximately 5000 BC the gulf was approaching its current level, and that sea levels appear to be stable from about 4000 BC to present.

Archaeological evidence shows sites began to spring up around the shores of the Gulf starting in 7000 BC and increase in frequency around 5500 BC. These sites display evidence of people fishing, cultivating date palms, keeping livestock, public architecture, and permanent structures. There are discoveries of domesticated cattle, sheep, and goats at sites by around 5500 BC, and evidence of irrigation along the northern shore by about 5000 BC. (Rose, 2010)

Other ancient near eastern scholars provide evidence that the Sumerian city of Ur was beachfront property on the Persian Gulf around 4000 BC, which means that the waters of the Gulf were approximately 150 miles farther inland compared to today. The explanation given for changes to the current landscape is that the silting up of the delta from river deposition has gradually filled in the head of the Persian Gulf and this is one the reasons (along with the Euphrates and Tigris changing their lower courses numerous times in the past) that many important southern city-states in Sumer lost relevance and the power centers shifted elsewhere.

We now have more independent evidence that the four rivers in Genesis 2 were real rivers that are described by the biblical text. And now we also add the evidence that the "mist" or "flow" (both good translations of the Hebrew) rising from the earth was referring to subterranean aquifers bringing water up to "**water the whole surface of the ground**" (Genesis 2:6). What we also have are constraints on the dating of the garden and when Adam and Eve may have dwelt there. Although Dr. Rose is making the case in his model for humans existing around 100,000 years ago or earlier in this dry basin, we are given more corroboration for our thesis that the Garden of Eden was in fact a historical place. Which I now propose was in the then-dry Persian Gulf downstream of the four rivers.

While the consensus among scientists seems to be that a shift from hunting and gathering to agriculture probably began to take place somewhere between 12,500-10,000 BC in different parts of the Fertile Crescent (of which Mesopotamia is part), Mesopotamia is often referred to as the cradle of civilization because this is where a lot of agriculture, irrigation, domestication of animals, the building of cities, commerce, religion, writing, and a dramatic population explosion are first seen. All these data put bounds on when in history the garden could be situated in this dry flood plain below the confluence of the four rivers prior to the Persian Gulf filling with water. When then is the most plausible timeframe for God to create the garden and Adam to be placed in there? We will return to this question in later chapters.

The question we should keep in the back of our minds at this point is, "How could Moses know the environmental conditions surrounding the garden of Eden, given that this was at least a few thousand years before writing was invented and the Persian Gulf had been under water for at least 3500 years when Moses was writing?" And before someone rolls their eyes about Adam and Eve, "Why are there no documents in the rest of the ancient near east like this one?" I think we are then left to conclude that Moses had knowledge of the geography, environment, and hydrology that dates millennia before even the first city-states in Sumer. Although the confluence of the Euphrates and Tigris today form the Shatt-Al Arab river and farther downstream the Karun River joins them, the Wadi al Batin River was dry by about 3000 BC. And even when it was still reliably running as a river, the hydrology of the region would not have shown a river with four confluences after 5000 BC because of the level of the Persian Gulf and how much farther inland it was. If fact, even the Euphrates and Tigris did not have a confluence during the Ubaid or Sumer periods. In other words, the confluence of the four rivers and the subterranean aquifers watering the surface of the ground could not have been observed after about 7000 BC. How does Moses know this information? We will back up this date in later chapters with a more thorough examination of the dating of the garden of Eden and the flood.

After having examined the first two chapters of Genesis we have seen that Chapter 1 is a unique genre of literature and we argued it was not meant to be interpreted as historical narrative. We have not yet discussed the broader implications of such a view but need to leave that until we have looked at a few more topics and how it all ties together for an orthodox Christian theology. We have also explored how Genesis 2 (and the rest of the book) are clearly historical narrative and should be interpreted as such. We will circle back to Genesis chapter 3 and the Fall. And then look more closely at Adam and Eve. But as this is also a book on the Christian faith and science, we will now turn our attention to modern day science and the claims it makes.

Chapter 8

Science on the age of the Universe and Earth

We have already seen that we do not need to force a concordant interpretation on Genesis chapter 1 to make it fit with contemporary scientific theories, because the type of literature in Genesis 1 is not to be read as historical narrative. So, although the dating of the garden and the subsequent genealogies can give us estimates of the amount of time from Adam and Eve until today, the age of the universe and Earth are not given to us by God. It was an indeterminate time "In the beginning".

But what about science? What are the specific claims, data, and theories about how old our universe and Earth are? Anyone who is a Christian and believes in a young Earth knows how daunting education can be to your faith. I was and continued to be a young Earth creationist during my time in college pursuing my degree. I was put on the spot more than once by professors who tried to make me feel ignorant and insecure. There was a tension in my mind while attending classes on geology and biology. I learned the material and ended up getting good grades in the courses, but I always kept the theories and their implication at arm's length, so as not to let them get the best of my doubts. After I told one of my professors that I was a Christian, he told me in a not-so-thinly veiled put down something along the lines of, "But you seem so smart and are doing well in your classes." The obvious implication was that Christians and their backward thinking were not "so smart". Let us then patiently wade through the scientific material together to see how strong the case is from science.

The fields of astronomy and cosmology claim that the age of the universe is around 13.8 billion years old. Geology and physics claim that the age of the Earth is about 4.5 billion years old. First, a brief lesson from the history of science to see how we got to today's thinking.

Historians say that the Greeks were among the first documented thinkers to realize that the Earth was a sphere and not flat. While certain passages in the Bible can be used to infer that the Earth is a circle, it doesn't comment on the shape of the planet. Many textbooks credit Pythagoras (570-500 BC) as the first to indicate that the Earth was a sphere.

While not long after Pythagoras, Xanthus (c. 480 BC) suggested that the presence of fossils in the mountainous regions was evidence that they were once covered by the sea. (McCann, 2008)

But this is in no way to denigrate the accomplishment of earlier cultures. The Mesopotamians and Egyptians had mathematics and astronomy millennia before the Greeks. And the Greeks themselves had a fondness and fascination with Egypt. Some scholars argue that the Greeks borrowed and then further developed their philosophy from the Egyptians, but I don't how tenable that suggestion is. Other scholars don't find abstract and philosophical ideas in ancient Egyptian writing. But you can't build the pyramids of the Old Kingdom, Obelisks, and temples in Egypt, and cities, temples, and ziggurats in Mesopotamia without math. And you can't develop calendars to know the times and seasons of the year with studying the stars. Although the evidence for more complicated mathematical understanding is clearer in Mesopotamia than Egypt.

One of the greatest minds and influences on western culture for the next two millennia, Aristotle (384-322 BC), confirmed the findings of Pythagoras that the Earth was a sphere by observing things like the Earth casting a curved shadow on the moon during a lunar eclipse. It could be argued that Aristotle was one of the first to describe and use something akin to the scientific method. He was quite meticulous about observations and reasoning. Among his voluminous works in many fields, he proposed a theory that the Earth is constantly changing, and that the rate of change is so slow that we cannot observe these changes within a human lifespan.

Aristarchus (310-230 BC) thought that the Earth rotated on its axis and revolved around the sun, but his ideas were essentially lost for over 1500 years. He should get much more credit for these very early ideas than he does.

Claudius Ptolemy (100-170 AD) introduced the geocentric model (the Earth fixed in the center with the celestial bodies revolving around it) in 150 AD. And this was pretty much the prevailing astronomy for the next 1400 years, with ad hoc additions of epicycles added to account for observations until the whole model was extremely complicated and unwieldy.

In 1543 Nicolaus Copernicus (1473-1543) rediscovered the ideas first put forth by Aristarchus, and even though it appears he knew of Aristarchus's work, he did not credit him in his published work. He proposed the heliocentric model (the idea that the sun was at the center of the solar system with the planets revolving around it) of the solar system.

Galileo Galilei (1564-1642), considered by some to be the father of modern science and a brilliant man who developed many fundamental ideas in classical physics, confirmed and defended Copernicus' model. And for this drew heat from the Catholic Church. He was questioned under the Roman Inquisition and denounced. He later wrote the *Dialogue Concerning the Two Chief World Systems* in 1632 that discussed some of Galileo's scientific ideas and their rationale versus the objections of his day. Some say that the character in the book who holds the Ptolemaic view of astronomy was meant to be interpreted as a slam on the Pope and/or the other "simple minded". Galileo was tried and found guilty of heresy and spent the rest of his life under house arrest. It is interesting that many people claim that this is a paradigmatic example of the danger of religion and close mindedness regarding science. But no decent interpretation of scripture would conclude that the Earth is at the center of the solar system. And the practice of punishing scientists for their ideas has more to do with the problem of religion seeking power and being yoked to politics than any Christian view of the interaction between the Kingdom of God and the state.

Leonardo da Vinci (1452-1519) had a keen mind and was an astute observer of nature. He hypothesized that fossil shells found on dry land were from organisms living in the sea that had been deposited on the sea floor after they died. He rejected the popular view of his day that the biblical flood explained the shells of the marine organisms on the grounds that if the flood had strewn them about then we would not find them in the regular layers in which they are found. (McCann, 2008) We then see that the idea of Noah's flood being a global flood was being questioned by a few individuals in the western Christian world by the time of the Renaissance based on the geological evidence.

Nicholas Steno (1638-1686) developed the law of superposition which was the realization that all rock (sedimentary, metamorphic, or igneous) is originally in a liquid state and is laid down horizontally if left undisturbed, so that the older layers are on the bottom and the younger layers are on the top. He realized that this is why fossils could be found in solid rock. And why we find the oldest fossils in the oldest rock layers.

Try to visualize this with a few examples. In the bottom of the ocean or sea, layers will be laid down gradually by sediment and the remains of dead marine organisms floating to the bottom. Or where a river meets the sea, all the sediment carried downstream by the river gets deposited in the delta and on the seabed. These layers build up over time with the youngest being on top. If an animal dies on the riverbank or in the delta, once the river covers the animal or its bones in sediment, they will continue to get buried if the conditions are right and then this sediment hardens into rock either through pressure or cementation. When we find this animal's bones inside of solid rock, we can therefore deduce that the fossil is of the same age as the rock it is entombed in, the rock under it is older, and the rock above it is younger. Steno postulated that fossils were a picture of the history of life.

James Hutton (1726-1797), often referred to as the father of modern geology, is most remembered for his ideas of uniformitarianism. Which states that the physically observable processes that we see working on the Earth today have generally been the same over long periods of time. During Hutton's time the young Earth creation view was still widely held, and sedimentation and its effects were largely attributed to the global flood of Noah. In 1795 he published his *Theory of the Earth* where he laid out the cycle of land creation and erosion and hypothesized that the Earth must be very old. He also offered explanations using the immense pressure and heat from inside the Earth as the catalyst for chemical reactions that create different metamorphic rocks.

Geologists today recognize that catastrophic processes often play a role in interpreting the evidence from the planet, although the hesitancy to interpret data based on possible hypotheses about catastrophic processes is still criticized by many young Earth Christian scientists. I will note in passing though that paradigms and scholarly consensuses are often hard to break out of in considering alternative hypotheses.

William Smith (1769-1839) advanced the idea that the different strata of rocks contain similar fossils laid down in similar order and could be used to correlate the dates of those strata between locations.

Charles Lyell (1797-1875) further developed Hutton's ideas on uniformitarianism and ideas about volcanism building up the land gradually (think Hawaii). Lyell was a contemporary of Charles Darwin and both had an influence on each other's work. During the 1800s geologists worked out how to relatively date the strata of rock formations.

Alfred Wegener (1880-1930) in the early 20th century developed his theory of continental drift based on the observation that the puzzle-like configuration of the continents looked like they should fit together in a single land mass. He could not come up with a mechanism to describe this drift, but in 1929 Arthur Holmes further expanded on it by proposing that the mantle of the planet is undergoing thermal convection, and this could explain the mechanism for the drifting of continents. This would later become known as plate tectonics.

In 1862 Lord Kelvin (1824-1907) estimated that the age of the Earth was between 20 and 400 million years old. He reached this conclusion by trying to calculate how long it would take for the Earth to cool if it started out as a molten mass.

At this stage in history all that really could be hypothesized was that the processes of the rock cycle pointed to an Earth that was much, much older than the 6000-12000 years given by many Christian theologians. Also, rocks could be dated relative to each other and to the fossils they contained, but absolute ages could not be derived from this information. With the discovery of radioactivity in the late 19th and early 20th centuries this all changed. And this was a monumental leap in the way things were dated by geologists.

With the work of scientists such as Ernest Rutherford, Marie Curie and her husband Pierre Curie, the knowledge was advanced that certain elements decay into other elements with mathematical precision and that they put out a substantial amount of heat in the decaying process. This heat would then contribute to the heat of the entire planet. This opened the possibility that the Earth was much older than previously thought, because people had assumed that the planet was cooling uniformly since its creation.

Carbon-14 dating was invented by Willard Libby in the 1940s and is still used today to date dead organisms up to 50,000 years old. Living animals take in carbon-14 from other plants and animals while they are alive, and when they die the carbon-14 begins to decay into nitrogen. By measuring the amount of carbon-14 remaining, we can get an estimate as to when the organism died. It has its assumptions and limitations, but the process is well understood and calibrated. It is usually checked against other non-radiometric age indicators for validation.

There are now over 40 other types of radiometric dating methods. These methods have been tested repeatedly over the years to determine the exact decay rates for the different isotopes used. An isotope is an atom of the same element but with a different number of neutrons in the nucleus (ex. C^{12}, C^{13}, C^{14}, are all carbon atoms but they have a different number of neutrons). The radioactive atom that you start with is called the parent element and what it decays into is called the daughter element. Multiple dating techniques are often used together to acquire the age of a rock. Different radiometric dating techniques can also be used independently to give us stronger confidence in their reliability. But because rocks are constantly undergoing changes with plate tectonics and erosion, most of the very oldest rocks have cycled between igneous, metamorphic, and sedimentary. Radiometric dating of rocks on Earth have yielded some that are 3.7-3.8 billion years old. But with the assumption that the Earth shared a common origin with other rocks in the solar system, scientists looked to meteorites that were gravitationally bound in the solar system and that had fallen to Earth, as they would not have coalesced into planetary bodies which were molten at their beginning. Dating these meteorites have yielded dates around 4.5 billion years old.

Here are a couple additional methods Earth scientists use for determining the ages of trees and ice sheets. Scientists claim these show that the age of the Earth is much older than 6000 years.

1. Counting tree rings. In each growing season a tree makes a new ring within its trunk, and we can determine the type of growing season by the

thickness of the ring. The oldest living trees have more than 4000 rings. We can also measure the relative thickness and thinness of the rings, as well as make comparisons between alive and dead trees, and can come up with a chronology of up to 12,000 years using trees alone.

2. Drilling through ice sheets and counting the layers. In the polar regions, ice builds up year after year with the snowfall and these often have layers of dust and ash from volcanic eruptions trapped in them from around the planet, which can then be compared with other data to verify how long it took to put down those layers. With this methodology, scientists claim to have 123,000-year-old cores from Greenland and 740,000-year-old cores from Antarctica.

Following from various dating techniques and lines of evidence, geologists and physicists estimate the age of the Earth is 4.5 billion years old. How about the age of the universe? What were some of the advancements in astronomy and cosmology since the time of Copernicus that led to the current understanding of the universe?

In 1842 Christian Doppler (1803-1853) presented his explanation of the phenomenon that would later bear his name, the Doppler effect. This principle says that the frequency and length of a wave depends on the speed and direction of the source of the wave in relation to the person observing it. These hypotheses were later verified by other experiments using both sound and light waves. In the mid-1850s scientists were able to explain this principle and how it relates to stars moving away from us. The idea, known as redshift, is the idea that as visible light from a star is moving away from us observing it here on Earth, the light waves are stretched out to a longer frequency and therefore move closer to the red end of the visible light spectrum. So, these stars appear more red to our eyes. This method was used to measure how fast stars were receding away from the Earth.

Jacobus Kapteyn's (1851-1922) work published in 1922 suggested that the Milky Way was about 40,000 light years across, and our sun was near its center.

Edwin Hubble (1889-1953) estimated that stars in the Andromeda Nebula (which he proposed was a galaxy made of stars, not just a cloud of dust and gas located in our own Milky Way galaxy) were much farther away than the estimated size of our Milky Way galaxy. Which eventually led many astronomers to the conclusion that the universe consists of a multitude of galaxies. He took the previous work done on redshift and by 1929 had observed that all the galaxies in the sky were moving away from each other. And that there was a direct linear correlation between how far galaxies were from us and how fast they are expanding away (the more distant a galaxy is from us, the faster it appears to be moving away). Think of 10 beads on an elastic string all bunched together when the string is not being stretched. Then taking each end of the string, stretch it and assume that all the beads are equally spaced once the string is stretched. From the perspective of an observer on a bead at the end, the farthest bead appears to have now receded away by 9 times the distance compared to the closest bead. This discovery of the expanding universe was resisted by many scientists because the implication seemed clear: If it is expanding now, it must have had a beginning if we wind the clock in reverse. This smacked too much of religious and theological undertones for many in the scientific community and possibly suggested, "In the beginning God".

These astronomical observations came as a surprise to many scientists, including Albert Einstein (1879-1955), who had assumed a static (unmoving) universe based on the prevailing observational astronomy of his day when he published his general theory of relativity in 1915.

But prior to these observations by Hubble, Dutch astronomer and mathematician Willem de Sitter (1872-1934) had published papers proposing an expanding model of the universe in 1916 and 1917 based on general relativity. By 1917, after being in correspondence with de Sitter, Einstein published a paper describing a static universe, one that people have termed "Einstein's world". He added a constant with a hand-picked value to his field equations for general relativity, which has been called the cosmological constant. Although there seems to be a lot of debate out there for why he used it, it seems most likely that Einstein thought that it was required to achieve a consistent relativistic model of the universe after trying other options in his theoretical cosmology. He also didn't appear to appreciate the instability of his model. But it is pretty hard to criticize Einstein, one of the greatest minds in history.

In 1922, Russian physicist and mathematician Alexander Friedmann developed an expanding model of the universe. He had models for open, closed, and flat spatial geometries. And I think should be given credit as the father of big bang cosmology.

Belgian astronomer, cosmologist, and priest Georges Lemaître (1894-1966), who was unaware of the Russian Friedmann's work, developed a similar theoretical big bang model of the expanding universe in 1927 in the framework of general relativity.

With Hubble's observational evidence of the expansion of the universe, Einstein removed the constant from his equations that he now thought was unnecessary. It is said that he expressed regret for this "blunder".

In 1964 astronomers Arno Penzias and Robert Wilson accidentally discovered the cosmic microwave background (CMB) radiation with a radio telescope they were working on for Bell Labs. Their discovery showed that a uniform radiation pervaded all of space. This was seen as a major empirical validation for the theory of the Big Bang because in the 1940s George Gamow, Ralph Alpher, and Robert Herman hypothesized that in order to create the elements hydrogen, helium, and lithium in the early universe it would require an extremely hot environment and that the remnants of this radiation should fill the universe and be observable today. This later refined theory of nucleosynthesis made predictions about the relative amounts of these light elements and has been confirmed multiple times by observations. Finding the CMB offered support for their theory that the early universe was once a hot plasma with free electrons not bound to nuclei, which prevented photons from freely traveling through space. Once the universe cooled sufficiently, the electrons were bound to nuclei forming neutral atoms, and thereby letting the photons travel through space. This redshifted radiation is what we observe in the remnants of the CMB. Later work clarified that the earliest light emitted from the universe and seen today in the CMB happened just 380,000 years after the initial moments of the universe. This visible light has been stretched according to the Doppler effect into what we now observe, which is microwave radiation pervading all observable space. This radiation is extremely uniform with tiny variations in temperature that are hypothesized to correlate to density differences in space that later became structures including stars and galaxies. Though it brought up other problems regarding the initial conditions of the universe.

The COBE satellite in 1989 and later the WMAP satellite in 2001 confirmed and refined the data on the CMB. By calculating how the expansion rate has changed with time, the WMAP satellite team in 2012 calculated the age of the universe at 13.772 billion years and the Planck satellite team in 2013 calculated it at 13.82 billion.

In 1998, two teams of astronomers discovered that not only was the universe expanding, but the expansion was accelerating. This led to reconsidering Einstein's constant and using it in the equations of general relativity to represent the vacuum energy, or the constant energy density of empty space. If space itself is expanding, the total energy pushing everything apart would increase with time. Cosmologists still don't know if the vacuum energy is actually a constant, but they propose this energy as the prime candidate for dark energy, the elusive form of energy that cosmologists claim make up almost ¾ of the matter and energy in the observable universe.

All of this is very interesting, and we can envision "rewinding" the universe, if we assume that the rate of expansion is constant over time (really it isn't but has been theorized to be explained by the different forms of matter and energy dominating at different stages during the evolution of the universe). But there are still discrepancies in the data and interpretations of how fast the universe is currently expanding.

One common misconception needs to be clarified when talking about the universe according to cosmologists. Cosmologists tell us that the first distinction to make is between the *observable* universe and the universe as a whole. In principle, we can only observe a subset of the universe because of the speed of light and the expansion of the universe. So, there is more out there than we can see. Many cosmologists would argue that based on the data that we currently have about the geometry of the observable universe, it points to a flat global universe. And one of the implications of this is that the universe is probably infinite in size and has always been so. Either that or inflation has essentially produced a flat geometry and we can't tell the difference. In other words, all the galaxies, stars, dust, gas, matter, and energy from our observable universe used to be closer together when we rewind the clock to the beginning, but the parts of the universe we cannot observe were also closer together. So many of them would warn us not to think of the entire universe as starting out as a tiny dot and then expanding, but that it is probably more accurate to think of an infinite universe where every point is at Planck density and then expands and cools. I know it's mind bending, but the important point is they don't claim to know for sure if the universe is infinite in extent. And even if it is, it doesn't contradict God creating the universe out of nothing. We should maybe even expect an infinite God to create an infinite material creation to display His nature and character. But we don't know.

What then are the implications for God creating the universe out of nothing? The first thing to recognize is that the data from cosmology and astronomy are consistent with God creating everything out of nothing. We have what appears to be a beginning to the universe. Many cosmologists would say it doesn't *have* to be a beginning, but it sure could be. The second thing is that honest cosmologists don't claim to know what happened at the beginning, but many would be leery to call it a beginning. The reason they are leery is because many of them believe the physics community will come up with a self-consistent and self-contained theory that explains the whole shebang. Or that what we think of as the beginning may just be a moment in the evolution of a universe that stretches farther back in time, maybe even infinitely far.

As it stands now, the general theory of relativity doesn't give us any explanatory power when we reach density greater than Planck density. It gives us meaningless answers, so cosmologists know the equations are inadequate for the job of describing anything before Planck density. Which is just a tiny, tiny, tiny fraction of a second from the start. Therefore, physicists and cosmologists have been busy for many years now trying to come up with a theory of quantum gravity to try to explain the mechanisms and physics of the first moments. It has also been in vogue for some time now to propose models of a multiverse. The skeptical among the religious and philosophical crowd see the multiverse theories as a way to use the law of large numbers from statistics to avoid the implications of a beginning and fine tuning. If there are many, maybe infinitely many, universes out there, then it would be no surprise that we find ourselves in a finely tuned universe conducive to life and observers. The skeptical among the scientific crowd usually complain that the theories don't make any definite and unique predictions and that there are no observations, even in principle, to validate such a model. Therefore, many of them don't even see it as science. Many cosmologists who believe in the theory of inflation object and claim that if inflation happened, then the multiverse is a necessary consequence of most inflationary models.

But here is a very important point regarding theoretical physics. Just because a physicist comes up with a model that is mathematically consistent, it doesn't mean that it corresponds to reality. It takes more than mathematical consistency to describe nature. It takes predictions that are unique to sort it out from the crowd and observations to confirm them. Some physicists spend a great deal of time and energy looking for solutions to Einstein's field equations, but even if they find one, it doesn't mean it corresponds to reality.

Although I am no expert, after studying this for some time and trying to understand cosmological models of the multiverse, those with no boundary conditions, cyclical models, and such, I think two logical options still hold; either the universe is eternal, or it had a beginning. However esoteric we try to make the definition of time. Although the concept of time is fascinating in its own right. Einstein taught us that time and space are not absolute, but Christian theologians have been saying this for millennia. That God is not bound by time nor space but is transcendent above both. He is eternal and His eternal being is not defined by a succession of events and their relationships. He knows the end from the beginning and causes all contingent events to work for His ends. So, the physicists who believe in determinism may turn out to be right after all, but it finds its ultimate cause in the free will of God, not the necessity of energy and matter. But I think the Bible leaves us with mystery and tension in these matters.

But I also think that just talking about how the universe began misses much of the improbability of the naturalist explanation. When we only consider how the universe started, we are leaving out many other issues that get brushed aside to make it seem like the only thing to account for in a naturalist explanation of the universe is what happened at the beginning. But there are many other aspects of the early universe that I believe make the naturalist explanation vanishingly improbable. Such as the low entropy in the early universe, the initial conditions, the probability and likelihood of the inflaton field, why it started to inflate, why it exited and decayed at just the right scale factor to make our universe possible, why not faster, why not slower, why are the laws of physics the way they are, how the four fundamental forces broke out from one unifying force, the values of the free parameters and constants that make this universe possible, and the probabilities of all these things being just right to produce our kind of universe.

Some physicists have proposed things like a wavefunction for the entire universe to account for the initial conditions, but I don't know what evidence they have for such proposals or how seriously the physics community takes them. It seems to me that this requires a many worlds interpretation of quantum theory and that every time a measurement takes place or a quantum system decoheres that a new branch of reality exists. Which seems like a stretch to save determinism and explain the initial conditions of the universe. Other scientists have come up with theories that the universe came from nothing without invoking supernatural intervention by appealing to quantum physics. But after reading much in this area from distinguished physicists, I am convinced that most physicists don't buy it. They believe that it is easy to mislead the public on what quantum physics is capable of. For instance, virtual particles are called virtual for a reason and are simply a heuristic to explain Feynman diagrams. They produce no effects in real spacetime. That there is no "quantum foam" that magically borrows energy and causes effects in real spacetime, much less produces a universe from nothing. That quantum tunneling may be interesting, but as an explanation for a universe from nothing, not so much.

Although the acceptance of inflation is definitely mainstream, other cosmologists and physicists are skeptical of the theory of inflation and argue that there is no inflationary theory that can start with generic initial conditions and then predict our universe. They complain that there are many models of inflation out there that can be changed to produce any result. But it seems to me that the cosmologists who believe inflation is correct (which is most of them) must still assume that this scalar field existed, that the conditions were just right for it to expand, that it decayed and exited at just the right time and with just the right expansion rate, etc. And I don't know if they just vary the amount of energy in the field, its potential, or how it decays to make post-dictions (instead of pre-dictions) to match our universe, or if they derive these properties from first principles. I really don't understand it good enough to know. And the homogeneity problem, which is one of the problems that inflation is supposed to solve, may yet turn out to be an incorrect assumption based on previously known large cosmic structures and data possibly showing the same coming from the James Webb space telescope.

So, in my view all these theories are still left with the following problems: how the universe began, the fine tuning of the initial conditions, the low entropy, inflationary theories with a lot of assumptions, the fine-tuning of the laws of nature, constants, and free parameters, the symmetry breaking being just right, matter-antimatter asymmetry, and last but not least human beings with desires to understand it all.

As Christians then, we don't need to be ashamed to say that we believe the best explanation for the material creation of the universe and its purpose is God created it out of nothing, by His will, for His glory, and for our everlasting joy.

I don't suggest that scientists quit trying and say, "God did it." But that they keep looking, trying to understand the physics and mechanisms, and then finally give glory to the Creator. Many will claim that God is not a good hypothesis, but I disagree. He is the best One.

So, let us summarize so far. Using our interpretation of Genesis 1 and what we have seen from the Bible's own teaching so far (more to come), the Bible makes no claims on when the universe and the Earth were created. It has been an indeterminant amount of time "In the beginning". Scientists claim, and have some compelling arguments, that the universe is 13.8 billion years old, and the Earth is 4.5 billion years old. For those readers who may still be uneasy about my tentative conclusions so far and whose minds are jumping to different sections of the Bible, please try to be patient. We will attempt to pull it all together.

Many Christians are suspicious and skeptical of scientific claims. And I should note here that sometimes that skepticism is warranted. Some, but by no means all, scientists can be very bold in their assertions. Leaving out open questions, what they assume in their theories, and inserting philosophical claims into their scientific conclusions. And there are some very intelligent Christian scientists who have proposed their own models to account for the data. Many of them point out unwarranted assumptions like uniformitarianism in geology and that catastrophes can explain a lot of observations. Or that the assumption of the homogeneity of the universe on the largest scales is unwarranted and even doubtful considering the observational evidence.

And they are right to question the assumptions. But modern geologists have come a long way since Lyell, and they acknowledge that catastrophes do happen and can lead to relatively fast changes. And modern cosmologists sometimes admit that their assumptions could be wrong. But a model still must account for most (if not all) of the data. Some creationist models assume things like the decay rates of radioactive atoms, or the speed of light, have changed over time but without providing enough evidence to convince secular scientists. I think scientists should point out anomalies that do not fit their models, but sometimes human nature just does not want to be shown it is wrong. So, they have biases like the rest of us. But proper science should be a collaborative effort and correct errors as it progresses.

One thing that we often hear from scientists or the public defending the objectivity of science is that many scientists are by nature "disruptors". They like to find things that do not match the current theories so they can have the notoriety of disproving a current paradigm, advancing their own theories to better account for the observations, and collect their Nobel Prize. They say that if the age of the universe or Earth were not highly validated through the years by theories, observation, and data, someone would have come forward with evidence to disprove them. While I think there are certainly scientists out there that match that personality profile, I think people in general don't like to rock the boat and be mired in controversy. Usually when new data or theories are presented that challenge the current paradigms or consensus, we find establishment experts are usually quite adamant that something in the data or interpretation must be wrong. Sometimes it gets even uglier with colleagues belittling, harassing, or affecting the careers of others. But from a Christian perspective, this should not surprise us. People are fallen sinners, whether highly educated or not. Sometimes it takes many, many years to overturn group thinking. But that is not always a bad thing. Sometimes radically proposed changes point to easier explanations like bad data or faulty interpretations. But other times they can lead the way to better theories. This is not to say that the theories are either infallible or just guesses; we have argued that science is always provisional and that theories get to be accepted because they have stood the tests of time and account for the majority of data. But if some data are discovered that modifies or even overturns them, the Bible is still mute on the subject.

But the real tragedy is that many of us Christians either deny the evidence, saying that the sciences of geology, physics, astronomy, and cosmology have all been wrong, or we experience a blow to our faith and have less trust in the word of God because we thought it was giving us an absolute age of the universe and Earth of around 6000-12000 years old. I think this is a false dichotomy and a choice we don't have to make. Next, we turn to the even more contentious issue of biological evolution.

Chapter 9

Should we even consider evolutionary theory?

It seems like we can't go through more than a couple days without watching a television show or reading an article that makes reference to biological evolution (if you are boring, or more charitably curious, and watch and read the stuff I do). We are surrounded by the theory, and it has become ubiquitous in all the disciplines of life from biology, psychology, sociology, political science, and even ethics. Any show we are likely to watch on animals, animal behavior, or human nature will most assuredly discuss evolution. It is in public education, from late elementary or middle school all the way to graduate school. And it is the one theory from science that I had most resisted, along with the rest of the Christian church, ever since its broad acceptance after Charles Darwin published his *On the Origin of Species by Means of Natural Selection, or the Preservation of Favoured Races in the Struggle for Life* in 1859.

While the idea of evolution had been around for some time before his publication, Darwin proposed a mechanism by which evolution could occur. His contemporary Alfred Russel Wallace actually sent Darwin a manuscript of the theory with the same broad outline prior to Darwin publishing his book. But Darwin had been working on his book for many years prior to the correspondence with Wallace and had been editing it and adding evidence to support his claims. Many historians believe the reasons that Darwin waited so long to publish were that he wanted as convincing a case as he could make by compiling vast evidence and reasoning, and was probably quite nervous about its reception, considering the implications of the work and the prevailing worldviews of his time in England. But the cultural worldview of Victorian England had changed quite a bit over the preceding couple decades and had become more receptive. Initially the work was slow to gain acceptance and there were many challenges and objections to it. But as some of the objections were answered, it eventually gained widespread approval.

I know the feeling all too well. Somebody starts discussing evolution and we feel the emotions rise up in us that we need to be on guard or be on the defensive to bring our counterattack. This is how I felt for almost my whole life and still do to some extent because usually the people that I have encountered in my life promoting evolution have not been Christians. And many, but not all, were hostile to Christianity. I am not the most sociable person, so my sample size of Christians is relatively small and is probably not very accurate. I now know that there are a large group of believers who believe that evolution occurred and who trust in Jesus Christ alone for their salvation. God has revealed to them by the Spirit the beauty and majesty of Jesus Christ by lifting the veil from their eyes. They have confessed their sins, turned from them, and embraced by faith the finished work of Jesus on the cross. But how is it that these people have embraced a theory of biology that seems so opposed to the clear teaching in the word of God and people being made in the image of God? There are several ways in which people have tried to reconcile their beliefs. But some people cannot see this as possible; I was one of them. I completely understand that feeling and viewpoint, and apparently do many Americans as well.

In a 2014 Gallup Poll, 42% of Americans said they believed God created human beings in their present form 10,000 years ago or less, 31% said that humans evolved but that God guided the process, and 19% said God had no part in the process. (Newport, 2014)

Now with any poll that only allows us to choose from some preset number of choices, people's true views are going to get pigeonholed into something that they would probably like to elaborate on. There are a variety of nuanced views out there among Christians and non-Christians alike. But the overall view of American culture in 2014 was that God plays a part in human origins.

But this emotional response of us Christians to evolutionary theory is by no means surprising. We have our interpretation of the Bible, the interpretations of Christian leaders that we trust and respect, and the social pressure from our families, Christian friends, and our churches that give us negative feedback and consequences for learning about the theory. Many Christians give us counsel to just shut our minds to hearing anything about the theory, while others come up with alternative theories and then try to present these theories to an educated world that knows the principles of biology very well. I personally don't think that this kind of approach serves the reputation of God's church well. And I think it brings about a kind of reproach to the Gospel and brands Christianity as anti-scientific or anti-intellectual.

But don't get me wrong. It is important to build walls against false teaching and God's word warns us repeatedly about such things. There are schools of Christianity that see open-mindedness as a virtue, but in the end many of these same people end up with a religion that looks very little like biblical Christianity. It is tempting for believers to capitulate to the spirit of our age, as we think the secular world will not take seriously our message as they group the whole church as anti-intellectual. So, I can sympathize with these believers. But I don't think the options for Christians are mutually exclusive. And that is one of the purposes of this book.

I am also not saying that the message of the Gospel is made palatable to people because we become concordists and try to squeeze the Bible into a scientific textbook. Both Jesus and Paul were quite clear that God has been pleased to reveal His Gospel to those with childlike faith and often uses it to confound the wisdom of men. When Paul was giving his sermon in Athens on Mars Hill to the philosophers of the day, he knew that Jesus Christ and Him crucified was foolishness to the minds of the professed wisemen of his day. And he did not compromise the message of the Gospel to fit with the prevailing theories and philosophies of the likes of the Stoics, Plato, or Aristotle, so that the power of the cross would be to no avail.

But I think all of us would be well served to listen to Augustine on Christians trying to dismiss the knowledge of men based on their experience, knowledge, and reason. Augustine makes the points that non-Christians know something about astronomy, biology, and geology. They can perceive and reason about such areas of study. He warns Christians against talking to unbelievers about subjects they don't understand and then to proceed to talk nonsense about these things, especially when we claim that our information is coming from an infallible interpretation of God's word. He says that the unbeliever will laugh at us and come to believe that the inspired authors and God Himself held such views. And that it is also counterproductive in terms of evangelism, for if they think the Bible teaches such things, why would they listen to us about salvation, the resurrection of the dead, the kingdom of God, and heaven itself? Augustine pleads with us to let the wiser Christians who can properly interpret scripture be the ones to defend it and not give unbelievers wrong interpretations. And because the Bible is the word of God, unbelievers should not be led to believe that it is full of falsehoods. (Augustine & Taylor, 1982)

I believe there is wisdom in heeding Augustine's warning for us believers. If we claim to know what we are talking about in regards to the sciences in which we have not been trained, or at least don't have a solid grasp of the concepts and mechanisms on which these disciplines are founded, we should do a great disservice to the likes of the apostle Paul, who was a very learned man trained under Gamaliel (Acts 22:3), a speaker of Hebrew and Greek, and accused of going mad due to his great learning by Festus (Acts 26:24).

This is especially true now versus when Augustine lived. During the 4th and 5th centuries AD, a person may reasonably expect to learn much of what was known by a lifetime of study. This is simply not the case today. Even if we lived many lifetimes, we would still only know a fraction comprehensively, and could not keep up with the new information and knowledge being developed. I am personally very curious and like learning, but not everyone does. We shouldn't necessarily expect to become experts in every subject before we tell people what God's word says on an issue or to have an opinion. But I think we should not attempt to debate with experts (or even educated laypeople) in their fields without having some knowledge of the field ourselves. And especially if we are making claims for the word of God relating to that field of study that make it look foolish. I am looking in the mirror during this little mini lecture because I have been guilty of this many times and still can be. We should try to maintain a humble spirit in all things but especially when discussing science with someone who really understands their subject matter. By the end of this book, I hope we will see that we have nothing to fear from the study of science and its findings. But what about biology?

I personally never felt like I could defend my position when presented with evidence from evolutionary theory. I felt like the world was against me as a young Earth creationist who thought that all humans had descended from Adam and Eve about 6000 years ago and that all the animals that were on the ark were responsible for filling this planet with the diversity of life that we see. The main reasons for my obstinacy and fear were two-fold. The first and most important was that it seemed to me that the idea was clearly antithetical to the Bible's teaching. Evolution was saying that we had a common ancestor with apes, and the Bible was saying that we all descended from Adam and Eve. How could both be true? I thought that the logical consequence of these seemingly opposing views was rejecting one or the other. The other reason was that it seemed like a slippery slope. Even if I took a more figurative interpretation of the story of Adam and Eve and saw Noah's flood as a local event, I could not seem to reconcile how Jesus himself and the rest of the Bible talked about Adam and Eve, the Fall, original sin, and people being made in the image of God. I also saw the philosophical implications of holding a view of evolution with all the random mutations and no purpose or goal to life. And that it might reduce Christianity to nothing more than a positive psychology that was fine, "If it helped get you through the day." I have since come to believe that these are false choices. But I still had many, many questions.

I could not work out how all the animals on Earth came to be where they are if they had to start from Mesopotamia. How would they get to North and South America, Australia, and the Hawaiian Islands. And if they somehow did, why did they look so different from each other? I know some young Earth Christians believe that the landmass was once a giant continent like Pangea, and then the continents moved apart, and the mountains rose because of the flood. But this still did not make sense to me, as the animals would not leave the ark until after the ground was dry. I couldn't work out why many of the fossils were of creatures that were no longer living. I didn't know what to do with dinosaur fossils. Why did they never find dinosaurs fossils together in the same layers as human ones? And why did the different rock layers tell a relatively consistent story of the history of life? Why were all the oldest rocks and sediments dated by scientists prior to about 550 million years ago associated with simple organisms and not complex ones? Why did our Y-chromosome DNA not lend support to the idea of descending from a man in the Middle East 6000-12000 years ago or our mitochondrial DNA from a woman of this time? I could not find an intellectually satisfying way to answer these questions from a young Earth perspective.

But the real sticking point for me was that is what I assumed the Bible taught. This kind of conflict in a person's mind is often experienced when a young professing Christian goes off to college and is constantly challenged about the Bible and the rationality of their faith. And many end up throwing in the towel on God or they just refuse to look at where the data from science leads. Or like me, they will just leave the two competing beliefs to hopefully be resolved later. I am very sympathetic to this plight. I have been there, and I imagine many of you reading this have been too.

After I came to the conviction that Genesis 1 was never intended to be interpreted as historical narrative, I felt more at ease with seeing the universe and Earth as being very old, but I still could see no way of allowing for common descent with modification for mankind. All I saw at the time was another tool in the repertoire of the naturalist to invoke their godless philosophy. I could only think of people like Richard Dawkins arguing that prior to Darwin an atheist could have logically held the position that God is not a good explanation for biological complexity but did not have a theory to account for his position. And therefore, it would have probably been an unsatisfying explanation. But when Darwin came along, the atheist could now feel intellectually satisfied with his position. (Dawkins, 1986)

I got some solace from faithful Christians who often pointed out that although hostile secular people many times proclaim science as the answer, that many aspects of this created universe point to a Creator. That there actually is design at every level of creation, from the beginning and fine tuning of the universe, to the tilt of our planet and size of our moon, to the complexity of biological systems and their interactions with the ecosystems, to the psychology and spirituality of man, their morality, and their need for meaning. Why a creature like man, that was a part of the natural world, should feel so alien to that world and be incensed at its injustice and imperfection? That overall, the God of the Bible is still the far superior explanation to materialism. Not to mention the historical evidence of the Old and New Testaments, the ministry and witnesses of Jesus, Jesus rising from the dead, our personal experience with the Savior, His church, and the living word of God. It will help to keep this in mind as we go through this book and then tie together the science, the Bible, and theology.

So, there was the *apparent* conflict that I saw as irreconcilable. But what if a large portion of us Christians are misinterpreting what God is telling us about the history of mankind? What if the view that the created world was a place of moral and physical perfection was never taught in Genesis? What if we had been conditioned to a view of the early Earth as one of idyllic beauty and harmony without death and disease? One that we are reading into the text of the Bible instead of reading that meaning out of it. What if our view of T-Rex and lions eating only grass and plants around 6000 years ago is not from the Bible but our own interpretation of what we think it means for the consequences of the Fall and the creation being unwillingly subjected to futility? What if these ideas are not taught by the word of God and we are subjecting ourselves to the ridicule and mockery of the world unnecessarily? I think we may be doing just that. We will get into these issues much more in depth in chapters 12 and 13 when we look at the image of God, the Fall, and original sin. But keep in mind that we have already argued that Adam and Eve were historical people miraculously created by God and that the garden of Eden was a real place, just as the Bible teaches. We will not abandon these ideas and will come back to them in chapter 11 after discussing the science.

If evolutionary theory provides compelling evidence of the descent of the animal kingdom from earlier creatures, what kinds of evidence should we expect to find? Does evolution prove that everything is random chance, that God is not in control, and that the materialists can rejoice in dethroning the King of Kings? I hope we will see that the implications are not as dire as some would like us to believe. And when we see the natural world operating in all its amazing adaptability, we begin to see how the personification of nature or natural selection with its anthropomorphic will and intention has strained credulity.

Is there even enough evidence to consider evolution? Isn't it just a theory? We can surely agree that when people have kids (unless they are identical twins), that all their children look different from one another. Someone may say, "You have your mother's eyes or your father's nose." We all realize that we inherit traits from both of our parents, and these traits get combined in a way that makes each of us unique looking. This uniqueness is because of variation in our parents' DNA and the way that the germ cells use crossing over and recombination to make haploid sex cells, how specific sex cells come together to make a zygote, and then the gene expression of chromosome pairs in the zygote.

But I think most Christians would be fine acknowledging that. I ask that we try to be patient as we work our way through the evidence. And if you have been opposed to the idea of evolution, just remember that I was once there too. I think the best place to start talking about the evidence for the theory of biological evolution is also one of the most familiar and less threatening aspects of organisms evolving, artificial selection.

Artificial selection

Artificial selection is when people selectively grow plants and breed animals to get the ones that have the traits that they desire. Early people noticed when they started cultivating crops and raising animals that they could choose which pairs they wanted to cross, and they could develop the traits they desired through each successive generation. The ancient farmers in Mexico saw that when they planted their corn some of the cobs were bigger and some tasted better, so they simply took the kernels from the ones they liked and replanted for next season's crop. Year after year, these desirable traits would get more pronounced. Later they realized they could hybridize different varieties of crops to get a better combination and a more resilient plant. Cabbage, broccoli, cauliflower, brussel sprouts, collards, and kale are all descendants of a wild mustard known as Brassica oleracea. These plants look quite distinct from each other, and this was achieved by people selecting for certain traits they deemed desirable.

This same process of artificial selection was done with animals for livestock. If a farmer had certain cattle that produced more meat or milk, he would only breed the animals that already displayed these traits. Eventually he would get a herd of cattle that were more productive in the traits he desired, such as larger bodies for more meat or higher producing milk cows. Essentially all the food we eat today is the result of artificial selection. Huge-breasted chickens, chickens that lay many eggs, cows, sheep, and goats that produce loads of milk, leaner pork, more steaks and hamburgers from a cow, wheat and corn with larger heads and kernels, more oil from canola, and the list goes on. The same idea is applied for animals like sheep, where we may want a different type of wool or to adapt them to a certain pastureland or ecology.

The example most familiar to many of us is that of dogs. There are more than 400 different breeds of dogs and all of us know that the differences in the physical appearances of these canines are vast. From a Chihuahua to a Mastiff, Shih Tzu to a Saint Bernard, many of these breeds are the product of humans selecting certain traits over time. In fact, dogs have the most morphological (body type) variety of any species on Earth. Dogs now have anywhere from 2 to 20 genes for an enzyme called amylase which helps break down the starch in carbohydrates. Which means that dogs can eat far more grain-rich diets compared to a wolf who has only two copies of this gene.

But artificial selection does not only have to do with physical appearance. Many animals have been selected for traits relating to behavior. In 1959, Dmitri Belyaev and Lyudmila Trut began an experiment in Russia to see if wild foxes could be domesticated and to test the hypothesis that certain physical traits were linked to domestication. They took wild foxes and then in each generation only let the tamest foxes mate to produce the next generation. Within less than 10 years, some of the foxes were acting like dogs, wagging their tails and happy to see people. Nowadays, almost the whole population from this experiment are tame and display many of the physical traits of domestication.

Sometimes selecting behavioral traits is done for the purpose of increasing productivity, as in the case of egg laying hen chickens. Selection experiments have demonstrated that just selecting the hens that lay the most eggs is not necessarily the best way to get the greatest production out of the chicken coop. Sometimes productive hens are also aggressive and end up harassing, killing, and eating other chickens. Some experiments have selected the most social and gentle birds instead of the highest laying individuals. And this ends up being a good strategy for maximum egg production, because fewer birds are killed, and the hens are more relaxed, less stressed, and therefore more productive.

But variation within a species is usually accepted by conservative Christians and is usually called microevolution. Animals have variation within their kinds. Which may lead some at this point to be thinking, "That is all well and good, but they are still dogs, foxes, chickens, cattle, sheep, and corn. This is just microevolution; I don't see them turning into other kinds or whole new organs developing." And this is correct. Although someone just looking at different types of dogs may be tempted to classify them into different animals if they didn't know better. But the broader point is that humans can breed animals and plants and cause them to change in their traits over time.

Fossil record

One of the things people have noticed for ages were the remains of organisms in the ground. We all know about dinosaur digs and the vast number of dinosaurs that scientists have dug out of the ground and have been able to reconstruct. Dinosaurs scattered all over the American West, Canada, China, Argentina, and Mongolia. Surely, we recognize that these creatures existed at some point in the past and no longer do today. Among these are a veritable plethora of animals that once lived and are now extinct. We also have fossils from wooly mammoths, saber-toothed tigers, giant sloths, camels, and monkeys in what is now the United States. Apes that were 10 feet tall and over 1000 pounds, huge flightless birds with carnivorous jaws, 50-foot-long snakes, and a centipede-like creature over 8 feet long. And the list goes on for wonderful creatures dug up from around the world that would seem very bizarre to us today. Over 1000 distinct species of dinosaurs have been named and described. About 15 skeletons, with 2 of them being almost complete, have been discovered for T. Rex. Scientists have uncovered around 250,000 different species of animals as fossil remains throughout the world. The vast majority of which are species that are extinct, for one reason because fossils are generally defined as being at least 10,000 years old.

What we find are that the ones that are found in the youngest layers of rocks most closely resemble the species we see on Earth today. And we observe that the oldest fossils usually look very different than the plants and animals that we see around us now. Another general observation is that fossils from layers of sedimentation that are adjacent to one another are more closely related than the fossils from distant layers of sedimentation. Also noteworthy is the fact that the fossils of a given continent are generally more closely related to the living animals on that land. From this information we can infer that most species eventually go extinct, the ones in the younger rocks more closely resemble species today, and the oldest rocks contain many organisms that generally look less like the animals we see around us now.

Although there are exceptions to these rules, as in the morphology of single-celled organisms and some so-called living fossils that look like they haven't changed for tens to hundreds of millions of years (ex. Stromatolites, Tree Ferns, Moss, Gingko trees, Horseshoe Crabs, Alligators, Crocodiles, etc.). Biologists do not claim that species *must* evolve or that the morphology of a species must change. If the selection pressures on a species are low and species are not isolated from one another, there will not be any incentive for novel traits to be favored, and therefore traits may get fixed in a population.

Any honest evolutionary biologist will admit that the fossil record is incomplete. This observation of the large gaps in the fossil record without the predicted gradual line of "missing links" was used by opponents of Darwin's theory after he published the *Origin of Species*. And it is still being used today by those opposed to evolution. Darwin himself argued that evolution always occurs during long periods of time with small, gradual changes in the population of an organism. But I would argue that the overwhelming pattern we actually see in the fossil record are long periods of stasis (organisms not changing the way they look, also known as their phenotype) followed by relatively sudden (in geological time) changes in morphology.

But most biologists will contend that this is simply due to the amazing incompleteness of the fossil record. And if all species were fossilized when they died, we would see a gradual change of the species over time with each successful subsequent change in morphology conferring a beneficial adaptation making them more fit for their environment and more successful at reproducing. I see this defense largely based on a desire to defend the gradualism of evolution against other proposed mechanisms of speciation like symbiogenesis, hybridization, horizontal gene transfer, whole genome duplication, and polyploidization. These mechanisms may have significant roles in speciation events. But we will find biologists on both sides of this debate. And most of them are probably somewhere in the middle, seeing evidence for both slow, gradual change and periods of punctuated equilibrium (relatively quick).

One line of evidence that biologists use to illustrate why the fossil record is incomplete is that some species like the lampreys and hagfish have not left any discovered fossils between 540 million years ago to the present, we have fossils of them from before that time, and we can watch them swimming around today. Another example is the coelacanth fishes thought to be extinct for 70 million years until one was found in the Indian Ocean in 1937.

What does it take to get fossils in the first place? First, we usually need to have water around for the process of sedimentation. Most fossils are found in sedimentary rock, but some are found in other types of rocks or materials. Animals may also be fossilized due to being buried in lava flows, covered in sand, stuck in tar pits, or encased in amber from tree resins. But for the most part animals must either live in the water, or they must be near water when they die. Somewhere the sediments can cover them before they are removed from the area by predators, or their bodies are converted back into organic matter and soil. This is the reason that the largest portion of fossils that we have are of marine organisms. The soft parts of animals are only preserved under very rare conditions, so a lot of the fossils we have discovered are of hard body parts, like bones and skeletons, that get covered by sediment which later turns into sedimentary rock through compaction or cementation. In these body fossils, the bones and skeletons are replaced by minerals in the sediment and essentially turn into rock. But these are not the only types of fossils. Trace fossils are not the remains of the organism itself, but impressions or traces that it leaves behind. Such as the impression of a leaf in shale or footprints and animal trails in sandstone. These trace fossils, plus molds and casts, can also give us information about the organism and its environment.

Once this unlikely process preserves a land animal and the processes of geology get this sedimentary rock closer to the surface, we must rely on the effects of erosion through water and wind to erode the layers of sediment, mass wasting due to gravity, or chemical weathering so that someone can find the fossil. Most fossils are still buried in the earth out of sight or too deep to find them successfully. Because of these factors a large portion of the fossil record is comprised of organisms that lived in the water. And of those, mainly in shallow seas. A substantial percentage of this record is composed of microscopic organisms like bacteria and pollen and small animals with shells.

It is estimated that between 1 and 4 billion species (obviously a large degree of uncertainty) have lived on Earth since the planet formed. There is thought to be somewhere between 7 and 10 million species living on the planet today, of which an estimated 80% or more have not been described. So, the 250,000 species or so that they have discovered in the fossil record only represent a very small portion of the life history on this planet according to biologists.

To interpret the data when finding fossils in rock, you must be able to date the rock layers. It will do us no good to find fossils and try to classify them and put them into a coherent picture of life history if we can't determine when the sediments formed around the fossils to preserve them. The commonsense idea of different layers of sediment laid down one on top of each other through time was codified as the principle of superposition by Nicolaus Steno in the 17th century. This states that as sediments are laid down the older layers are at the bottom and the younger layers are at the top. He developed three other principles of geology, including that the layers are originally deposited horizontally and then can be tilted at an angle or folded by geologic forces, a principle described by mechanisms like uplift, earthquakes, volcanoes, and plate tectonics. He developed the idea of an intrusion of magma into sedimentary rocks where one could deduce that the magma intruding into the sedimentary rock had to be younger than the surrounded rock into which it cut. He also realized that the layers could be laid down in a marine environment until something blocked their deposition or until erosion began its process. These principles laid the foundation for people digging through rock to give them *relative* ages.

With this knowledge they could say that one layer was older than another based on the deposition of sediments. Then people noticed that types of fossils seemed to appear to be grouped into rocks of similar composition, and they would generally find the same types of fossils together. People took this and thought that if we could find rock layers that contained the same type of fossils then we could generally say that they were of the same age of rock. This may be true for a local area where it is easier to show that the sedimentation and rock types are the same, but it initially appears like poor logic based on comparing rock strata that are widely separated by distance. This was accomplished in some instances with what is known as an index fossil, one that only occurs in a small thickness and specific type of sedimentary rock. The presence of certain index fossils correlated well with specific rock types and men like William Smith and

Georges Cuvier used this information to produce the first geologic maps in Britain around the turn of the 19th century.

Geology progressed a long way from there, and they could confidently say that certain layers of rock along vast distances belonged to a certain geologic age. Using rough estimates (and several assumptions) for the laying down of sediments over time, geologists in the 1800s came up with estimates for the age of the planet anywhere from 3 million to 6 billion years. Not exactly a statistician's dream! But science needed another tool to avoid circular arguments using fossils and rocks. This came in the way of *absolute* dating.

As we have previously seen, from the work of scientists such as Ernest Rutherford, Marie and Pierre Curie in the late 19th and early 20th centuries, the knowledge was advanced that certain radioactive elements decay into other elements with mathematical precision. These elements can be found in rocks and minerals. Different dating methods are chosen based on what element is present to measure, and the conditions under which the rocks formed or the mineral crystallized. Some are useful for younger rocks and others for older ones. They can be checked against the logic of the relative rock strata and with other methods of radiometric dating to confirm the absolute date. These methods did not come without their growing pains, like any scientific tool, but have been honed and improved over the last 100 years to be very reliable today.

Those skeptical of these dating methods have a harder time explaining why the ages obtained from the various radiometric methods will yield similar results. These five methods using different radioactive isotopes have been used on meteorites to come up with similar ages: Rb-Sr, Sm-Nd, Pb-Pb, Re-Os, and Lu-Hf. It is true that sometimes these methods have used assumptions that look like circular reasoning, (i.e. they infer the decay rate based on an Earth that is 4.5 billion years old), but they have taken pure samples of the radioactive isotopes into the laboratory and let them decay into the daughter element for some fixed amount of time and then measured the amount of parent to the daughter. From these controlled experiments one can extrapolate the half-life for each radioactive element. There are limitations for which dating method one uses based on different factors. It depends on the type of rock or mineral we are trying to date, what types of conditions the rock was under when it solidified, and the physics and chemistry of different methods applied to different rocks.

Without further belaboring the science behind absolute dating using radioactive isotopes, since the 1940s we have had a way to give absolute ages to rocks. With this knowledge they could assemble the different rock layers according to geologic time and correlate the fossils that they found within those layers.

According to most biologists, the first 2 billion years of life consisted only of prokaryotes, which are single-celled organisms like bacteria and archaea that lack a membrane-bound nucleus and other membrane-bound organelles of the cell. Eukaryotes, which contain all these features in the cell, are thought to have evolved around 2 billion years ago and are believed to be a result of the process of symbiogenesis (or endosymbiosis), where two types of prokaryotic cells merged or engulfed one another. The mitochondrion organelle is believed to come from an evolved aerobic bacterium and the chloroplast organelle in plants is believed to originate from an evolved photosynthetic bacterium.

Most biologists seem to believe that multicellularity evolved independently many times. First for prokaryotes by around 3.5 billion years ago and then for eukaryotes by about 1.5 billion years ago or earlier. In eukaryotes, they think it occurred for animals, plants, red algae, green algae, brown algae, and fungi. Evidence for truly complex eukaryotic multicellular organisms show up in the fossil record around 1 billion years ago. The first jawless fishes and chordates in the Cambrian explosion 500-550 million years ago (mya), diversification of metazoan families 440-480 mya, first land plants 430 mya, amphibians and fishes with jaws 350-400 mya, reptiles 300-350 mya, reptiles diversify 250-290 mya, dinosaurs and mammals 200-250 mya, dinosaurs diversify and first birds 150-200 mya, primates and flowering plants 90-150 mya, and mammals dramatically diversify 2-60 mya.

So, does this mean that science has ended the discussion on interpreting the fossil record and there is nothing left to explain? Not really. There continues to abound many questions relating to the fossil record and the mechanisms by which evolution proceeded. Starting with the origin of life itself. This is a monumental mystery that science is aware of. Some have claimed that it is easier to envision and explain the complexity and diversity of life today from a prokaryotic bacterium than it is to posit the beginning of organic life from non-living substances. There is a general consensus that it would have to be the simplest of a self-sustaining replicating biomolecule capable of being acted upon by evolution. Because there is evidence for stromatolites dating back to around 3.5 billion years ago, the reasonable inference given evolutionary theory is that they descended from earlier life. Some have suggested that extremophile bacteria could have been capable of living in the hostile environment of Earth prior to this time and that has been a lot of the focus on finding the origin of life. The most widely held hypothesis is that of a self-replicating RNA molecule.

But critics point out that there are many problems with the RNA hypothesis. Among some of the problems are the following: Which came first, proteins or RNA? We need RNA to build proteins, but we need proteins to decode RNA. RNA is inherently unstable. RNA is too complex a molecule to come from non-living sources. Catalytic activity is limited and is only observed in long RNA sequences. And you must have a molecule that contains information and can reproduce itself. Others describe the main problem for the origin of life is where the information comes from to construct DNA or RNA. Because they contain information in a code that tells the machinery of the cell how to operate, you need the information (software) *prior* to the organism (hardware) for it to operate on. Another current hypothesis being investigated is the "metabolism first" one. I am not sure if this is a solvable problem for science or not, but scientists are actively at work on experiments and developing theories.

The second problem for the fossil record, which we touched on earlier, is the discontinuity of the fossils. With relatively few exceptions, the fossil record shows distinct species with little evidence in the form of transitional fossils, especially the kind of gradual slow change in morphology caused by genotypic changes predicted by Darwinian evolution. Although Darwin didn't know about genetics.

Ernst Mayr gives us a glimpse into this dilemma from Darwin to at least the early 1980s in his book *The Growth of Biological Thought* in which he describes that one of the most serious objections many had to Darwin's *Origins* was the incompleteness of the fossil record and why we don't find more transitional links indicated by the theory. Darwin argued that the lack of preservation of fossils is due to the less-than-ideal nature of rocks and fossilization and tried to vindicate this with many examples. Mayr goes on to explain that even up to the present day the fossil record is incomplete and discontinuous and can lend skepticism for gradual evolution by natural selection. Even leading some paleontologists in his day to hold to the belief that genetic changes were often large, fast, and led to new species within generations, versus Darwin's slow and gradual changes and speciation. (Mayr, 1982)

Another line of reasoning used by biologists on the sparsity of the fossil record is that the history of life from the fossil record appears to be one of extinction and diversification repeated several times. Furthermore, once the descendent species developed traits that made them well adapted to their environment and reproductively successful, the less suitable antecedents would quickly lose out and go extinct. Which would decrease the likelihood of them being fossilized and found. Many view speciation events happening more often in small, isolated populations that would also decrease the probability of finding these founding groups in the fossil record. Whatever the argument, there is still the reality of the record as we have it. But there are some cases that biologists point to for evidence of transitional forms.

One example is the horse family. Fossils from around 50 million years ago show a mammal about the size of dog living in the forests of North America. These transitional horse fossils dating in sediment from 35, 15, 8, and 1 mya show the horse growing in size, the forefeet gradually going from spread-out toes to the hoof of the anatomically modern horse, and the teeth become larger and more durable to accommodate an herbivore's diet.

Another example is the whale. Biologists claim that the fossil record shows the evolution of whales from a land mammal. Whales are mammals; they are warm-blooded, they give birth to live young, and they produce milk with which to feed their young. Whales have vestigial hind legs that are not attached to their skeleton. The claimed fossils in this progression are the Pakicetus 50 mya, Ambulocetus 50 mya, Kutchicetus 43-46 mya, Rodhocetus 47 mya, Dorudon 43 mya, Dorudon 40 mya, Odontocetes (toothed whales) and Mysticetes (baleen whales) 40 mya.

The discovery of Tiktaalik in 2004 by Neil Shubin of the University of Chicago heralded what biologists claimed to be a transitional form between fish and amphibians. The interesting thing about this discovery was the prediction that was used to look for the fossil. Biologists had predicted that the limbs of tetrapods (four-legged vertebrates) would have evolved from the fins of fishes. So, they looked at the fossil record and determined that where they needed to concentrate their search would be in sediments somewhere between 360-390 million years old. Neil and colleagues eventually found this fossil that shows traits of both fishes and tetrapods in the rocks of Ellesmere Island dating to around 375 mya.

The most famous of these examples is the discovery of Archaeopteryx in Germany in 1860. Naturalists had long noticed the anatomical similarities between reptiles and birds and had hypothesized that birds may be descended from reptiles. Archaeopteryx had features of dinosaurs, reptiles, and birds.

The Cambrian explosion (545-485 mya) also initially appears to present quite a difficulty for biologists to explain. The vast number of body plans that came into existence and left fossils in certain shale deposits within a relatively short geological time frame. It did not seem to uphold the idea of gradual, slow changes predicted by Darwinian evolution. Most of these species are shown to go extinct equally as fast at the end of the Cambrian. Many biologists claim that this is simply illusionary for the fact that many of these lineages appear before the Cambrian, and many persist past the end of it. Certain hypotheses like those developed by Stephen Jay Gould and Niles Eldredge in 1972 attempted to explain these phenomena.

It has long been known that marine fossils and shells have been discovered high in the mountains and this seemed like straightforward evidence that Noah's flood had covered all the land on Earth at one time. But once the geological processes of uplift and plate tectonics for building mountain ranges came to be understood, it was easier to see how sediments with fossils of marine organisms could be uplifted during mountain building at plate boundaries. To be fair, many intelligent Christians have models where the mountains rise closer to their present heights as a result of the flood, so they don't have to suggest that Noah's flood covered modern Mount Everest or all the Himalayas.

But it is always quite a bit easier to try to dismiss a theory based on its weak points than it is to propose a better one based on the available evidence. I think we Christians should be humble as to how we would better explain the fossil record. Let's think about what we would expect to see in the fossil record if the world was 6000-12000 years old and all species except the ones living in water were represented in Noah's ark. We would reasonably expect that if God filled the Earth with creatures great and small and the flood was global, they would all be represented in the Ark, except for the marine organisms. Or at the very least, there could be a few thousand kinds that had the diversity in their genome to produce a lot of variation within their kinds. Two of the cat kind, two of the dog kind, two of the monkey kind, etc. From these kinds we could get all the cats, all the dogs, all the monkeys, etc.

We are never told that God commanded Noah to fill the Ark with plant species, so we could reasonably assume that most plant life would die in the flood in the duration of 1 year and 10 days. Assuming all of that, the sediment laid down after the flood would first have to be compressed or cemented to create rock and the rock types on the surface would be mainly composed of sedimentary rock, as opposed to metamorphic or igneous. This is not really an issue though as most of the rock at the Earth's surface is indeed sedimentary rock. And many geological models of Christian scientists have a lot of plate tectonics and volcanism occurring during and after the flood, along with the vast erosive power of the water receding back into new, deeper ocean basins. So, they purport to explain igneous and metamorphic rock at the surface through these mechanisms.

But it would seem reasonable to infer that most of the living species would be buried in this sedimentary rock and would be of the same age and found in the same layers. We should expect to find all kinds of fossils from simple to complex in all these layers of sedimentary rock, and we would not expect the fossils to look that much different than the varieties that we have on Earth now. Certainly, some microevolution causing variety within kinds may be expected, but it would seem to strain credulity to expect them to be vastly different than the samples from the Ark within a few thousand years. But the other problem that I see is that there would not be enough time for the processes of permineralization and lithification to turn bones into fossils.

We could postulate that God created different forms of life than those around Mesopotamia which boarded the Ark, and those creatures appear as long extinct species. But it does seem like special pleading and reading too much back into the text. We are never told that God created new kinds of plants and animals after the flood to refurnish the local ecologies around the planet. We could adopt the gap theory and explain the fossil record that way, but that is reading a whole lot back into the biblical text and it still doesn't explain the distribution of animals after getting off the ark, much less the evidence from genetics.

This problem of how the species distributed themselves around the globe when the land masses were separated by oceans seemed to me to be an intractable one. I know in some creationist models that plate tectonics resulted from the flood and moved apart the giant supercontinent, but this physical process would surely have ceased by the time Noah and the animals exited the Ark. And even if the animals spread out by land throughout the Earth in the next several thousand years, to me it still couldn't explain the distribution of species and the presence of them on islands.

As a young Earth creationist myself for much of my life, I can sympathize with trying to come up with models and solutions for some of these problems. And I have heard some pretty ingenious ones from smart, educated Christians.

Geographical distribution (biogeography)

Biogeography is the study of why plants and animals are living where they are. Take oceanic (volcanic) islands for example. Oceanic islands are distinguished from continental islands because they were never connected to another land mass, but instead arose in the ocean by volcanism (think the Hawaiian Islands). What we see over and over is that these islands do not have mammals, reptiles, or amphibians (unless introduced by man sometime in the past). None of these species are native to these island environments. We mainly see species of birds, insects, and plants. The variety of species living on these oceanic islands differs from those on continental islands, as the species that reside on continental islands closely resemble those of the nearest land mass or continent. The reason for these distributions became clearer when scientists began to realize that the continents move due to plate tectonics. Or more accurately, the plates on which the continents sit do the moving. When we examine the history of the continents due to plate tectonics, we can begin to understand how continental islands that were once connected to the larger mainland would have species of mammals, reptiles, and amphibians that are closely related to those of the mainland. But they may have evolved in their separate populations, depending on how long the island has been separated.

We also observe that species in certain ecological zones (like the deserts, tropics, or temperate forests) from different parts of the globe are often similar in their phenotypes (how they look) but can be quite different in their genotypes (what their sequence of DNA looks like). In fact, they more often resemble these other species separated by vast distances compared to species on the same continent that live in different ecological zones. To put it more simply, why do we find plants and animals with similar body plans and traits, but they differ significantly in their DNA?

Biologists account for this by what is known as convergent evolution. It is the theory that populations of animals who face the same types of selection pressures from similar environments will have certain heritable traits that converge with one another as a result of the animals being best adapted to their environment. So, if two animal species live far apart but both are in regions covered by snow and ice, the animals that make fur colors that do not match the surroundings will be killed more often by predators generation after generation until only animals with genes that make fur camouflaged to the environment will be left. And then those traits get fixed in a population. Examples of convergent evolution are sprinkled throughout the tree of life. Some traits that are proposed to have evolved independently many times include eyes, the ability to fly, a streamlined body for swimming, opposable thumbs, echolocation, meat-eating plants, sex determination in chromosomes, and many others.

Another thing we find in most cases is that the fossils that are in the youngest rocks closest to the surface resemble the animals that live in that region. Could we not interpret this to mean that God acted after the flood to make special creations of animals for each place? Apart from no biblical support for this idea, this idea of special creations for each little niche in God's creation was an idea that was widely held in Darwin's day.

Ernst Mayr relates that Darwin himself still held to this belief while he was aboard the Beagle doing his long excursion of sailing and naturalism prior to developing his theory of evolution by natural selection. That the idea was common in Darwin's day and Darwin himself thought it was a good explanation for why certain animals were distributed the way they were. (Mayr, 1982) But Darwin changed his mind before he wrote *Origin of Species*. He essentially laid out the groundwork for the study of biogeography without having the knowledge that would come about by plate tectonics.

Plate tectonics, which explained the mechanism for continental drift, made the biggest difference in recognizing the interconnectivity of species and where they came from and dispersed to. People had previously recognized that the continents looked like they fit together in a jigsaw puzzle. Prior to this, some theories had been proposed that postulated willy-nilly land bridges springing up to connect almost anything together to describe the distribution of species. A lot of these theories were developed because the scientists did not think that insects, birds, and plants could be dispersed over the vast distance of oceans, but the latter has shown to be the case.

There seems to be no other way that an oceanic island like Hawaii could get populated by these groups unless God made special creations on the island. Which then begs the question of why there would only be birds, insects, and plants like the other oceanic islands? Certainly, feral mammals like the mongoose and pig have done very well in Hawaii. It has long been agreed upon that continental islands like Great Britain, Ceylon, and the Sunda Islands once had connections to the larger continents. As well as the Bering Strait connecting North America to Asia and the Isthmus of Panama separating the Pacific Ocean from the Atlantic around 3 million years ago. When sea levels dropped, this newly exposed land bridge allowed for the passage of animals like the opossum, armadillo, and porcupine into North America from South America. While bears, cats, dogs, raccoons, and horses all made their way south over the bridge. According to the theory of plate tectonics, most of the land masses were joined together in a supercontinent called Pangea about 270 mya and began to break up around 200 mya. Austaulia, Antarctica, South America, Madagascar, India, and the Arabian Peninsula started breaking apart around 180 mya and continued until about 30 mya.

Jerry Coyne explains how predictions can be made based on fossils, living animals, and plate tectonics. Such as how marsupials ended up in Australia. He describes how it would seem implausible from the current day geography to figure out how they got to Australia, given that the oldest fossils of them are from North America around 80 million years ago. Their fossils had also been found at the tip of South America and dated to around 40 million years ago. And based on the theory of plate tectonics, the theory had been worked out that the supercontinent of Gondwana began breaking up prior to this. But the southern tip of South America, Antarctica, and Australia should have still been connected. So, scientists should be able to find fossils of marsupials in Antarctica dated to around 30-40 mya. And that's what they found when they went to Antarctica and looked. Many fossils of marsupial species dating to the right time. (Coyne J. , 2009)

Surely, we must give due credit to the predictive power of this hypothesis pulling together different disciplines. We should ask ourselves how another model would explain such evidence as that of biogeography.

Homologous structures

Homologous structures are body parts that are structurally similar between species but used for different purposes. The idea of the study of these structures was to show that the animals shared a common ancestor. One often used example is the forelimb in different vertebrates like human, cat, bird, bat, whale, and lizard to show the similarity between the different types of bones required to build the forelimb. Anatomists compared the bones and noticed that all shared the upper arm bone (our humerus), the two bones of the lower limb (our radius and ulna), the smaller bones that make up our wrist (our carpal bones), and then finally to the bones that make up our fingers or similar appendages on other vertebrates (humans, cats, bats, frogs, and lizards all have five).

This basis of taxonomy (how to group and classify living things) based on physical form was first developed by Carolus Linnaeus in the early 18th century. His basis of organizing creatures was mainly based on how they looked and the function of body parts. He grouped species mainly on morphology, but later scientists like Ernst Mayr separated species by the fact that they were reproductively isolated from one another. This is one of the hallmarks of the current evolutionary paradigm. That speciation cannot occur unless populations are reproductively isolated. Otherwise, the genetic information would get recombined into the gene pool of the population and no divergence of species could occur. But in the current understanding of evolution, it is more accurate to say that most biologists think of reproductive isolation as either prezygotic or postzygotic. These fancy terms just ask what limits possible mating from producing fertile offspring between populations? The first are factors that limit mating opportunities, like physical separation. The second are factors that prevent the zygote from developing or being fertile itself.

This taxonomy developed first by Linnaeus and later by Mayr has been upgraded over the years and biologists assert that outward appearance alone can be deceiving, as mechanisms such as convergent evolution can make organisms look alike but their evolutionary history may be quite different. There is still no universal consensus among biologists regarding taxonomy and some seem to think some outlier animals and plants can't be neatly placed in a simple group. Even the concept of what defines a species is still not agreed upon between biologists.

Embryology

Embryology is the study of sex cells, fertilization, and the development of embryos and fetuses. A human embryo is defined by medicine as the time from the fertilization of the egg to the end of the 8th week of gestation. It is the time when the function of totipotent cells differentiate to form different cells, tissues, organs, and the whole body.

Ernst Mayr relates how Darwin felt like his data from embryology was his strongest argument for evolution in *The Origin of Species* and was dissatisfied that it seems to be glossed over by most reviewers of his book. (Mayr, 1982) So, Darwin thought that the data he had gathered from the study of embryology was his strongest point in favor of his arguments for common descent by natural selection. Darwin was rather conservative about adopting certain implications for the similarities in embryos of different animals. This was not the case for a scientist by the name of Ernst Haeckel, who developed a theory about the development of embryos and what they could tell us about common descent and how it unfolded. Haeckel essentially thought that we could view the evolutionary past by viewing the different stages in the development of the embryo to tell the lineage of descent through different species. Put differently, he claimed we could watch a human embryo develop and view the stages of chordates, fish, tetrapods, mammals, primates, apes, and humans and clearly see the evolutionary history of man. He proposed that a human embryo would have traits of the adult version of less complex organisms as it developed in the womb. This theory was widely held and popular for at least 40 years beginning in the 1870s. Haeckel was later found to be deceptive in his drawings to make the embryos fit into his theory of recapitulation.

Later biologists thought what Haeckel got wrong was that he was expecting to see the adult forms of the previous ancestors in the development of the human embryo (human embryo would look like an adult fish, amphibian, reptile, ape, etc.), but they still maintained that the human embryo goes through stages of development recapitulating the proposed ancestors. What is really happening is that there are common embryotic features of all vertebrates that differentiate and develop into adult structures. There is no real sense of human development showing ancestral lineage. Although you still see this idea pop up now and again. I could have sworn it was still in a biology textbook for a class I had in college in the late 1990s.

Many biologists make similar assertions about different structures of embryos that they say only make sense in the light of evolution. Some point to why young baleen whales develop teeth or why the more complex vertebrates should have a notochord. Or regarding the homologous structures discussed in the previous section like the forelimbs of humans, cats, birds, bats, whales, and lizards and how they develop during the embryonic stages.

This study of similar features and how they develop and differentiate has led to the field of evolutionary development (or evo-devo). This field is relatively new and has been the result of advances in developmental biology, molecular biology, genetics, and epigenetics. Some of this work has concentrated on some of the master control genes, like HOX and PAX, that coordinate many genes during development. Biologists have noted how some master control genes are often very similar in different organisms. And these master control genes work on a different set of genes for each organism to produce the body plan and parts. So, master control genes can use genes that differ across organisms to build the same kinds of structures in different species. And there seems to be a modularity to some of these control genes that conserve them for similar functions but can produce different morphologies that lead to animals and plants looking very distinct. Many biologists see this similarity of master control genes across species as further evidence for common descent. But it does bring up questions of how random mutations really are.

Human evolution

In 1856, just three years before the publication of Darwin's *Origin of Species*, the discovery of a Neanderthal skullcap (now known as Neanderthal 1) was discovered by limestone quarry miners in a German cave located in the Neandertal valley. Although two other fossil remains had been previously discovered in western Europe in 1829 and 1848, this was the first to be described by an anatomist to be an early human fossil, in this case Hermann Schaaffhausen. It consisted of 16 bone fragments including skullcap, a complete right-side humerus, complete right radius, five ribs, pelvis, among others. The classification allowed scientists to realize the two earlier discoveries in 1829 in Belgium and in 1848 in Gibraltar were also Neanderthals. In 1886 two nearly perfect skeletons were found in Belgium, in 1906 a nearly complete skeleton was found in France, 9 incomplete skeletons were found in northern Iraq from 1953-1957, remains from 12 Neanderthals were found in 1994 in a cave in Spain, and remains from over 400 individuals have been found to date. In 2010, the entire genome of a Neanderthal (which is complicated and rarely accomplished with ancient fossils) was sequenced showing that 1-4% of the DNA in the genome of non-Africans may have come from the Neanderthals, which scientists claim shows that they interbred with Homo Sapiens. Although more recent studies show that even Africans have a small percentage of Neanderthal DNA.

A skullcap and thighbone were discovered by amateur anthropologist Eugene Dubois on the island of Java, Indonesia in 1891. He was dubbed Java Man (now known as Trinil 2) and later classified as Homo erectus. Dubois' claim that this was our earliest ancestor was largely disregarded for about 40 years, until more discoveries of Homo erectus were found in China and the East Indies. In 1921 a Swedish geologist named Gunnar Anderson discovered a single molar tooth which an anatomy professor recognized as from a human ancestor and as a result excavations at the site ensued and produced bones from 50 individual Homo erectus individuals. In 1960 a partial cranium was found in Africa, and in 1984 a nearly complete skeleton was recovered, later dubbed Turkana Boy. Several other finds of Homo erectus have been found from Africa to the Republic of Georgia (north of Turkey) to China to Indonesia.

In 1907 a jawbone was found and later described by the German anthropologist Otto Schoetensack, which he termed Homo heidelbergensis, as it was found close to the town of Heidelberg, Germany. Among other fossils found with the jawbone were mammals who lived 500,000 years ago. At least 7-11 fossils have been found that have been classified as Homo heidelbergensis, although specialists often debate the anatomy and classifications.

In 1924 anatomy professor Raymond Dart described a skull found in a limestone quarry in South Africa. He gave the specimen the classification of Australopithecus africanus, although it was popularly known as Taung child. He classified it as an ape to human intermediate species and said that it most likely walked upright based on where the skull was situated over the spinal cord. This finding looked like evidence confirming an intermediate missing link species between humans and apes but was largely rejected for various reasons. Among which was the previous claim of an intermediate species that came to be known as Piltdown Man, which was discovered and described between 1910-1912, and was claimed to be the link between apes and humans. But Piltdown Man was later shown to be a deliberate hoax. Between 1936 and 1948 Robert Broom discovered various Australopithecus skulls in South Africa.

Donald Johanson in 1974 discovered the famous Lucy. It was classified as Australopithecus afarensis. It was about 40% of a complete skeleton and the most complete one to date. Others have been found in Ethiopia, and all the fossils discovered so far have only been found in East Africa.

In 2010 Lee Berger discovered two partial skeletons that he named Australopithecus sediba.

In 2008 a tiny finger bone was found in a cave in Russia. DNA analysis revealed a close genetic relative of Neanderthals. Scientists named the previously unknown group Denisovans based on the name of the cave in Siberia in which it was found. Other fragments were found and analyzed as well. In 2019 a mandible was found in a cave in China and sequenced. It was also classified as Denisovan. They are thought to be another Homo species alive at the same time as Neanderthals and Homo Sapiens.

Scientists claim they have early "human" fossil remains from more than 6000 individuals as of the date of this writing. They tell us they have thousands of fossils documenting the process of species getting more and more "human-like" after the branches split from the apes. Although scientists disagree on where every fossil fits in the lineage and which can be considered direct lines of descent or just extinct cousins, the generally agreed upon sequence goes something like this:

At least 7 million years ago (mya), the lineage leading to the Homo genus separated from the apes. The species that Lucy was a part of (Australopithecus afarensis, 3 mya) are believed to be the earliest fully upright walking ancestors. We can tell this by the anatomy of their pelvis and legs, as well as where the brain stem goes into the skull. Apes who walk on all fours most of the time have their leg bones splayed out, angled away from the inside of the pelvis which makes them waddle awkwardly when walking upright. Homo Sapiens on the other hand have their leg bones angled to the inside which places their center of gravity in the middle, so they don't move side to side while walking. Lucy's pelvis resembled that of Homo more than apes. Changes from apes are also seen within the spinal column of Australopithecus afarensis.

Homo habilis (2.8 mya) is considered to be one of the first to master stone tool technology. Homo habilis is thought to have existed with Homo erectus for more than a half million years.

Homo erectus (1.8 mya) evolved in Africa, its forehead has less of a slope and its teeth are smaller. Anthropologists tell us that Homo erectus lived between 1.89 million– 110,000 years ago, was the first of the genus Homo to migrate out of Africa almost 2 mya, then spread to Asia, and perhaps parts of Europe. There was a lot of variability among individuals, probably due to how long they were around and how vastly they had spread out geographically. Their brain size ranged from slightly larger than a chimpanzee to close to the size of the average modern human. They were the first of the Homo species to make hand axes out of stone. They may have used controlled fire, especially the more recent Homo erectus, but scientists aren't sure.

Homo heidelbergensis lived between 700,000-200,000 years ago, had a brain size slightly smaller than modern humans, possibly used controlled fire, and lived across Africa and Europe. They were the first Homo species to use wooden spears and routinely hunt large game. Some claim they were the first to build rudimentary dwellings to live in.

Neanderthals (Homo neanderthalensis) lived between 400,000-40,000 years ago, and fossils have been found in Europe and western Asia. They had a brain size that was slightly larger than modern humans on average, used controlled fire, had an advanced toolkit, regularly hunted and butchered game, wore clothing, and lived in shelters. There is evidence that they buried some of their dead and possibly even decorated some burial sites. The evidence is rather weak, but ambiguous enough that some experts claim that they produced symbolic art, including cave paintings and jewelry. Others claim that the paintings were produced by Homo sapiens or copied by Neanderthals mimicking what they had seen Homo Sapiens produce, and that the jewelry was traded and was originally made by Homo sapiens. Whatever the truth, there is little doubt that they were not the dumb, knuckle-dragging, cave men of popular imagination.

Some anthropologists and linguists believe that Neanderthals had language. And some believe that even Homo heidelbergensis could communicate through language. They point to various arguments and lines of evidence compared to Homo Sapiens, including anatomical ones, cooperation in hunting and living, brain size, genes for language, and others. I'm not sure if it is an answerable question, but it wouldn't be like our speech if they did. Homo Sapiens had vastly different material culture than Neanderthals, which would indicate differences in brain function and language. This material culture of Homo Sapiens exploded around 50,000 years ago.

Most anthropologists believe that the evidence points to Homo Sapiens, Neanderthals, and Denisovans descending from or sharing the common ancestor Homo heidelbergensis. We have now seen various lines of evidence that are used to support evolutionary theory. In the next chapter we will examine the most recent line of evidence, genetics.

Chapter 10

Genetics and evolution

If we were to ask 100 biologists what is the most important thing that Darwin didn't understand about evolution, 99 of them would probably answer genetics. It is debated whether Darwin knew of the work of Gregor Mendel, but if he did, he did not appreciate its significance. Besides the lack of examples of natural selection at work and to be observed in a human lifespan, Darwin's critics also pounced on the fact that Darwin did not have a coherent mechanism for inheritance. Darwin believed in the inheritance of acquired characteristics, such as the theory put forth by Jean-Baptiste Lamarck, which was widely believed in Darwin's day. Darwin hypothesized that traits and changes to an organism during its lifetime were transferred to the germline cells by something he called gemmules. Darwin also believed, which was popular at the time, that traits would be blended from parents to offspring. When it was argued compellingly that this would quickly eliminate variation in a population within several generations, Darwin seemed to understand this limitation.

This all began to change in the early 20th century. Experimentation and reasoning eventually won the day, and the idea of Lamarckian inheritance of acquired traits was gradually abandoned. Although the concept of acquired traits has been renewed in modern biology in a much more limited context based on the study of epigenetics and gene expression. The work of Gregor Mendel was rediscovered, and he was eventually given the credit he deserved and had not experienced during his lifetime. And with the rise of genetic sequencing technology, the stage was set for modern biology and new evidence for evolution and common descent.

But this is not to say that biology now leads to naturalism. I am sympathetic to some of the arguments from the intelligent design camp that see the explanations of evolutionists for speciation, complex biochemical structures, and novel, beneficial adaptations as many times a black box and a lot of hand waving. The claims of some biologists often start with assertions of evolution as chance, random, and unguided. The first important thing we need to realize about such assertions are that they are not scientific statements, they are philosophical ones. So, a scientist trying to argue from random genetic mutations and variance in a population to evidence that God is not involved has made a leap in logic without justification. Natural processes and mechanisms do not preclude God doing it, as we have seen in our chapter on skepticism and God's ordinary providence.

Many evolutionary biologists and biology professors will often use words like random and purposeless when speaking about mutations in DNA. One of the central tenets of the Neo-Darwinian Modern Synthesis theory of evolution for the past 100 years or so is that mutations, as the source for novel variation, occur randomly. But it is important to understand that biologists will describe mutations as random but the process of natural selection as non-random. A mutation that happens in a coding section of DNA may produce a novel protein, it may produce a similarly functioning protein with the alteration of one amino acid, or it may render the protein non-functional. It may also occur in a duplicated sequence that is not yet being expressed, or it may happen in a non-coding region. When this novel or different protein is finally expressed, it will most likely be maladaptive or neutral for the organism, but occasionally it will be beneficial. But the selection pressures on that new gene are not random. If the mutation(s) leading to the protein makes the organism better adapted to its environment and more reproductively successful, then it is more likely for this allele to spread in a population. If the protein makes the organism less likely to survive and reproduce (which is more likely), then these mutations will be lost more quickly. These selection pressures are acting non-randomly. They are sorting out which genes make for a more adaptive organism from the maladaptive ones. So, biologists make the distinction between the mutations themselves being random and natural selection being non-random.

Many resources on biology and evolution go out of their way to state that the mutations are neither "good" nor "bad". They say they are random and the cell containing the genome does not "know" how to adapt to its environment for its beneficial end. You will find that natural selection, sexual selection, and genetic drift are usually considered the main mechanisms to explain why we have certain genes in a population. But that is just kind of sidestepping the question of how the beneficial genes arose to begin with. Selection pressures operate on organisms and can explain the variation and distribution of genes in a population, but it does not explain how the traits developed. For instance, one category of selection pressures, natural selection, is an explanation for why some organisms survive and reproduce compared to others, but it does not tell us how the fitter genome was changed to confer that benefit to begin with. Perry Marshall in his book *Evolution 2.0* makes the case that random mutation is noise and noise always degrades, it never improves. He argues that random mutation + natural selection + time = extinction, not evolution. (Marshall, 2015) Whether we think that is a good argument or not is one question. Most biologists would not think it is a good argument because they would point out that a mutation in a single base often leads to the same protein. This happens because the codes for making amino acids are redundant. We can get a total of 64 combinations of four bases chosen three times, but there are only 20 amino acids. Therefore, most amino acids can be coded for by between two and four combinations of the four bases, A, G, C, and T. But it is usually only a position change in the third base that does not affect the amino acid. But even if a mutation happens in the first two bases of the codon, a different amino acid may not affect the shape and function of the protein. But it often does.

Take sickle-cell disease for example. A change in a single amino acid alters the shape and function of the protein which makes it worse at doing its job. But being the condition is recessive, having just one copy of the mutated gene confers the benefit of resisting infection to malaria. Therefore, there is evidence that a mutation affecting protein shape and function can confer benefit to the organism, but also evidence that there is a cost to be paid by someone who has two copies of the gene. But I think it makes an important point that mutations are much more often maladaptive than beneficial to an organism. But is it really true that mutations are random?

It has been shown that mutations happen more slowly in very important genes. It has been shown in different studies that the environment can have a direct impact on how fast cells allow mutations to occur. This happens in our immune system when confronted with an infection as the immune cells are furiously mutating until they can come up with a solution to a foreign invader (or perceived invader in the case of autoimmune diseases). Many types of cells do this when they are stressed and are trying to survive. They may mutate faster seeking a solution to use a food source when nothing else is available. These mutation races are not always fruitful and many times the organism dies. But the idea, although controversial, is that the cell has a "will" to survive.

Now to be fair, many experiments have been done that show it is more likely that a mutation existed prior to a new environmental condition. And other experiments have been done that show single point mutations conferring stepwise beneficial adaptations. We also know of mechanisms like single gene duplication, whole genome duplication, gene transfer, and others that allow a potential new gene or genes to mutate without affecting the fitness of the organism. But the more biology learns about mutations, epigenetics, gene expression, gene conservation, and the like, the less dogmatic the claim of randomness appears to me. And the more we learn about genetics and biology, the less convincing it looks for a simple explanation of random mutations producing and expressing beneficial genes correctly. But here is the crucial point, random does not imply atheism. Some biologists seem to think that explaining a mechanism implies materialism, or that God is no longer required in their hypotheses. This again is an error in logic.

But in my opinion, random here is really just a byword for ignorance. Genetic experiments which have used radiation to induce random mutations have failed miserably to invoke beneficial attributes to the organism. If cancer can change and evolve so that treatments no longer work, if bacteria can develop resistance to antibiotics quickly, if a cell develops a novel way to use a new food source, or a yeast can change its genes to survive in a lethal salt environment in as little as 25 generations (Bell & Gonzalez, 2009), provoking random mutations throws up a red flag. Or at least it should be qualified with humility instead of asserted with certainty.

I think this is one reason why so many people find evolution so hard to believe. Because we see so many plants and animals that are so exquisitely adapted to their environment. While many would say it is the illusion of design, some of the body plans, phenotypes, and behaviors of organisms beggar belief if mutations are truly random. How long would it take to produce genes randomly and in the correct sequence, that are then correctly expressed, to make insects that look exactly like the plants and leaves they live on? What sequence of events, including mutations to produce the proteins and then the correct expression of those proteins in specific cell types, to form a caterpillar that produces pheromones that causes ants to take the caterpillar back to its home and feed it until it matures? Or a plant that looks and smells like a bee to trick the bee into mating with it and dispersing its pollen and fertilizing another plant of the same species? Or having attributes that can only be exploited by one another in their local spheres. Can this adaptive ability really be attributed to a stochastic (random probability) process? Does the process of random mutations, with most being either deleterious or neutral, explain the diversity and adaptation of plants and animals around the globe? Does the loss of function from some genes like vitamin C synthesis for apes and humans, eyes for cave dwelling fish, fins for bottoming dwelling fish, or positive functioning genes for digesting starch or lactose possibly point to environment affecting genes directly versus randomly produced alleles?

I don't claim to know the answer. And we need to be careful with god-of-the-gaps arguments. As we have previously touched on, this pejorative refers to theists using the lack of scientific understanding in some areas to place God in there as an explanation. Said another way, we don't understand how something works, therefore God must have done it. And for the thesis of our book, it doesn't ultimately matter. Because we believe that the Bible teaches that God is at work in all processes and mechanisms of His creation. I just think that biologists should be more careful in expressing absolute certainty and by not making the illogical leap of apparent random processes to stating there is no purpose in nature. That is a philosophical position, not a scientific one.

Scientists for and against intelligent design (ID) go back and forth arguing about irreducible complexity and the evidence for and against evolutionary mechanisms. ID folks point to complex structures that cannot function without all the parts there at the same time while evolutionists claim to bring forth examples of organisms becoming more adapted with each small genetic change that eventually leads through deep time to a new complex biological system. ID folks make arguments based on the improbabilities of random mutations producing beneficial, functional traits while evolutionists claim to provide evidence that these things have been observed in the lab using fast replicating organisms like viruses, bacteria, and yeast or they must have happened in the past by examining the genomes of different plants and animals. The evolutionists seem to have very few detailed explanations of the steps that may have occurred to create the complex biochemistry in various cells, new body plans, new organs, new body systems, or new species. They often respond by saying that just because you can't think of how some biological system would come about does not mean it didn't happen. While the ID proponents often seem to ignore the evidence of the resiliency of proteins, how they can be repurposed for other functions, how single point mutations can enhance adaptation, the creation of de novo proteins, and how new proteins can bind with one another without affecting the fitness of the organism. They both seem to arrive at the same conclusion that evolution happened and that the evidence of it can be inferred from the genomes of organisms and by the biodiversity that we observe. The ID folks seem to be saying that a designer is required to account for what random mutations and selection pressures are impotent to accomplish, while many evolutionists seem to be saying that the mechanisms are sufficient and therefore exclude a designer. In my opinion the ID people cannot conceive of an experiment or prediction to validate their hypothesis and the evolutionists have no logical warrant to reach their conclusion. Therefore, they don't seem to be really arguing about science. They appear to be arguing whether God is necessary or not. That is why I keep coming back to having a proper understanding of ordinary providence from a biblical perspective. God is working, but it doesn't necessarily have to be miraculous or divine intervention. So, whether genetic and evolutionary mechanisms can account for biodiversity, or we have a coherent explanation of how it unfolded, materialism does not follow.

Now, I think there is a different perspective on which ID can make an even more compelling case. If mutation is truly random, if it can affect any gene with equal probability, and most mutations are deleterious or neutral, then it is far more likely that a Designer has orchestrated biology and the environment for His ends to begin life, to prevent it from going extinct, and for producing mankind.

But with that little aside, let's quickly look at some genetic evidence for evolution with common descent.

The first thing many biologists mention is that all living organisms use the same code and molecule, DNA. From bacteria to barnacles, fungi to foxes, plants to peacocks, hippos to humans, all biological life uses the same molecule to store information. And that molecule is DNA. And the same nitrogenous bases adenine (A), guanine (G), cytosine (C), and thymine (T) in various sequences to store that information. Three of these bases together make a codon, which when translated by the tRNA in the ribosome organelle, attach the same kinds of amino acids across all organisms. A string of these amino acids makes proteins that do virtually all the functions in the cell, which leads to functions in the tissues, up to organs, and finally the organism as a whole. The fact that all life uses the same information molecule and ways of making proteins is one line of evidence for common descent.

When looking at proposed related organisms based on the hypothesis of common descent, biologists find that their genomes contain many sequences that are identical. This suggests that they had a common ancestor from whom they inherited these DNA sequences. As a side note, we have already discussed convergent evolution. This could be an alternative hypothesis because many proposed unrelated organisms have traits that are similar, i.e. wings, eyes, a streamlined body for swimming, echolocation, etc. But it is unlikely that the gene sequences for similar physical traits will be identical in these organisms, although they often use similar kinds of proteins for function. It then follows that having the exact same sequences in DNA is more simply explained by common descent. And this genetic evidence is then corroborated using other lines of evidence, especially among those species who are hypothesized to be related in the past. They also find that many sequences in non-coding regions of the genome are also alike among these organisms. If we have large portions of identical "non-functional" or non-coding regions of DNA among proposed related organisms, this is further evidence that these organisms had a common ancestry. Of course, it takes additional reasons and arguments to show that the regions they are comparing are valid comparisons to begin with.

Biochemistry and ancient preserved genes. All organisms use many of the same biomolecules for metabolism and development. And they also use many of the same master control genes for their body plans.

Since the advent and refinement of DNA sequencing technology, biologists have used this technology to group organisms into trees of relatedness, which most often agrees with classical taxonomy, but sometimes produces surprising results. They claim that by using such phylogenetic trees, that humans are most closely related to chimpanzees and bonobos and are more distantly related to gorillas and orangutans.

Scientists have developed methods for estimating genetic clocks, or how fast areas of the genome mutate. And then by using this, they estimate how long ago two species diverged from one another. Some assumptions go into these techniques and are debated among experts, but it will suffice for our purposes to be aware they exist and are used by biologists to make estimates and predictions about the tree of life.

So, at this point I'm sure some of you are feeling depressed and deflated. I know the feeling. I have read through so much science on evolution over the years and felt depressed when I was done. I was wondering where God was in all of this, and it made the world feel so mechanical and dead. But we are not going to leave it there. The Christian faith is one of hope but also of truth. We will therefore turn our attention back to the word of God and examine Adam and Eve more closely.

Chapter 11

Adam and Eve

I have heard it said that somewhere in Scandinavia they have a saying to show someone is being naïve and childish. It goes something like, "You don't still believe in Adam and Eve, do you?" That is unfortunately the attitude and condescending tone you get from many people today when you tell them you are a Bible-believing Christian. But is this belief irrational? Are there no evidence or good reasons to believe that Adam and Eve were indeed historical persons and miraculously created by God, apart from a blind leap of faith in the Bible? While the infallible word of God is enough reason for us Christians, I don't think God hardly ever requires a blind leap of faith contrary to reason and evidence. And we have already shown that the garden of Eden is firmly rooted in history.

Let's first take a Bible-wide view of Adam and Eve and see what God's word proclaims about them. On a side note, when I refer to Moses telling us something, I am referring to the fact that Jesus and the rest of the Bible's own testimony attribute the first five books of the Old Testament, known as the Pentateuch, to the writing of Moses. We have seen that there are good reasons to maintain the conservative view that Moses was the original human author of the first five books of the Bible.

Adam in the Old Testament

Genesis 1:26-31

26 Then God said, "Let Us make man in Our image, according to Our likeness; and let them rule over the fish of the sea and over the birds of the sky and over the cattle and over all the earth, and over every creeping thing that creeps on the earth." 27 God created man in His own image, in the image of God He created him; male and female He created them. 28 God blessed them; and God said to them, "Be fruitful and multiply, and fill the earth, and subdue it; and rule over the fish of the sea and over the birds of the sky and over every living

thing that moves on the earth." 29 Then God said, "Behold, I have given you every plant yielding seed that is on the surface of all the earth, and every tree which has fruit yielding seed; it shall be food for you; 30 and to every beast of the earth and to every bird of the sky and to every thing that moves on the earth which has life, I have given every green plant for food"; and it was so. 31 God saw all that He had made, and behold, it was very good. And there was evening and there was morning, the sixth day.

Although we have argued that Genesis 1 was not meant to be interpreted as historical narrative, we also saw that this does not mean there are no truths to be learned from the text. We just have to be careful to interpret it in a way that it was meant to communicate. From this text we know that God made man in His own image, God created them male and female, the man is a single person and humanity in general, God blessed them, God commanded them to reproduce and fill the earth, God gave them dominion over the earth and animals, and God pronounces it all very good.

Genesis 5:1-2

This is the book of the generations of Adam. In the day when God created man, He made him in the likeness of God. 2 He created them male and female, and He blessed them and named them Man in the day when they were created.

Although this chapter begins a new section about the generations preceding from Adam and Eve and may have been a preexisting source and book (as verse 1 implies), it ties the historical man miraculously created by God to the more general man and mankind created in the image of God from Genesis 1. And it sets up the genealogy of Adam through Seth as the chosen line in which the purpose of the image of God is to be realized.

Genesis 2:7-8

7 Then the Lord God formed man of dust from the ground, and breathed into his nostrils the breath of life; and man became a living being. 8 The Lord God planted a garden toward the east, in Eden; and there He placed the man whom He had formed.

Genesis 2:18-25

15 Then the Lord God took the man and put him into the garden of Eden to cultivate it and keep it.

Genesis 2:22

22 The Lord God fashioned into a woman the rib which He had taken from the man, and brought her to the man.

God fashions Adam from the dust of the ground and Personally breathes into him the breath of life, animating his body. While we understand that this may be an anthropomorphic way of describing God forming and breathing, it none the less conveys the personal way of working with this son of God. We have seen previously that God is making a human into His image and likeness, versus the pagans fashioning gods with their own hands to become animated in the likeness of the gods. We have also seen that God places His image bearer in the temple-garden to be a priest in this sacred space. God miraculously creates Eve from Adam's side and the two are in the garden to serve and guard it.

Genesis 3:23-24

23 therefore the Lord God sent him out from the garden of Eden, to cultivate the ground from which he was taken. 24 So He drove the man out; and at the east of the garden of Eden He stationed the cherubim and the flaming sword which turned every direction to guard the way to the tree of life.

After the Fall (which we will come back to), God drives Adam and Eve out of the garden, most likely to the east, as that is where He stations the cherubim. The man and his wife are driven from the idyllic, sacred space of the garden out to the common and profane, to work the very ground from which Adam was formed. Instead of cultivating the holy ground and reaping blessings, Adam will cultivate the common ground and it will frustrate him. The cherubim now guard the way to God's presence and the tree of life.

Genesis 4:1-2

4 Now the man had relations with his wife Eve, and she conceived and gave birth to Cain, and she said, "I have gotten a manchild with the help of the Lord." 2 Again, she gave birth to his brother Abel. And Abel was a keeper of flocks, but Cain was a tiller of the ground.

Adam and Eve have their first male child; Moses does not tell us whether any females were born before Cain. Abel was the second male child born. We know when we get to Chapter 5 that Adam and Eve also had daughters. It is reasonable to infer that Adam and Eve had no children when they were in the garden from this account.

Cain was a farmer of the land and Abel was a keeper of sheep. Moses does not tell us how Adam learned to farm initially. He does not tell us if and how Adam taught Cain this profession. He does not tell us how Abel obtained his livestock or gained the knowledge required to keep domesticated animals. But we are not told a lot of things that we may have questions about. It is not the purpose of the story to discuss mundane details, but rather about the broader theological points in relation to God, His promises, and His kingdom.

Genesis 4:25-26

25 Adam had relations with his wife again; and she gave birth to a son, and named him Seth, for, she said, "God has appointed me another offspring in place of Abel, for Cain killed him." 26 To Seth, to him also a son was born; and he called his name Enosh. Then men began to call upon the name of the Lord.

Adam and Eve have a son named Seth. Seth and his line of descendants will be important for the rest of the book. After seeing the ungodly line of Cain, his descendants, and the wickedness that seems to be propagating, we are given hope that Seth and his line will worship the Lord and crush the head of the serpent.

Genesis 5:1-5

5 This is the book of the generations of Adam. In the day when God created man, He made him in the likeness of God. 2 He created them male and female, and He blessed them and named them Man in the day when they were created. When Adam had lived one hundred and thirty years, he became the father of a son in his own likeness, according to his image, and named him Seth. 4 Then the days of Adam after he became the father of Seth were eight hundred years, and he had other sons and daughters. 5 So all the days that Adam lived were nine hundred and thirty years, and he died.

Here is where the allegorical or metaphorical interpretations of Adam and Eve as non-historical persons really come into question for the first time since the creation account. This genealogy in Genesis 5 traces Adam's descendants through the line of Seth down to Noah. There is debate whether the genealogies are complete or not but even if they are not, I think it would be prudent to say any alleged gap would maybe add a couple thousand years at the most. But it should be noted that many scholars think most of the biblical genealogies are incomplete or stylized for symmetry to make theological points or to highlight the important descendants in question. More to the point though, these genealogies are to be read as real, historical persons. Just like the rest of the genealogies in the Bible are meant to communicate truth about ancestry and family history, while often being stylized or compressed for theological emphasis.

Adam and Eve's son Seth is said to be in the likeness and image of Adam. We see that the image of God is not gone after the Fall but is somehow passed on from Adam to Seth. It is curious that it is not Cain or Abel who is said to be in Adam's image, but one of the points of this chapter and the genealogy is to show that Seth is the promised child and that the godly line of Seth will include Enoch, who is taken by the Lord and does not taste death, and Noah, who will be righteous and a type of savior. These descendants of Adam give us hope that the image of God will influence all the earth.

We will return to the amazing age of Adam and the people included in these genealogies later.

1 Chronicles 1:1-4

1 Adam, Seth, Enosh, 2 Kenan, Mahalalel, Jared, 3 Enoch, Methuselah, Lamech, 4 Noah, Shem, Ham and Japheth.

Same exact genealogy as Genesis 5.

Hosea 6:7

7 But like Adam they have transgressed the covenant;

There they have dealt treacherously against Me.

We will return to this idea of Adam transgressing the covenant in later chapters.

Adam in the New Testament

Matthew 19:3-6

3 Some Pharisees came to Jesus, testing Him and asking, "Is it lawful for a man to divorce his wife for any reason at all?" 4 And He answered and said, "Have you not read that He who created them from the beginning made them male and female, 5 and said, 'For this reason a man shall leave his father and mother and be joined to his wife, and the two shall become one flesh'? 6 So they are no longer two, but one flesh. What therefore God has joined together, let no man separate."

Mark 10:2-9

2 Some Pharisees came up to Jesus, testing Him, and began to question Him whether it was lawful for a man to divorce a wife. 3 And He answered and said to them, "What did Moses command you?" 4 They said, "Moses permitted a man to write a certificate of divorce and send her away." 5 But Jesus said to them, "Because of your hardness of heart he wrote you this commandment. 6 But from the beginning of creation, God made them male and female. 7 For this reason a man shall leave his father and mother, 8 and the two shall become one flesh; so they are no longer two, but one flesh. 9 What therefore God has joined together, let no man separate."

In these two Gospel accounts Jesus quotes either Genesis 1:27 or Genesis 5:1-2 and ties it into Genesis 2:24 when responding to the Pharisees who are trying to test Him. As we touched on earlier, Jesus shows that marriage between a man and a woman was rooted in the original plan of God for marriage and that Adam and Eve were that prototype. And we know from New Testament theology that the marriage of the man and woman are an ectype of the archetype of Christ and the church. Jesus therefore acknowledges the union of Adam and Eve as historical and the reason for the covenant of marriage is because Eve was taken from Adam and man and wife are to become one flesh.

Luke 3:23-38 (………. inserted to shorten genealogy)

23 When He began His ministry, Jesus Himself was about thirty years of age, being, as was supposed, the son of Joseph, the son of Eli,................., the son of Shem, the son of Noah, the son of Lamech, 37 the son of Methuselah, the son of Enoch, the son of Jared, the son of Mahalaleel, the son of Cainan, 38 the son of Enosh, the son of Seth, the son of Adam, the son of God.

I will not go into the differences between the genealogies of Matthew and Luke, but what is important for us here is that Luke traces the genealogy of Jesus back to "Adam, the son of God" confirming the account from Genesis 2 and 5 that Adam had no earthly parents who begot him but was created from the dust of the ground by God. I don't think we are warranted to take liberty of ascribing any others in this list to a metaphorical or allegorical meaning and therefore we should not do the same for Adam himself.

Romans 5:10-21

10 For if while we were enemies we were reconciled to God through the death of His Son, much more, having been reconciled, we shall be saved by His life. 11 And not only this, but we also exult in God through our Lord Jesus Christ, through whom we have now received the reconciliation.

12 Therefore, just as through one man sin entered into the world, and death through sin, and so death spread to all men, because all sinned— 13 for until the Law sin was in the world, but sin is not imputed when there is no law. 14 Nevertheless death reigned from Adam until Moses, even over those who had not sinned in the likeness of the offense of Adam, who is a type of Him who was to come.

15 But the free gift is not like the transgression. For if by the transgression of the one the many died, much more did the grace of God and the gift by the grace of the one Man, Jesus Christ, abound to the many. 16 The gift is not like that which came through the one who sinned; for on the one hand the judgment arose from one transgression resulting in condemnation, but on the other hand the free gift arose from many transgressions resulting in justification. 17 For if by the transgression of the one, death reigned through the one, much more those who receive the abundance of grace and of the gift of righteousness will reign in life through the One, Jesus Christ.

18 So then as through one transgression there resulted condemnation to all men, even so through one act of righteousness there resulted justification of life to all men. 19 For as through the one man's disobedience the many were made sinners, even so through the obedience of the One the many will be made righteous. 20 The Law came in so that the transgression would increase; but where sin increased, grace abounded all the more, 21 so that, as sin reigned in death, even so grace would reign through righteousness to eternal life through Jesus Christ our Lord.

Here Paul makes unmistakably clear that Adam was a real, historical person. Because Adam transgressed God's covenant by eating the fruit from the forbidden tree, death spread to all men, even to those who did not sin in the likeness of Adam. If Adam was not a real person, then sin and condemnation could also be viewed as metaphorical. Which would lead to the inference that we don't need an actual savior to justify us. But because Paul surely knows that Jesus Christ is a real person and a real Savior, the analogy demands that Adam was a real person and his sin had real consequences.

I believe Paul here is talking about spiritual death and not physical death. Why do I say this? The first reason is the context itself. The context is about contrasting spiritual death in Adam with spiritual life in Christ. The second reason is the way that Paul talks about Adam in 1 Corinthians 15. The third reason is how Adam was created from the ground to begin with. And the last reason is that Adam and Eve did not immediately physically die when they ate from the tree. Life in the Bible is about much more than being physically alive. It is defined more in terms of living in communion with God and in the light of His countenance, as the following verse demonstrates. I will deal with some of these issues as we go, and the doctrine of original sin in a later chapter.

John 6:51-53

51 Truly, truly, I say to you, if anyone keeps My word he will never see death." 52 The Jews said to Him, "Now we know that You have a demon. Abraham died, and the prophets also; and You say, 'If anyone keeps My word, he will never taste of death.' 53 Surely You are not greater than our father Abraham, who died? The prophets died too; whom do You make Yourself out to be?"

Adam was type of Christ to come. I think Paul makes clear that in the same way Adam's guilt of the first sin is imputed to us because he is our representative head, so also is Christ able to satisfy the judgement of God on behalf of His people, by bearing the wrath of God in our stead and imputing His righteous life by perfectly fulfilling the law to the account of those who believe in Him through faith. The main point for our chapter here is that Paul acknowledges Adam as a historical person whose sin had real consequences for those he represented.

1 Corinthians 15:19-23

19 If we have hoped in Christ in this life only, we are of all men most to be pitied. 20 But now Christ has been raised from the dead, the first fruits of those who are asleep. 21 For since by a man came death, by a man also came the resurrection of the dead. 22 For as in Adam all die, so also in Christ all will be made alive. 23 But each in his own order: Christ the first fruits, after that those who are Christ's at His coming,

In the beginning of this chapter of 1 Corinthians Paul is describing the fact of the resurrection of Jesus Christ. Here again Paul affirms the historicity of Adam and that by his sin, death came to all men. Again, we must say that if Adam was not a historical person, then Paul's analogy and contrast with Christ doesn't make sense.

What kind of death is referred to here? A straightforward reading of the text would lead us to assume physical death is being described here because Paul is contrasting the resurrection of our bodies with death in Adam. And I think that is correct. Physical death did come through Adam as the word of God affirms. But what conclusions can we draw from this? Are we bound, like most orthodox Christian theology, to see this death as something new introduced into the creation? Let's first look at the other relevant passages from 1 Corinthians.

1 Corinthians 15:42-50

42 So also is the resurrection of the dead. It is sown a perishable body, it is raised an imperishable body; 43 it is sown in dishonor, it is raised in glory; it is sown in weakness, it is raised in power; 44 it is sown a natural body, it is raised a spiritual body. If there is a natural body, there is also a spiritual body. 45 So also it is written, "The first man, Adam, became a living soul." The last Adam

became a life-giving spirit. 46 However, the spiritual is not first, but the natural; then the spiritual. 47 The first man is from the earth, earthy; the second man is from heaven. 48 As is the earthy, so also are those who are earthy; and as is the heavenly, so also are those who are heavenly. 49 Just as we have borne the image of the earthy, we will also bear the image of the heavenly. 50 Now I say this, brethren, that flesh and blood cannot inherit the kingdom of God; nor does the perishable inherit the imperishable.

Paul quotes Gen. 2:7 and is again affirming the historicity of Adam and his miraculous beginning. But what is interesting, and I think missed by many expositors of scripture, is that Paul is contrasting our natural bodies with our resurrected bodies, and how Paul applies this to Adam. And most tellingly to Adam ***prior*** to the Fall. He tells us that Adam was from the earth (Gen. 2:7), and that our natural bodies must first be changed in order to inherit the kingdom of God. In other words, we need to be made of a different substance in resurrected bodies to live embodied lives in a heavenly state. So, Adam was made from the ground, in a natural body of flesh and blood prior to the Fall. In other words, mortal. Adam himself would have needed to be transformed from a natural body to a spiritual body to inherit the kingdom of God.

The way that Adam brought physical death to all through his disobedience was because he did not pass his probation in the garden, nor expand the boundaries of the garden. And the way back to the tree of life was inaccessible to his posterity and contemporaries because of him. So, it is not that Adam's guilt and corruption somehow brought foreign physical death into a world which had never known it, but instead all died in the physical sense because the way had been shut to receive life. But his sin is imputed to those in Adam (Romans 5) in the way of spiritual death. We are getting a little ahead of ourselves, but I needed to bring this up now while we were looking at these verses. Much more to come.

1 Timothy 2:13-14

13 For it was Adam who was first created, and then Eve. 14 And it was not Adam who was deceived, but the woman being deceived, fell into transgression.

Paul again reveals by the Spirit the literal, historical account of the miraculous creation of Adam and Eve and the reality of the Fall as an event in space and time.

2 Corinthians 11:3

3 But I am afraid that, as the serpent deceived Eve by his craftiness, your minds will be led astray from the simplicity and purity of devotion to Christ.

Paul once again affirms the historical reality of Eve, the garden, and the Fall.

Jude 14

14 It was also about these men that Enoch, in the seventh generation from Adam, prophesied, saying, "Behold, the Lord came with many thousands of His holy ones,

Jude references the genealogy of Adam as truth.

Let's bring this together. We argued the literary analysis of Genesis 2 (and the rest of the book) is historical narrative. It may have theological themes, typology, and symbolism, but none of these are incompatible with history. The Bible presents history all the time with different types of literature, literary devices and structure, and with theological viewpoints. But we have seen that all of scripture and Jesus Himself affirms the historicity of Adam and Eve. We have also briefly argued that Adam was created mortal, with flesh and blood and a proclivity to die, just like us. His disobedience had wide-ranging effects, but I will argue not like many of us usually think. We will take a closer look at that sin and the context in which occurred.

The Fall and the immediate environment of Eden

We last saw Adam and Eve as a married couple of one flesh naked and unashamed, placed in the garden-temple as priests to serve and protect it. And the good Lord had provided everything they needed for delight and flourishing, as well as walking with them in intimate communion. And now we encounter an ominous scene.

Genesis 3:1

Now the serpent was more crafty than any beast of the field which the Lord God had made. And he said to the woman, "Indeed, has God said, 'You shall not eat from any tree of the garden'?" 2 The woman said to the serpent,

"From the fruit of the trees of the garden we may eat; 3 but from the fruit of the tree which is in the middle of the garden, God has said, 'You shall not eat from it or touch it, or you will die.'" 4 The serpent said to the woman, "You surely will not die! 5 For God knows that in the day you eat from it your eyes will be opened, and you will be like God, knowing good and evil."

We encounter a crafty, sly serpent who can speak. The New Testament identifies him as Satan and the devil (Romans 16:20, Revelation 12:9, Revelation 20:2). He seems out of place in this idyllic environment, especially when he begins to speak. What are we to make of this talking snake? While I see no reason to interpret the serpent as metaphorical from the text, the rest of the Bible gives us good reasons to see the physical snake as a conduit and mouthpiece of the devil and Satan, the deceiver and accuser. So, the devil is doing his tempting through this snake.

He begins by questioning what God has commanded by twisting God's words into something He never said. God had told them they could eat from *any* tree in the garden except one. Eve seems to defend God and tell the serpent they can eat from any tree except the one, but also adds a prohibition against touching it and lessens the severity of the original consequence from surely die, to just die. I think we can see here that Eve is already succumbing to the subtleties of the serpent. The serpent proceeds to then contradict God directly by telling Eve she will not die. For that is what the father of lies does (John 8:44). The serpent then impugns the motives and goodness of God to make his command against eating seem to be restrictive and selfish and joins it with a half-truth to entice her. The sad and ironic thing is that Adam and Eve were already in the image of God. They were already like Him to begin with.

There is the impression that Adam and Eve were holy, righteous, and good. They did not know evil or have any selfish desires or motives. They were pure and walked with God.

Genesis 3:6-7

6 When the woman saw that the tree was good for food, and that it was a delight to the eyes, and that the tree was desirable to make one wise, she took from its fruit and ate; and she gave also to her husband with her, and he ate. 7

Then the eyes of both of them were opened, and they knew that they were naked; and they sewed fig leaves together and made themselves loin coverings.

Disastrously, the woman is deceived, falls for the temptation, and believes the lie. And now we see that the lust of her eyes and the lust of her flesh desire the fruit of the tree. She takes and eats and gives to her husband and he takes and eats. We also notice that Adam was with her during the temptation. He does nothing to stop the deception but instead partakes in it. The eyes of them are opened. Not physically, but to an experiential reality of sin, and they now feel guilt and shame. They cover themselves in the areas of their bodies meant to be enjoyed in marriage. The corruption of sin has led to their experience of shame.

Genesis 3:8-13

8 They heard the sound of the Lord God walking in the garden in the cool of the day, and the man and his wife hid themselves from the presence of the Lord God among the trees of the garden. 9 Then the Lord God called to the man, and said to him, "Where are you?" 10 He said, "I heard the sound of You in the garden, and I was afraid because I was naked; so I hid myself." 11 And He said, "Who told you that you were naked? Have you eaten from the tree of which I commanded you not to eat?" 12 The man said, "The woman whom You gave to be with me, she gave me from the tree, and I ate." 13 Then the Lord God said to the woman, "What is this you have done?" And the woman said, "The serpent deceived me, and I ate."

We recognize that this may be another anthropomorphism of God walking in the garden, but it conveys the intimacy that they had with God. They hide themselves and it is not that God doesn't know where they are, but He is going to call them to account for their actions. The guilt of sin has led to hiding from God for fear of punishment. After asking whether they broke His commandment, the man starts blaming the woman and, by extension, the Lord God Himself. While the woman blames the serpent. So, the tragedy has unfolded. The covenant, promise keeping God and Creator, the One who is both transcendent and immanent, who gave Adam and Eve the best possible environment and always had their best interests in mind has been sinned against because of unbelief and the lusts of the flesh. And no one wants to take personal responsibility for what has happened. So, God pronounces the curses for breaking His covenant on the serpent, Eve, and then Adam.

Genesis 3:14-19

14 The Lord God said to the serpent,

"Because you have done this,

Cursed are you more than all cattle,

And more than every beast of the field;

On your belly you will go,

And dust you will eat

All the days of your life;

15 And I will put enmity

Between you and the woman,

And between your seed and her seed;

He shall bruise you on the head,

And you shall bruise him on the heel."

16 To the woman He said,

"I will greatly multiply

Your pain in childbirth,

In pain you will bring forth children;

Yet your desire will be for your husband,

And he will rule over you."

17 Then to Adam He said, "Because you have listened to the voice of your wife, and have eaten from the tree about which I commanded you, saying, 'You shall not eat from it';

Cursed is the ground because of you;

In toil you will eat of it

All the days of your life.

18 "Both thorns and thistles it shall grow for you;

And you will eat the plants of the field;

19 By the sweat of your face

You will eat bread,

Till you return to the ground,

Because from it you were taken;

For you are dust,

And to dust you shall return."

For most of my Christian life I believed that a lot of physical things of cosmic significance happened in these verses. From earthquakes to tornadoes, from predation to viruses, from rust to physical death. But can we read these global effects out of the curses God pronounces on Adam, Eve, and the serpent?

The first one to receive the curses is the serpent. While many are tempted to read this as snakes are finally forced to crawl on their bellies, I believe it is better to read as a humbling pronouncement on Satan himself, using the body and mouthpiece of a serpent. Crawling along the ground with a mouthful of dust is a curse of humiliation. And the enmity which is to be between the seed of the woman and the seed of this serpent surely points us to the surrounding context of what is to follow. That is, that the elect line of Seth will carry on the promise to bruise or crush the serpent's head.

The next is the woman and pronouncements on her. God says that her pain in childbirth with be multiplied. I see no reason not to assume this is literal, as the Hebrew word is for conception and pregnancy. But I don't think it is all physical pain. We will see that her first two male children bring enormous emotional pain and suffering with murder. She will desire her husband, but he will rule over her. The same kind of language is used by God in the next chapter to describe sin's desire for Cain and that Cain must master it. So, this is not a good kind of desire and mastery. Eve will want to assert her control and manipulate Adam, but Adam will dominate Eve and force her to submit. Their relationship has lost the original intention God had for them. Instead of being naked, unashamed, and helpers to one another who fulfill and complete each other, they are now riddled with guilt, shame, and want to dominate and manipulate each other. Instead of being fruitful and multiplying with little to no pain in childbirth, now Eve will suffer and be frustrated in her mandate to be fruitful and multiply.

The last curses belong to Adam, who we are taught later in scripture (Hosea 6:7, Romans 5, 1 Corinthians 15) is the covenant head of natural humanity. Although we can infer his covenant headship from the immediate context based on expulsion from the garden and the cancer of sin spreading through his descendants. God curses the ground because of Adam and tells him that he will toil in his work as a farmer, weeds will frustrate his plans for desirable crops, and he will work and sweat to produce cereal grains and turn them into bread until the day he dies. Adam will return to the ground from which he was taken and formed. Adam has also been frustrated in his mandate to exercise dominion and to expand the kingdom of God.

Here is a good time to discuss prophecy in the Bible. There are many things going on here that are immediately fulfilled, but that is usually not the end of the story with prophecy from God. Many, if not most, prophecies have both near and far fulfillments. And our story here is no different. Yes, the serpent has been humiliated, the woman has been cursed with pain in fulfilling her mandate, the promise of the seed of the woman is given to Eve, and the curse and enmity fall on the serpent in the garden. But the rest of the Bible plays out the children of God and the children of the serpent in enmity. And while Seth is the appointed child and heir of the promise, his descendants ultimately led to the seed of the woman who will crush the head of the serpent.

Revelation 12:1-2

12 A great sign appeared in heaven: a woman clothed with the sun, and the moon under her feet, and on her head a crown of twelve stars; 2 and she was with child; and she *cried out, being in labor and in pain to give birth.

This woman, symbolizing Israel and the church, is in labor pains to give birth to the Messiah. The serpent still has enmity with the seed of the woman and tries to devour the One who will crush his head.

Revelation 12:3-4

3 Then another sign appeared in heaven: and behold, a great red dragon having seven heads and ten horns, and on his heads were seven diadems. 4 And his tail *swept away a third of the stars of heaven and threw them to the earth. And the dragon stood before the woman who was about to give birth, so that when she gave birth he might devour her child.

Who is this child that the dragon is so intent on destroying?

Revelation 12:5

5 And she gave birth to a son, a male child, who is to rule all the nations with a rod of iron; and her child was caught up to God and to His throne.

This is the Messiah, Jesus the Christ. (Psalms 2:9).

And who is this dragon?

Revelation 12:9

9 And the great dragon was thrown down, the serpent of old who is called the devil and Satan, who deceives the whole world; he was thrown down to the earth, and his angels were thrown down with him.

The serpent of old, who is the devil and Satan, knows that the seed of the woman has come. Satan and his reprobate angels are thrown down from heaven, no longer to accuse the children of God. He wages war against the church knowing that his time is limited. The kingdom of God that Adam was supposed to expand to the ends of the earth and that the serpent foiled, has finally come. All rule and authority have been given to this Seed and those who follow Him will overcome by the blood of the Lamb. The pain of labor and childbirth have come to fruition and now even the creation itself waits, suffering the pains of childbirth, to be redeemed according to the children of God (Romans 8:18-22).

Back to the immediate context and implications of the curses.

In orthodox Christianity through the ages this has been interpreted as the entire planet has changed from one of abundance to one of frustration. While I agree that the ground has been cursed and will frustrate Adam and his descendants, I think this is more of a local curse and phenomenon as we will see.

But the same idea of being frustrated in the original plan and mandate for the couple is present. Instead of having dominion, ruling, and subduing, Adam will now be frustrated with his calling and vocation. We know from the previous chapter that God formed Adam somewhere outside of the garden and then placed him into it later.

But the scene is definitely not one of triviality. The couple have disobeyed God and paradise has indeed been lost. But the greater tragedy is the loss of communion with God Himself. We see hints here that the mandate of the couple is in jeopardy. What will it take to restore fellowship with the Lord and to expand the kingdom?

Genesis 3:20-21

20 Now the man called his wife's name Eve, because she was the mother of all the living. 21 The Lord God made garments of skin for Adam and his wife, and clothed them.

Here we get the proper name of the woman, Eve, which means life or living. She is called the mother of all the living by Adam, and we get the first hints of faith exercised by Adam in the promise of God that the seed of the woman will bruise or crush the head of the serpent. He also appears to believe that he and Eve will be fruitful and multiply despite the judgement and curses. God makes them garments of skin and clothes them. Here is the beginning of the covenant of grace within history. Instead of exercising justice toward His creatures He created and provided everything for, which God had every right to do, He performs grace. He takes their meager fig leaf loin coverings and replaces them with proper garments. Many note that this implies the killing of an animal to make the clothing. While not explicit in the text, the idea of a sacrificial atonement for sin would be clear by now to the Israelites. But we have our first glimmers of hope and promise after the Fall.

Genesis 3:22-24

22 Then the Lord God said, "Behold, the man has become like one of Us, knowing good and evil; and now, he might stretch out his hand, and take also from the tree of life, and eat, and live forever"— 23 therefore the Lord God sent him out from the garden of Eden, to cultivate the ground from which he was

taken. 24 So He drove the man out; and at the east of the garden of Eden He stationed the cherubim and the flaming sword which turned every direction to guard the way to the tree of life.

We first notice that as Adam and Eve were priests in the garden to serve and guard the holy garden temple, now the cherubim and flaming sword guard the way to the tree of life. Adam and Eve are presented prior to the Fall as holy, righteous, and good. They had personal communion with the Lord God and did not need a mediator to stand between themselves and the Lord God. He walked and talked with them. And they have lost this. How will we get it back?

Revelation 21:3-4

3 And I heard a loud voice from the throne, saying, "Behold, the tabernacle of God is among men, and He will dwell among them, and they shall be His people, and God Himself will be among them, 4 and He will wipe away every tear from their eyes; and there will no longer be any death; there will no longer be any mourning, or crying, or pain; the first things have passed away."

What are we to make of God's pronouncement that man has become like one of Us, knowing good and evil? In chapter 4 when we examined the myths of the ancient near east, I commented on several myths like Adapa and the Epic of Gilgamesh where people try to say this language in Genesis 3 in an adaptation of those older stories. We discussed there why those arguments fall short and are nothing like the consequences of the Fall. So, what is going on here?

God knows good and evil in every sense of the word knowing, except knowing evil experientially. There is nothing impure in Him and the rest of the scriptures testifies clearly that God hates sin and evil. So, in what way does Adam know good and evil like God? Some scholars suggest that knowing good and evil here represents making judgements in regard to justice. Or knowing evil in the sense of moral distinctions. But we need to first look at the tree of life to see if that makes sense. What does Revelation say about the tree of life in the new Jerusalem?

Revelation 2:7

He who has an ear, let him hear what the Spirit says to the churches. To him who overcomes, I will grant to eat of the tree of life which is in the Paradise of God.'

Revelation 22:1:-3

Then he showed me a river of the water of life, clear as crystal, coming from the throne of God and of the Lamb, 2 in the middle of its street. On either side of the river was the tree of life, bearing twelve kinds of fruit, yielding its fruit every month; and the leaves of the tree were for the healing of the nations. 3 There will no longer be any curse; and the throne of God and of the Lamb will be in it, and His bond-servants will serve Him;

Revelation 22:14-15

14 Blessed are those who wash their robes, so that they may have the right to the tree of life, and may enter by the gates into the city. 15 Outside are the dogs and the sorcerers and the immoral persons and the murderers and the idolaters, and everyone who loves and practices lying.

Revelation 22:18-19

18 I testify to everyone who hears the words of the prophecy of this book: if anyone adds to them, God will add to him the plagues which are written in this book; 19 and if anyone takes away from the words of the book of this prophecy, God will take away his part from the tree of life and from the holy city, which are written in this book.

We can then see that the tree of life will be freely offered to those who are in Christ. And only those who are covered in the righteousness of the blood of the Lamb will have access. The sinful and impure cannot touch or eat from this tree. We can infer from our text in Genesis that Adam and Eve are barred from the garden and the tree because they have sinned and are impure. Many recognize that there was no explicit prohibition given to the couple about not eating from the tree of life, so it is implied that they could freely eat from it, along with the other trees in the garden.

I therefore conclude that the tree acts as a sign and seal of the sacrament symbolizing their communion and life with God, not that it had magical properties of eternal life. Therefore, knowing evil, both in a sense of moral distinction, like God Himself, and experientially, something totally foreign to God, God could not allow them to partake of the tree that signified eternal life in communion with Himself. By eating from the tree of life after being defiled by sin, Adam and Eve would have had no chance at redemption. They would be approaching the sacrament of life with unclean hands and would therefore be confirmed in their spiritual damnation. God not allowing them to enter the garden after their disobedience and defilement is another act of grace.

The garden priests, Adam and Eve, have defiled themselves in the temple-garden and the Lord God cleanses his temple by banishing them from the garden of Eden to cultivate the very ground from which Adam was taken. God sends them out and stations cherubim at the east of the garden, along with a flaming sword to guard the way to the tree of life. Cherubim played a prominent role in the later tabernacle and temple. Cherubim had their wings covering the mercy seat on the ark of the testimony (covenant). But more applicable to our verses, they were embroidered on the veil that separated the holy of holies from the holy place in the tabernacle and in both the inner and outer sanctuaries of Solomon's temple. (Exodus 26:31-34, 1 Kings 6:23-30) The way to God's immediate presence has been shut.

We can then summarize the Fall. Satan, using the body and form of a serpent, lies and tempts Adam and Eve to break God's commandment. They failed to exercise their duties as priests to cast the serpent out of the garden by not properly serving and guarding it. Instead, they let the unholy into the temple-garden and succumbed to the temptation by not trusting God's word through unbelief, by impugning God with selfish motives, and by wanting to determine right and wrong on their own terms. Adam and Eve lose their original purity and now know good and evil, both mentally and experientially. Now their original creation mandate to be fruitful and multiply, to fill the earth, and to have dominion over the creatures and the land has been frustrated and twisted. Their relationship with God, relationship with each other, and relationship with God's creation has changed. God pronounces curses on them for breaking His covenant (we will discuss how far-ranging these effects were in short order) and then banishes them from the garden. They have defiled the sacred and clean space of the garden and are cast out into the common and profane. The way is shut to the tree of life and personal communion with the Lord God. There are hints that not all hope is lost though. Adam exercises faith in the promises of God by calling his wife Eve.

How will man once again be able to walk with God in the garden? If a man and his wife living in an ideal environment and created directly by God in true knowledge, holiness, and righteousness cannot obey God, is there any hope for mankind? The rest of Genesis follows the descendants of Adam to find out. But we still have a few questions to answer. What was the creation like outside the garden? When in history is the most probable timeframe for the garden? What to make of Cain and his descendants? Seth and his? The flood? We won't try to exposit all the verses in chapters 4 through 9 but we will talk about the most pertinent ones germane to our discussion. And then wrap up with a general summary and place the garden squarely within history.

Genesis 4-9: What kind of world is this?

As we pick up in Genesis 4, Adam and Eve have been cast out of Eden. Adam and Eve have their first male child, Cain. And their second male child, Abel. Cain was a farmer and Abel a shepherd. The chapters switched from the name God (Elohim) in chapter 1, to Lord God (Yahweh Elohim) in chapters 2 and 3, and now just to Lord (Yahweh). Contrary to critical scholars who see evidence of different authors and sources with the divine name changes, it is extremely important and deliberate by Moses for the theology of these first few chapters to link the creator God with Israel's self-existent, immanent, covenant-keeping God.

Both Cain and Abel bring offerings to the Lord. Cain brings not the first fruits of his land, but just some fruit of the ground. Abel, on the other hand, brings the first fruits of his flock. This would make sense to the Israelites who have just been given the law at Sinai. Abel is obedient and faithful while Cain does not give the first and best of his land. Abel is acting in faith, while Cain is giving a token offering not combined with faith (Hebrews 11:4). The Lord has regard for Abel's offerings but not Cain's. Cain is emotionally disturbed by this and ultimately murders his brother Abel. We can already see the effects of sin and the Fall increasing rapidly. Now we have cold-blooded, premeditated murder between kin.

Not only is the ground cursed because of Adam, but Abel's blood is crying out from the ground. The sin of man affects the creation as well. We pick up with God pronouncing judgement on Cain.

Genesis 4:10-17

10 He said, "What have you done? The voice of your brother's blood is crying to Me from the ground. 11 Now you are cursed from the ground, which has opened its mouth to receive your brother's blood from your hand. 12 When you cultivate the ground, it will no longer yield its strength to you; you will be a vagrant and a wanderer on the earth." 13 Cain said to the Lord, "My punishment is too great to bear! 14 Behold, You have driven me this day from the face of the ground; and from Your face I will be hidden, and I will be a vagrant and a wanderer on the earth, and whoever finds me will kill me." 15 So the Lord said to him, "Therefore whoever kills Cain, vengeance will be taken on him sevenfold." And the Lord appointed a sign for Cain, so that no one finding him would slay him. 16 Then Cain went out from the presence of the Lord, and settled in the land of Nod, east of Eden. 17 Cain had relations with his wife and she conceived, and gave birth to Enoch; and he built a city, and called the name of the city Enoch, after the name of his son.

We have something curious here. For the first time since the beginning of the book we have reference to someone not mentioned in Adam and Eve's family. After Cain kills Abel, the ground will not be productive for him in his farming, he is driven away to be a vagabond and wanderer, and Cain is fearful that anyone finding him will kill him. Cain is expelled from God's presence, which is symbolic of being spiritually lost, and he literally navigates further east of Eden. He foreshadows what will happen to Israel when they are later exiled for breaking the covenant.

Assuming Adam and Eve are the first human beings, then Cain is fearful of his own family taking vengeance for his murder. While possible, and even probable, based on laws regarding the blood avenger in the Sinai covenant (Numbers 35:16-21, Numbers 35:30-34, Deuteronomy 19:11-13), it need not be the case. But this has been the prevailing interpretation taken by conservative Christians for millennia. Interpreters of Genesis have noticed this problem of Cain being worried about being found and killed for a long time, not to mention Cain's wife, and offered various proposals. One of which was the suggestion that other people were around during the time of Adam and Eve. And I argue that seeing Adam's family as the only people around doesn't make the best sense within the broader context.

We have argued that the creation narrative is not to be taken as historical narrative. That there is darkness and the sea as part of the finished ordered creation, implying forces hostile to life. And that this is contrasted with the new heavens, new earth, and new Jerusalem. We have also argued that the garden of Eden is a temple-garden, or sacred space, within the larger environment of a common or profane world. We have argued that the pronouncements of the goodness of creation don't refer to moral goodness, but functional goodness. That is, that God has so ordered his creation to be set up with fructifying life and in such a way that the son of God, Adam, can fulfill his mandate as God's vicegerent, image bearer, and holy priest to expand the kingdom of God.

I therefore propose that the garden of Eden was a paradise temple into which the miraculously created Adam and Eve were priests to serve and guard and have fellowship with the Lord. And out of this base of operation, they were tasked with expanding the boundaries of the garden and extending the kingdom of God into the broader world. They were to accomplish their subduing and dominion by being in close communion with God and producing godly offspring that would continue the expansion of the kingdom of God. Back to the text.

God shows his mercy towards Cain in that the Lord gave Cain a sign so that no one finding him would kill him. Cain settles in the land of Nod, named after his wandering, east of Eden. There is already a broader context to the land surrounding Eden. We are now introduced to Cain's wife. No mention about where she is from or if she is related to Cain. I will argue later that this makes sense in the context of the judgement leading to the flood. They have a son, and they name him Enoch. We are told that Enoch builds a city, and it is named after him. Although a "city" in this time could be any place with a wall around it, it makes the most sense within the context of a broader population that is already east of Eden.

The rest of chapter 4 goes through ten generations of Cain's family. It isn't given the same amount of attention as the line of Seth, and for good reason. Seth is the chosen line who will keep the promise alive of the seed that will crush the head of the serpent and by whom his descendants will once again be tasked with expanding the kingdom of God. One of the literary purposes of the descendants of Cain is to show the increasing brutality and propagation of sin. Whereas God said anyone who kills Cain will be avenged sevenfold (Genesis 4:15), the polygamist Lamech boasts about killing a man for wounding him and saying that he should be avenged seventy times seven. The Israelites could already see that eye for an eye justice is needed in the covenant given at Sinai so that people do not take disproportionate revenge. And this contrasts starkly with Jesus's teaching about how many times we are to forgive when a brother sins against us, also seventy times seven (Matthew 18:21-22). But there is grace even toward the reprobate line of Cain. For they too are blessed with children to be fruitful and multiply. And they produce life-enhancing technology and culture through their generations. Some of their descendants dwell in tents and have livestock, others are musicians, while others are metalworkers in bronze and iron. These references to bronze and iron are interesting.

Genesis 4:22

22 As for Zillah, she also gave birth to Tubal-cain, the forger of all implements of bronze and iron; and the sister of Tubal-cain was Naamah.

Many scoff at this point because if true bronze and iron is meant in the sense of smelting and producing alloys, we are talking thousands of years after Eden before we have evidence of such things from archaeology. As we will see later when we discuss the dating of the garden of Eden, we are talking about thousands of years before the bronze or iron age. But the text is simply describing the far descendants and peoples coming from Cain's offspring and the types of arts, trades, and professions that came to characterized them. **20 Adah gave birth to Jabal; he was the father of those who dwell in tents and have livestock. 21 His brother's name was Jubal; he was the father of all those who play the lyre and pipe. 22 As for Zillah, she also gave birth to Tubal-cain, the forger of all implements of bronze and iron; and the sister of Tubal-cain was Naamah.**

Just a little tangent. But the purpose of the narrative is not necessarily technology. It is to show that even the ungodly line of Cain, who are getting progressively morally corrupted, experience the common grace of God.

We then read that Adam and Eve have another son and name him Seth. This is obviously not in chronological order. Seth was born long before the end of Cain's genealogy, but the purpose of Moses is not strict chronology. Whereas Eve had said that she had gotten a man from the Lord with the birth of Cain, now Eve shows her faith in the promise of God when she declares that God has appointed (Hebrew Shiyth) her another seed or offspring, Seth (Hebrew Sheth), in the place of Abel. The hope of the seed to crush the head of the serpent is still alive and Eve trusts the promise. Seth has a son named Enosh. We are not told about Seth's wife or where she came from. And then men begin to call upon the name of the Lord. The chosen line of Seth looks to and worships the Lord of their parents. We are given no such indication from Cain's line.

Chapter 5 starts with a new toledot section and could indicate that Moses is using an already existing written source. We are again reminded about the image of God and how it ties to Adam and Eve. This in an important introduction to the section, as it relates the godly line of Seth and his descendants.

Genesis 5:3-5

3 When Adam had lived one hundred and thirty years, he became the father of a son in his own likeness, according to his image, and named him Seth. 4 Then the days of Adam after he became the father of Seth were eight hundred years, and he had other sons and daughters. 5 So all the days that Adam lived were nine hundred and thirty years, and he died.

Seth is born in the likeness and image of Adam. We will discuss this further when we get to the chapter on the image of God. But it is important to see the image passing from Adam to Seth. Adam was in the likeness and image of God, and Seth is in the likeness and image of Adam. There is no indication in the text so far that explicitly tells us that the image of God has changed or been corrupted, but later scripture does indeed indicate that it is marred and defaced. And it is progressively restored again to those who are in Jesus Christ, the true image of God.

Adam is 130 years old when Seth is born and lives to the ripe old age of 930 years. What are we to make of the ages of Adam and his descendants in this chapter? As we have noted earlier, many people make comparisons between the Sumerian king lists and these genealogies. In the king lists, the reigns of the kings prior to the flood are in tens of thousands of years, then the flood swept over, and then their reigns decrease (but are still hundreds to thousands of years long). While we see the same general pattern here of very long lives prior to the flood, the flood, and then lives that are shorter (but still very long). But here we are dealing with hundreds of years, not tens of thousands. We have already argued that it is far more likely that the descendants of the sons of Noah, some of whom make up the land of Sumer, Akkad, and Babylonia, incorporated this shared tradition into their own exaggerated chronology, religion, and politics. As is usual for Mesopotamia, they take actual history and mythologize it.

Let's say my hypothesis is correct for the sake of argument. What then about these ages in Genesis? There is nothing in biology that says organisms cannot live for long periods of time. Some simple organisms and bacteria can live for hundreds of thousands, even millions, of years. Trees can live for thousands of years. The Greenland shark can live between 300-500 years, tortoises can live for 200 years, and so forth. But what about humans? My tentative proposal is that the chromosomes, genes, and biochemistry of cell division and metabolism of Adam and Eve as newly created humans were pristine and allowed them to reach incredible ages. With thousands of years of mutations, sexual reproduction with others outside of the garden, damage to DNA and other biochemical molecules, their subsequent generations slowly assumed the kinds of lifespans that we are used to. Why organisms actually age and why we have to get old is still a debated topic among scientists.

Of course, this is far afield from the text and any information that Moses gives us. But I personally don't see how the text can be interpreted any other way. The way the narrative is structured seems to rule out any other options in my mind. It states how old they were when they became the father of X (who the Israelites would know as the first-born son), how many years they lived after becoming the father of X, and then the total number of years they lived. Many interpret these lifespans as evidence that Adam and Eve were initially immortal, but we have seen why that isn't the best interpretation from various lines of evidence in scripture itself. I think we can safely assume that these people were made of flesh and blood like us, and they would need to be transformed into a different type of body to inherit the kingdom of God. Others say the numbers are symbolic, but I find them unconvincing based on the ones I have read so far. But I admit that is still possible. The critics claim that the Israelites simply borrowed and then adapted the idea of the antediluvian kings or sages from the king lists and other literature for their own theology and origin stories. My argument again (barring any evidence or persuasive arguments to the contrary) is that Adam's descendants lived lives we would consider incredibly long, and the Mesopotamians took this history and made it look even more fantastic for their own documents and myths.

What is theologically significant though, is the constant refrain that they died. In one sense Adam and Eve died when they ate from the forbidden tree. They died spiritually and no longer had close, personal communion with the Lord God. And they later died physically because the way had been shut to the tree of life. They and their descendants would need atonement for their sins from this point forward or they would not only taste physical death but spiritual death as well. Their hope was to be found in God and His promises about the woman's seed and the salvation He will bring. The narrative here in Genesis 5 then is hammering home the picture of both hope and the reality of the Fall.

We have ten generations here, like in Cain's genealogy. Both genealogies are compressed for theological reasons. We know explicitly that those in the line of Seth had other sons and daughters and can infer as much from Cain's as well. This would make sense to the Israelites hearing it because they lived in patriarchal societies where genealogies were traced through males in the household, and they knew from the other books of Moses about God's laws regarding inheritance and headship. The first-born male got a double portion of the inheritance and then the rest of the inheritance went to remaining male children. Daughters were given a dowry in marriage and then were under the care of their husband's family. The first-born son was also considered head of the household upon his father's death. So, laying out the genealogies by naming first-born sons makes sense within the Mosaic covenant and ancient near eastern cultures. What is also interesting is that Adam's first-born son Cain does not receive the blessing and double portion. Cain forfeits his first-born prerogatives because of his sin, Seth is appointed to Eve and chosen by God to receive the blessing, and he is named in the genealogy as if he is Adam's first born. This concept of the first-born son relinquishing their inheritance or God choosing a younger son to use and bless will show itself throughout the rest of the book of Genesis, and especially in the line of Abraham. Here Seth is the child of promise. Just like Isaac was, just like Jacob was.

In the seventh generation from Adam (Jude 14-15) we meet Enoch. And we learn that Enoch walked with God and God took him. We know that men have been calling on the name of the Lord in the godly line of Seth and here we encounter a righteous man walking by faith with God. The fact that the text seems to intimate he does not die gives us hope about this line of descendants (Hebrews 11:5-6). Lamech has a first-born son named Noah, and he is given that name because Lamech believes Noah will give them rest or comfort from their work and the toil of their hands working the ground that the Lord had cursed. Because the Lord God had cursed the ground on account of Adam for his disobedience. Noah becomes the father of three sons: Shem, Ham, and Japheth.

Noah's flood

Now we move into Genesis 6 and the beginning of the story of Noah's flood. As it is relevant to our discussion about the context of the garden of Eden in terms of geography, dating, human populations, the original state of the world outside the garden, the image of God, reasons for the flood, and some exegetical differences within the flood. Let's dive in.

Genesis 6:1-4

6 Now it came about, when men began to multiply on the face of the land, and daughters were born to them, 2 that the sons of God saw that the daughters of men were beautiful; and they took wives for themselves, whomever they chose. 3 Then the Lord said, "My Spirit shall not strive with man forever, because he also is flesh; nevertheless his days shall be one hundred and twenty years." 4 The Nephilim were on the earth in those days, and also afterward, when the sons of God came in to the daughters of men, and they bore children to them. Those were the mighty men who were of old, men of renown.

We just got done looking at the genealogy of the godly line of Seth and the wicked line of Cain. I proposed that there were other people around outside of the garden, and these other people are who Cain was worried about and where

Cain's and Seth's wives may have come from. Then what does it mean that daughters were born to men and the sons of God took them as wives?

In the immediate context I think one solid interpretation is that the sons of God are from the line of Seth and the daughters of men are from the line of Cain. Many commentators take the interpretation that the sons of God are angels using references from the New Testament in 2 Peter and Jude as evidence, but I don't think this interpretation of the New Testament verses is sound to begin with. Plus, this hardly fits the context. We are taught by Jesus that angels do not marry, and the judgement is against people, not angels. Others think it refers to kings or other royalty, but again this seems out of context to me and an untenable interpretation. By using our previous conclusion about people outside the garden we have a few options. One is that both the lines of Seth and Cain are marrying women outside the descendants of Adam and Eve. But we already proposed they were doing this in Genesis 4 and 5. The other option was the first one mentioned, that the descendants of Seth were marrying the daughters of Cain. The third is a combination of the two. I think that the sons of God clearly refer to the line of Seth because Adam was the son of God (Luke 3:38), Seth was born in the likeness and image of Adam, and people in Seth's line were calling on the name of the Lord. And both Enoch and Noah were righteous men. Therefore, I think the best interpretation is that men in the line of Seth were marrying into godless or idolatrous cultures and people without regard to the will of God. The Israelite audience would understand the prohibition against marrying pagans and the lure of their idolatry. The way that the sons of God chose wives in this passage is reminiscent of how Eve was tempted in the garden. Just as Eve saw that the fruit was good, pleasing, and then took and ate, so too these men see the women are good, pleasing, and then they take. The lusts of the flesh are overriding the will of God.

The Lord pronounces that His Spirit with not strive with man forever and therefore man's days will be 120 years. Many interpret this verse to be referencing a new lifespan given to man after the flood. But I don't think that is the best interpretation for a couple of reasons. The lifespan of people after the flood continues to be longer than this for thousands of years afterward, and many people live lives much shorter than 120 years after the time of Moses and Joshua. I think it is best understood as the time between the pronouncement of judgement and the start of the flood. God in His grace is giving people time to repent before judgement (1 Peter 3:20).

Who were these Nephilim? The Nephilim were an ethnicity of people who were very tall "giants" according to Numbers 13:33. It is important to recognize here that the Nephilim were around before the flood and also afterward according to our text (Genesis 6:4). And then they are again attested to in the book of Numbers. So, the first conclusion we can begin to tentatively draw is that people outside the ark survived the flood.

But back to the Nephilim. Many people have wild interpretations of this one. Half-angel, half-men demigods roaming the earth. Can we get that from the text? I don't think so. The best interpretation of verse 4 in my opinion is that the Nephilim were not the offspring of anyone in our story. Not of angels, tyrants, nor the sons of God (Seth's righteous line) and the daughters of men (women from pagan cultures). But it is simply pointing out that the Nephilim were alive at this time in history "and also afterward" (v. 4). In other words, it is a chronological marker and nothing else. Otherwise, besides being very strange in the context and the light of biblical theology, it would lead us to conclude that Caleb and the spies sent to spy out the promised land in Numbers 13 still saw these half-angel demigods wandering about. It makes much more sense to view them as giants or mighty men (think Goliath) which may give some in the audience an idea of the time in history prior to the flood.

Starting in verse 5, we see:

Genesis 6:5-8

5 Then the Lord saw that the wickedness of man was great on the earth, and that every intent of the thoughts of his heart was only evil continually. 6 The Lord was sorry that He had made man on the earth, and He was grieved in His heart. 7 The Lord said, "I will blot out man whom I have created from the face of the land, from man to animals to creeping things and to birds of the sky; for I am sorry that I have made them." 8 But Noah found favor in the eyes of the Lord.

Here we have one the most sweeping pronouncements of moral depravity in the entire Bible. Like a scourge, sin is multiplying and intensifying in mankind.

Man has gotten to the logical end point of the corruption of sin without the restraining grace of God. Every thought and intention were evil, resulting in evil behavior. Unlike the flood myths from Mesopotamia, the reason for God's judgement and bringing the flood is the sin and moral depravity of man. God is deeply grieved by sin, then and now. The morally perfect God who in His goodness and love made a couple without sin and put them in a perfect environment to expand the kingdom of God is now faced with a land teeming with sin and violence. This is indeed a turn of events from the beginning scenes of Genesis. God wills to bring a flood to not only wipe out mankind but air and land creatures as well. The sin of man affects the rest of creation. Lest we and the original audience despair, Noah finds favor in the eyes of the Lord. By God's unmerited grace and love, the line of Adam and Seth will continue with the hope of the seed of the woman that will crush the head of the serpent.

The narrative continues with another toledot section which relates the generations of Noah, his wife, his sons, and their wives through the flood. We see again that the earth was corrupt, and instead of being filled with godly offspring expanding the kingdom of God and ruling wisely and subduing it, they instead have filled the earth with violence. God gives instructions to Noah about building the ark and the details point to our location in southern Mesopotamia. Probably closer to the foothills of the Zagros mountains in Iran, as this would be east of Eden and have trees for constructing the ark. Bitumen naturally flows out of the ground in this area for waterproofing vessels. But we must remember that the climate and ecology of southern Mesopotamia was different back then. It was not nearly as arid as today, so gopherwood and other trees may have been more abundant and accessible to Noah.

God tells Noah that He will establish His covenant with Noah and commands him to take aboard two of every animal, male and female, along with his wife, sons, and sons' wives. The animals will come to Noah on their own volition and then he is to gather food for himself and the animals. Noah is obedient in all that God commanded him.

Starting in chapter 7, the Lord commanded Noah to enter the ark with his family, stating that Noah alone has the Lord found to be righteous in this generation. In His grace, God delivers Noah's family and the animals on account of the faith and righteousness of Noah (Ezekiel 14:12-20, Hebrews 11:7). This is the first explicit mention of clean and unclean animals in the Bible. Noah is to take seven clean animals, male and female, and unclean animals by two, male and female. The Lord tells him that in seven more days (seven is the number of completeness throughout the Bible) He will bring a flood upon the earth for 40 days and 40 nights (40 is a number associated with judgment and trial in the Bible, think wandering 40 years in the wilderness, Jesus in the desert 40 days before being tested by Satan, etc.) to blot out all life. We can infer 120 years have passed from God's judgement to flood the earth until the rains begin. We are given very specific information about when the rains started.

Genesis 7:11-12

11 In the six hundredth year of Noah's life, in the second month, on the seventeenth day of the month, on the same day all the fountains of the great deep burst open, and the floodgates of the sky were opened. 12 The rain fell upon the earth for forty days and forty nights.

The rains come, waters come from the earth itself (snow runoff, river flooding, underground aquifers, miracle?), the waters cover the earth, the ark floats on the surface of the waters, the waters cover the high mountains (or hills, Hebrew har) which were under all the heavens, the waters keep rising another 23 feet, and everything with the breath of life in it including mankind and all animals are killed. The waters prevailed upon the earth for 150 days.

What we have here is de-creation. All that God had made was destroyed, except for the people and animals in the ark. From an ordered world to one that is once again covered by the deep. And then the pivot in the story.

Genesis 8:1

But God remembered Noah and all the beasts and all the cattle that were with him in the ark; and God caused a wind to pass over the earth, and the water subsided.

Now we begin re-creation. We have the deep and God causes a wind (Hebrew Ruwach) to separate the waters. The same Hebrew word Ruwach for the Spirit in Genesis 1, except there the grammar implies God's Spirit.

The ark comes to rest on the mountains of Ararat. Where is this? First, we need to determine the scope of the flood. Was it global or local?

The first thing to bear in mind is that the Hebrew phrase (kol erets) is translated, "whole earth", "all the earth", "whole land", or "all the land", and is determined by the context of the passage. There is no doubt that this is a massive flood, but nothing in the text (except reading into the text the assumption of a global flood and landing on a 16,000-foot-high mountain) would preclude that the picture is one of water from horizon to horizon with all the foothills covered in water.

Acts 17:26

26 and He made from one man every nation of mankind to live on all the face of the earth, having determined their appointed times and the boundaries of their habitation

Paul tells the men of Athens that God is Lord of heaven and earth and all that it contains. We are told from verse 26 that from one man God made every nation of mankind to live on all the face of the earth. Who was this one man? Paul is either talking about Adam or Noah.

Genesis 9:18-19

18 Now the sons of Noah who came out of the ark were Shem and Ham and Japheth; and Ham was the father of Canaan. 19 These three were the sons of Noah, and from these the whole earth was populated.

We need to be mindful of not letting the word "earth" used by the translators to fix our minds on the planet Earth rotating in empty space, instead of the earth (dirt, soil, land) at our feet. Or to get hung up with the words "all" and "whole". We will see that the ancient near east is the area in view of many biblical passages including these. Think "land" by default for the Hebrew word (erets) in Genesis 9 and "land" for the Greek word (Ge) in Acts 17. If this is too ambiguous and not convincing, let's look at some other verses about what is intended by "the whole earth".

Genesis 13:8-9

8 So Abram said to Lot, "Please let there be no strife between you and me, nor between my herdsmen and your herdsmen, for we are brothers. 9 Is not the whole land before you? Please separate from me; if to the left, then I will go to the right; or if to the right, then I will go to the left."

Verse 9 uses the same Hebrew (kol erets) for "whole land". The translators could have used the phrase "whole earth" here, but the context makes it clear that they are talking about the land of Canaan.

Genesis 41:53-57

53 When the seven years of plenty which had been in the land of Egypt came to an end, 54 and the seven years of famine began to come, just as Joseph had said, then there was famine in all the lands, but in all the land of Egypt there was bread. 55 So when all the land of Egypt was famished, the people cried out to Pharaoh for bread; and Pharaoh said to all the Egyptians, "Go to Joseph; whatever he says to you, you shall do." 56 When the famine was spread over all the face of the earth, then Joseph opened all the storehouses, and sold to the Egyptians; and the famine was severe in the land of Egypt. 57 The people of all the earth came to Egypt to buy grain from Joseph, because the famine was severe in all the earth.

This is relating the account of Joseph in Egypt when the famine began in Egypt after preparing for seven years prior. Verse 54 tells us there was famine in all the lands except Egypt. What are "all the lands"? In verse 56, it is equated with "all the face of the earth". So, it is explicit here that "all the lands" are the same as "all the face of the earth" for this passage. The face of the earth here is at a minimum from Egypt to the land of Canaan (where Joseph's family was). And at a maximum the immediate surrounding nations. But not the entire planet. We are told by Moses in verse 57 that "all the earth came to Egypt to buy grain". Are we to read into this text that it means the whole planet? Moses is just relating how the Pentateuch often refers to "all the face of the earth". Verse 57 just reiterates this point again using the Hebrew (kol erets) for "all the earth".

1 Kings 10:24

24 All the earth was seeking the presence of Solomon, to hear his wisdom which God had put in his heart.

Again, it is clear that all the earth here is in regard to the people of the ancient near east. And probably more limited from Sinai to Arabia to Phoenicia to northcentral Mesopotamia.

Genesis 11:1,4,8-9

11 Now the whole earth used the same language and the same words.

4 They said, "Come, let us build for ourselves a city, and a tower whose top will reach into heaven, and let us make for ourselves a name, otherwise we will be scattered abroad over the face of the whole earth."

8 So the Lord scattered them abroad from there over the face of the whole earth; and they stopped building the city. 9 Therefore its name was called Babel, because there the Lord confused the language of the whole earth; and from there the Lord scattered them abroad over the face of the whole earth.

This is the account of the tower of Babel and God dispersing the people into the different nations and languages of Genesis 10 (Genesis 11 is first in chronology and then Genesis 10 explains the dispersal according to nations and languages). The text makes it clear that all the people speaking the same language here are the descendants of Noah's three sons. Many biblical commentaries suggest that the tower of Babel happened in the days of Peleg (Gen. 10:25), which would place it in the range of 100-300 years after the flood if there are not too many gaps in those generations. We would expect these people to still be speaking the same language as Noah and his family. For the topic at hand, we are concerned about "the whole earth" in verse 1. We are told repeatedly in this passage that God scattered them over the face of the whole earth after Babel and we can see from Genesis 10 that the area occupied by these peoples is still talking about the ancient near east. So, the whole earth in verse 1 is Shinar (Mesopotamia), and the place where they are scattered over the face of the whole earth is the ancient near east from modern-day Egypt to Turkey to Iran.

We then reach the conclusion that the waters did indeed cover the entire land under all the heavens, in the sense of water being everywhere from horizon to horizon that Noah and the crew could see. But it was not a worldwide flood. The text of the flood itself can be read faithfully as a local event. Which means that the ark did not come to rest on 16,000-foot-high Mt. Ararat of today. But the text doesn't say that. It says the mountains of Ararat.

Genesis 8:4

4 In the seventh month, on the seventeenth day of the month, the ark rested upon the mountains of Ararat.

We have seen that the Hebrew word (har), here rendered mountains, can also be translated hills or mounts. The word Ararat here seems to be a word of foreign origin and is simply transliterated. Some Hebrew lexicons give the meaning of the word Ararat as "curse reversed", which would make perfect sense in the context, because Noah was prophesied to bring relief from the curse of the ground. And the curse of the deluge itself has begun to recede. A competent Hebrew listener or reader would hear various wordplays in the text and the naming of people and place names to reflect their point in the narrative.

The tentative dating of Noah's flood depends on when in history we situate Adam and Eve in the garden, which we haven't concluded yet. But jumping ahead just a bit, my tentative date for the garden is around 8000 BC. Which would put Noah's flood at a tentative date around 6500 BC. The Persian Gulf was farther inland (approximately 150 miles farther) for sure by 5000-4000 BC, which pushed the delta flood plain north of its present location and could give us more northernly locations for the where the ark came to rest. But nothing as far north as eastern Turkey. But it appears from my research that in 6500 BC, the Persian Gulf was not as far north yet, so we are probably talking near the northern end of the present Gulf. Lorence G. Collins has ably argued that the flooding could have easily been as deep as 96 feet or more on the southern extent of the flood, and as wide as 100-200 miles, taking many months to drain. (Collins L. G., 2009)

I think it could have been even deeper and wider given the ancient coastline of the Persian Gulf during our proposed timeframe. But contrary to most interpretations, I don't think we are talking about the ark coming to rest on mountains in Armenia or eastern Turkey. Based on the ancient geography and faithfully interpreting the text, I propose that the ark came to rest in the foothills of the western Zagros mountains, probably closer to the river delta (northern present Gulf), and that the hills of Ararat are named for reversing the curse, not a mountain in modern-day eastern Turkey. On to the text.

The waters keep receding until the tops of the hills are visible. This makes sense because the bottom of the ark is approximately 22 feet under the level of the water. So, the ark would ground itself and two and half months later, the tops of the hills poke out of the water. Noah sends out a raven. Then he sends out a dove and it comes back to him because water was still covering the surface of the land, and he lovingly brings the bird back onboard. Noah waits seven days and sends the dove out again. She comes back with a freshly picked olive leaf, he waits seven more days, sends her out, and she doesn't return. Bruce Waltke points out that here again the biblical account is much more realistic than the Babylonian version. The raven is the stronger flier and could feed on carrion, so it makes more sense to send the stronger bird first than it does first sending out the dove. The standard Babylonian version has dove, swallow, and then raven. (Waltke, 2001) Of course, this detail with birds is very close to the flood account in Gilgamesh. I have already argued that those flood myths are faint and inaccurate recollections of Noah's flood. But the story containing the birds in Gilgamesh is only attested from the 7th century tablets found at Nineveh. Those tablets would be composed 600 years after Moses wrote Genesis. So, it is anyone's guess how this information came into the Gilgamesh flood and why it is not present in earlier flood myths from there. But it is still immaterial because I propose a date for Noah's flood around 6500 BC. Yet others are skeptical of the detail of the dove bringing back an olive leaf, as they say olive trees do not grow in Mesopotamia. That may be true today, but we are talking about the western edge of the Zagros foothills in Iran, and the climate was different and wetter in 6500 BC.

About a month and a half shy of a full year, the waters are dried up on the surface of the land. About two months later the land was completely dry.

What do we have so far? As various theologians have noted, we have the destruction of the ordered world and an echo of re-creation. We witnessed the de-creation of the world, including the death of mankind and animals, and once again the world was covered with the deep, formless and void. God remembers Noah, sends a wind, the clouds part, the rains stop, and the sunlight shines. The waters above the expanse have been separated from the waters below. The ark comes to rest in the foothills and after a while the tops of the hills appear. The waters are gathered into one place and the dry land appears. The dove comes back the second time with an olive leaf in her beak. The earth has sprouted vegetation. God tells Noah to depart from the ark with all his family, the birds, animals, and creeping things so that they can swarm abundantly on the earth and to be fruitful and multiply. The land is again populated with man and animals, and they are blessed. We therefore see that the re-creation is complete, harkening back to Genesis 1, and a second Adam begins again.

Noah builds an altar to the Lord and offers burnt offerings of clean animals and birds. This is the first mention of an altar in the Bible so far. Moses told Pharaoh that they needed to take sacrifices and burnt offerings for the Lord during the plagues in Egypt (Exodus 10:25). What is the burnt offering?

Exodus 20:23-24

23 You shall not make other gods besides Me; gods of silver or gods of gold, you shall not make for yourselves. 24 You shall make an altar of earth for Me, and you shall sacrifice on it your burnt offerings and your peace offerings, your sheep and your oxen; in every place where I cause My name to be remembered, I will come to you and bless you.

Exodus 24:4-5

4 Moses wrote down all the words of the Lord. Then he arose early in the morning, and built an altar at the foot of the mountain with twelve pillars for the twelve tribes of Israel. 5 He sent young men of the sons of Israel, and they offered burnt offerings and sacrificed young bulls as peace offerings to the Lord.

Exodus 29:18

18 You shall offer up in smoke the whole ram on the altar; it is a burnt offering to the Lord: it is a soothing aroma, an offering by fire to the Lord.

Leviticus 1:4,9

4 He shall lay his hand on the head of the burnt offering, that it may be accepted for him to make atonement on his behalf.

9 Its entrails, however, and its legs he shall wash with water. And the priest shall offer up in smoke all of it on the altar for a burnt offering, an offering by fire of a soothing aroma to the Lord.

Although we now have re-creation, sacrifice and atonement are needed in this new world. The garden has been washed away. There is a gap between man and God, but God's grace still endures.

Genesis 8:21-22

21 The Lord smelled the soothing aroma; and the Lord said to Himself, "I will never again curse the ground on account of man, for the intent of man's heart is evil from his youth; and I will never again destroy every living thing, as I have done.

22 "While the earth remains,

Seedtime and harvest,

And cold and heat,

And summer and winter,

And day and night

Shall not cease."

Contrary to the actions of the gods in the Mesopotamian flood myths that gather and swarm over the sacrifice like flies to satiate their hunger and thirst, God is pleased and reconciled by the atoning sacrifice of clean animals, He is remembered at this place, and Noah and his family are standing in proper relationship to the Holy One of Israel.

Many commentators think that the curse referred to here is the curse of the flood. But I think a better interpretation given the context is that the curse of the ground pronounced on Adam is removed. I believe this for a few reasons. Noah's father Lamech prophesied and named his son after the hope that he would give them rest or comfort from the toil of working the cursed ground because of Adam (Genesis 5:29). Another reason is that the Lord promises He will never again destroy every living thing. This clearly seems to be a second promise relating to the flood, not an elaboration of the first promise to "**never again curse the ground on account of man**". The third is that there seems to be no indication in the rest of Old Testament that the Israelites are frustrated with the lack of productivity from the ground, unless it is a curse from God as a result of breaking the Sinai covenant (law of Moses) and being unfaithful (Deuteronomy 28:15-68). Instead, the fields and crops are abundant due to keeping the covenant (Deuteronomy 28:1-14). And lastly, because the "world" has been destroyed and re-created with fresh blessings pronounced.

Both promises of Genesis 8:21 are tied in with the poem that ensues. Seedtime and harvest will not be thwarted by the curse of the ground. The ground again with yield crops with abundance. And the days and seasons will continue while the earth remains. But this is a new era unlike Adam and Eve. The corruption of sin has taken hold, even in the elect line of Seth through Noah and his sons. This is a type of new beginning and a type of second Adam, but it is unlike the first scenario. And yet, the promise of the seed remains, and God is a gracious, covenant keeping God.

God blesses Noah and his sons and tells them to be fruitful and multiply in any echo of Genesis 1. Now instead of naming the animals like Adam, the animals will fear Noah and his sons, and they can eat them as food, as long as they do not eat the blood. Because the life is in the blood and God will require life for life. There is a pronouncement of capital punishment for murder and the justification given is because man is created in the image of God. So, even though this new creation is different with a stark reality of man's depravity, sacrifices, atonement, and consuming the animals as food, the image of God remains. We will discuss this much more in the chapter on that subject. God again blesses them and tells them to be fruitful and multiply.

God establishes a covenant with Noah, his family, their descendants, and all the living creatures that come out of the ark. This is the first explicit mention of covenant in Genesis, but I hinted that there was an implicit one with Adam and Eve as well, even though the word covenant is not there. A covenant is generally an agreement between two parties that has conditions, what happens if the conditions are met, and what happens if they are not. Not all biblical covenants are the same and many are types of covenants that other ancient nations used as well. The educated Israelites would have been familiar with many of these. But God is always the One to establish His covenants and their terms in the Bible. But the idea of covenant can be a large topic and is not in the purview of this book.

This particular covenant could be called a promise covenant because God is not requiring any conditions of the other parties. The covenant is between God and Noah, his family, their descendants, and the animals. God promises to never flood the earth again to destroy all flesh that He created. And the sign of the everlasting covenant was to be the bow in the cloud. The Hebrew word denotes a bow, as in bow and arrow. When God saw the bow, He would remember His covenant with the earth and all flesh not to destroy it again with a flood. But the bow here is clearly the rainbow in the clouds. But the bow is pointed up, away from the earth and its life.

The sons of Noah were Shem, Ham, and Japheth. We get the parenthetical remark that Ham was the father of Canaan, which is explained in the next paragraph. From these three the whole earth is populated (or scattered). The whole earth here is also a subset of the entire planet. Noah plants a vineyard and becomes drunk by the fruit of its vine. While Noah is again taking up agriculture and subduing the earth in his own type of garden, the corruption of his nature displays itself in the sin of drunkenness and uncovering his nakedness. Noah's son Ham acts shamefully and instead of being discreet and honoring his father by covering his nakedness, he tells his brothers Shem and Japheth. These two brothers take a garment, walk backwards into Noah's tent, and cover their father without looking at him (Leviticus 18:7-17).

Noah awakes from his drunken stupor and proclaims a curse on Canaan, which may seem odd considering it was Ham who committed the inequity, but Canaan's descendants will be some of Israel's chief enemies, including the Egyptians, Philistines, and Canaanites. Noah gives blessings to Shem, from whom will come Abraham from Ur in Mesopotamia (along with the other Semitic tribes of Mesopotamia and Syria) and says let Canaan be his servant. The descendants of Jacob will dispossess the Canaanites to enter the promised land and the land will be fully given to them by the time of king Solomon. Noah blesses Japheth saying that he will dwell in the tents of Shem. The descendants of Japheth are largely the Gentiles of Asia Minor and Greece. And they will be blessed when the ultimate descendant of Shem, Jesus Christ, the seed of the woman who crushes the serpent's head, opens the kingdom of God and the repentance that leads to life for the Gentiles also.

Noah lives another 350 years after the flood for a total of 950 years, and he dies. This refrain, and he died, appears again as to end the genealogy started in chapter 5 with Adam. Despite the re-creation, blessings to be fruitful and multiply, man still in the image of God, and a promise of not wiping out mankind and all life again with a flood, the way to the tree of life is still inaccessible and Noah dies.

What does the account of Noah and the flood tell us about the kind of world we are in? We have argued that Genesis 1 is a type of literature that is not to be interpreted as historical narrative. This means, among other things, that it was never meant to be a scientific description of material creation and that it does not give us information about the age of the universe and Earth. We also discussed that elements hostile to life were part of the theology of the original creation, in contrast to the new heavens, earth, and Jerusalem. And thereby concluded that Adam's sin did not bring natural death, predation, entropy, earthquakes, floods, bacteria, viruses, disease, and other factors opposed to life into the profane area outside of the garden.

Starting in Genesis chapter 2 we argued that we have a real, historical place in the garden of Eden with two miraculously created human beings, Adam and Eve, who are born morally pure, righteous, holy, and having a true knowledge of God. They were priests in the temple-garden and their mission was to extent the boundaries of the garden, and hence the kingdom of God, by living in personal communion with God, producing godly offspring, being righteous and wise vicegerents in subduing the common and profane environment outside of the garden, and by reflecting God's image and glory to the people and Earth. If they did this, they could continue to eat from the tree of life and would be transformed into immortal and incorruptible bodies to dwell with God forever.

These human beings made in the image of God were created mortal and immediately died spiritually when they disobeyed God and ate from the tree of the knowledge of good and evil. God cleansed His temple-garden and by closing the way to the tree of life shut down the possibility of partaking of the sacrament that signified life with Him. Physical death for Adam and his posterity was now the reality. But God makes a promise that the seed of the woman would ultimately crush the head of the serpent and sin, evil, lies, temptation, and death will not have the last word. There will once again be a way to eat freely from the tree of life in the paradise of God. The chosen line of Seth calls upon the name of Lord, and although his descendants initially display faith and righteousness, they eventually also follow the lusts of their flesh instead of the will of God. And they look at the pagan women, see that they are desirable, and they take and marry, whomever they chose. Noah alone is righteous and finds favor with God.

So, this is our world up until the flood. How does it change with the flood? We have argued that the flood is historical narrative and really happened. But it is a local flood only, not global. And that it was in the delta flood plain of the Euphrates and Tigris rivers around 6500 BC. God does indeed destroy all living things He made, including mankind and animals. But didn't we argue that Genesis 1 was not literal in the historical sense? We did, but if we remember towards the beginning of the book, we argued that God is the Creator of the universe and the governor of all its processes and mechanisms. He gives life and takes it away. So, whether we are talking about the special, miraculous creation of Adam and Eve, and possibly the animal and plant life in the garden, all material creation is dependent on God for life and existence, regardless of mechanism.

All living things died in the flood, including the descendants of Adam and Eve (except Noah and his family), the animals in the garden, and all other people and wildlife in that geographic area. But how can this be? Why would God only judge the people of southern Mesopotamia, including Adam and Eve?

Let's zoom out for some perspective. God chooses people by the counsel of His own will to expand the kingdom of God, bring lost sinners into the fold of the children of God, and to display His glory throughout the entire Earth. Is this not what we see from the entire Bible? God creates Adam and Eve and commissions them to expand the garden, they sin and fail in their mission. God chooses Noah and saves him and his family through the judgement of the flood to continue the promise of the seed, despite the moral depravity of man. God choses Shem and through his descendants come Abraham. God chooses Abraham that he may become a father of many nations, to be a blessing to the nations, and that in him all the families of the earth will be blessed. The New Testament tells us that the ultimate seed of Abraham was Jesus Christ, the Messiah.

Galatians 3:8

8 The Scripture, foreseeing that God would justify the Gentiles by faith, preached the gospel beforehand to Abraham, saying, "All the nations will be blessed in you."

Galatians 3:16-17

16 Now the promises were spoken to Abraham and to his seed. He does not say, "And to seeds," as referring to many, but rather to one, "And to your seed," that is, Christ. 17 What I am saying is this: the Law, which came four hundred and thirty years later, does not invalidate a covenant previously ratified by God, so as to nullify the promise.

Galatians 3:29

29 And if you belong to Christ, then you are Abraham's descendants, heirs according to promise.

Romans 4:13-14

13 For the promise to Abraham or to his descendants that he would be heir of the world was not through the Law, but through the righteousness of faith. 14 For if those who are of the Law are heirs, faith is made void and the promise is nullified;

Romans 4:16-18

16 For this reason it is by faith, in order that it may be in accordance with grace, so that the promise will be guaranteed to all the descendants, not only to those who are of the Law, but also to those who are of the faith of Abraham, who is the father of us all, 17 (as it is written, "A father of many nations have I made you") in the presence of Him whom he believed, even God, who gives life to the dead and calls into being that which does not exist. 18 In hope against hope he believed, so that he might become a father of many nations according to that which had been spoken, "So shall your descendants be."

Out of all the nations on the Earth, Israel is to be holy like God is holy so that they would be sanctified, and the peoples of the pagan nations would seek God and be saved. But they too failed in their mission to expand the kingdom of God and are judged, chastised, and exiled. God chose the tribe of Judah and had grace on many undeserving people in this tribe leading to the birth of Jesus. Adam and Israel failed in their missions, but the God-man Jesus does not fail. He delights to do His Father's will and lives in perfect obedience to that will. He is the true image of God and reflects this glory to the world. He willingly lays down His life for His sheep, so that they would go into all the world expanding the kingdom of God, in order that God would bring many sons and daughters to glory. It is a rescue mission from start to finish with the goal of bringing the kingdom of God to the ends of the Earth. When Adam failed, God made a way. When Israel failed, God made a way. Jesus Christ did not fail, and the result will be the redemption of the entire creation after the sons of God. The nations have been given to the Son as His inheritance, and at His name every knee will bow, and every tongue will confess that He is King. And then when all the enemies are defeated, including death itself, all things will be handed over to the Father, so that He is all in all. Praise be to God!

Now back to the flood.

Creation is recapitulated during the flood. The spheres of creation are separated and filled once again. But the man Noah and his family are not starting out with the moral purity of Adam and Eve. There is no idyllic garden and personal communion where God will walk with them in the cool of the day, wipe away every tear, where He will be their God, and they will be His people. Now there is sacrifice and atonement for sin and man needs a mediator between himself and God. The curse of the ground is lifted, but moral corruption impedes the mission. God makes an everlasting covenant not to send a flood like this and in His everlasting covenant protects the earth and fructifies it to be productive. The sons of Noah are indeed fruitful and multiply and help to populate the ancient near east over the next few thousand years. Let us now look at the garden of Eden again and situate it within history and geography before moving on to a summary of Adam and Eve and to the implications for theology and science.

Dating of the garden

We know that Abel was a keeper of flocks (domesticated animals) while Cain was a tiller of the ground (domesticated crops). Everything in the context of the chapters of the generations of Adam and Eve points to a time of settled living with domesticated crops and animals. Unless we use special pleading for innovation and technology unique to Adam and Eve, this puts the earliest time of the creation of Adam and Eve at around 12,000 BC based on findings from archaeology and anthropology.

When we looked at the evidence from geography and hydrology for the location of the garden, we saw that the most recent time for the location of the garden would be around 7000 BC. After this, the slow incursion of salt water into the Persian Gulf would no longer fit the environment described by Moses.

The Samarra culture is attested around central Mesopotamia and dates to around 6000 BC. There is evidence of settlements from the Ubaid culture in southern Mesopotamia dating to around 6000 BC, which is firmly established around the shoreline of the Persian Gulf in southern Mesopotamia by 5500 BC and later. Noah's flood, and therefore the garden, need to be earlier in time to account for these settlements, as they would have been wiped out in the flood. And Noah's flood is at least 1000 years after the garden.

So, we now have various constraints for our dating. With the strict constraints of the local geography and hydrology, we have a window of 12,000-7000 BC. But the lower limit does not provide enough time for nations and cultures to spread from the sons of Noah. So, my tentative dating for the garden of Eden is around 8000 BC in the then-mostly dry river basin of the Persian Gulf.

One of the main challenges for the skeptic from this discussion of the ancient geography and hydrology of the area of southern Mesopotamia is they need to offer explanations as to how Moses would know about this dry river flood basin, these four rivers coming together in one confluence river upstream of the garden, along with the subterranean springs to water the ground, in an area that had been beneath the waters of the Persian Gulf for at least 4000 years by the time Moses penned Genesis in 1250-1200 BC.

The Persian Gulf was too high in Moses' day to accurately describe where the 4 rivers came together and flowed downstream to water the garden, much less to tie the location together with the underground aquifers. The coastline of the Gulf has been steadily moving southward due to silting from river deposits since around 5000-4000 BC. In 4000 BC it was approximately 150 miles farther inland than the present-day north shore, which would have put the Sumerian city of Ur on the beachfront. At this time in history even the Euphrates and Tigris did not have a confluence. They flowed separately into the Gulf. And when the Wadi

Al-Batin (Pishon) river was still running prior to around 3000 BC, the waters of the Persian Gulf would not have allowed the knowledge and description of the garden's geography and hydrology. Therefore, from about 7000 BC to today, the geography that Moses is describing would not make sense, even to an audience in Mesopotamia, much less the Israelites getting ready to enter Canaan. Although some of them may have been able to get the general geographic picture from describing the lands of Havilah and Cush.

Our thesis then produces a couple of testable predictions. There may still be the ancient remains of a garden with large, mature trees under the waters of Persian Gulf dating to between 9000-8000 BC. And there should be widespread, thick flood deposits in the southern delta flood plain of Mesopotamia dated to between 7500-6500 BC. The flood deposits should be below most, if not all, Ubaid cultural artifacts and may contain sparse artifacts and material culture from the population living in the 1000-1500 years from Adam to Noah.

As we saw earlier in the book when we discussed the mythology of Egypt and Mesopotamia, many scholars strain to make connections between them and the biblical account. We saw that most ancient near eastern scholars believe the flood myths from Mesopotamia were written first and were the inspiration and source material for Noah's flood. We did not deny the similarities, but we argued, based on several lines of argument, that it is more probable that the much earlier oral traditions of the historical event of Noah's flood were applied to locally impressive floods attested by archaeology in Southern Mesopotamia at the city of Ur around 3500 BC and at Kish and Shuruppak around 2900 BC. And they were then used for their own political and religious purposes in Mesopotamia. Said differently, there is evidence for real floods in southern Mesopotamia dated to 3500 BC and 2900 BC. But they are rather localized floods, sometimes not even extending between the close city-states.

There are also good reasons to believe that the various Mesopotamian flood myths and Sumerian king lists have kernels of truth in them. But we agreed with Professor Kitchen that the Mesopotamians mythologized history instead of historicizing mythology. (Kitchen, 2003) Many ancient near eastern scholars agree that at least some of the kings were real people ruling real cities and that real floods could have swept through these city-states. We argued that in contrast to the other flood myths, the biblical account is realistic in its details and paints a picture that rings true to reality. A boat that is sea-worthy and takes decades to build (not days), enough time, rain, and other sources of water to flood a large area (not seven days), the months it would take to drain such massive amounts of water (instead of days), sending birds in a logical order, and the incongruities of Mesopotamians offering sacrifices outside of a temple or ziggurat, etc. Therefore, I conclude that the descendants of Shem and Ham in Mesopotamia (and Japheth in the rest of the ancient near east) orally passed down Noah's flood story and tradition until finally committing it to writing somewhere in the 3rd-2nd millennium BC. And that finding widespread flood deposits in southern Iraq dating to around 7500-6500 BC would lend more credence to this hypothesis.

There is something else interesting about our thesis, the predictions, and how they could relate to Mesopotamian mythology. Our conclusion that the garden of Eden was lost long ago underneath the waters of the Persian Gulf brings back to mind one of the myths that we examined in chapter 4, The Epic of Gilgamesh.

After Gilgamesh learns from Utnapishtim that the gods will not give him eternal life, Utnapishtim tells Gilgamesh that there is a plant underwater that will restore his youth. Gilgamesh dives and finds the plant. He plans on giving it to the elderly in Uruk and to eat some himself. On the journey back to Uruk, Gilgamesh takes a bath, a serpent senses the presence of the plant, and takes it away. The snake sheds its skin and is gone. Which seems to explain why snakes shed their skin and maintain their youth. The moral of the story at this point seems to be that no matter how fervently Gilgamesh seeks immortality and youth, he must accept his mortality and learn to appreciate what he has.

But here we could have yet another example of the faint memories of actual events from the distant past, which have been adapted and skewed for the Sumerian, Akkadian, and Babylonian religions and politics. This time a memory of the garden of Eden and within it, the tree of life. We can almost imagine the biological or genealogical descendants of Shem and Ham in Mesopotamia losing the theological significance of these events over the many generations but remembering enough to create a story incorporating a legendary demigod, mythologized about an actual king, looking for a plant associated with restoring youth and loosely associated with a snake. To be found somewhere under the waters of the Persian Gulf possibly not far from the mouth of the rivers where Utnapishtim and his wife are said to dwell. As we have shown from chapter 4 and our study of the garden, the stories are completely different and the geography for the garden of Eden has much evidence for its historic truth. Quite unlike the pantheon of human-like gods, epic journeys, monsters defeated, trees that bear gemstones, lethal waters of death to cross, and eternal life granted by the same god who is angry that people survived the flood in Gilgamesh. But I find it interesting none the less. My conclusions are almost the exact opposite of most secular scholars. If borrowing and influence are coming from anywhere, they go from Adam, Eve, and Noah to the later Mesopotamians.

Now for the summary of what the Bible generally, and Genesis more particularly, says about the "first" man, Adam. He was created miraculously by God from the dust of the earth, he did not share common ancestors with the great apes or other animals, and he had no earthly parents. He was the son of God (Luke 3:38). God created him and Eve in His own image, they had a fully developed language that they did not have to learn, and both Adam and Eve were morally pure and without sin (Ecclesiastes 7:29). But they were both mortals, made of flesh and blood, even in their positive holiness (1 Corinthians 15:42-57).

God pronounced the creation of Adam and Eve as good in a functional *and* moral sense to represent Him and reflect His glory to the rest of the world and expand the kingdom of God. The world that Adam and Eve were put into already contained forces hostile to life: drought, famine, pestilence, earthquakes, volcanoes, hurricanes, floods, predation, disease, and death. They were placed in the sacred temple Garden of Eden as priests to serve and guard it, to walk in fellowship with God, and to expand the boundaries of the garden by fulfilling their commission and purpose. Adam and Eve were blessed to multiply their descendants upon the earth, to have dominion over the land and animals, to act as wise and righteous vicegerents, and to subdue the creation for the glory of God and bring citizens into God's kingdom.

Adam and Eve were real historical people, as the testimony of God's word throughout attests to. And they had real progeny that ultimately led to Jesus Christ, the son of God, the Messiah, the final Adam. God gave Adam a prohibition to test his faith and obedience and Adam transgressed against this command, doubting the word of God and questioning God's goodness. This led to spiritual death for all those whom Adam represented and physical death for himself, Eve, their offspring, and all they were commissioned to reach, because the way to the tree of life was shut. The sacrament of life for those walking in communion with God was now reserved for the new Jerusalem (Revelation 2:7, Revelation 22:2, Revelation 22:14), the final temple and holy of holies. Adam was the representative head for all humanity and as a result, we humans all suffer due to the consequence of this one sin. Adam was also a type of Israel, God's son whom He called out of Egypt (Hosea 11:1), and Jesus Christ (Matthew 2:13-15, Matthew 3:17, Mark 1:11, Luke 3:22, Romans 5:14, 1 Corinthians 15:45), God's beloved Son, the final Adam.

Adam and Eve were cast out of the temple garden into the profane and common, in which Adam was to work the ground by the sweat of his brow. God cursed the ground on account of his sin, but this curse was not global in scope. It did not change the laws of nature or how the world worked globally, it did not change an incorruptible Earth into a corruptible one, and it did not bring death to the people and animals in Adam's world in the way traditionally thought.

God brought forth the covenant of grace in history after the transgression of Adam and made gracious promises and provisions to Adam's line. The chosen line of Seth began to call upon the name of the Lord and worship Him, but even their faithfulness was short-lived. The earth eventually became filled with sin and violence, instead of with righteousness and peace. God floods and then recreates the world in this area of the ancient near east and starts again with Noah and his family. But they are not starting out in innocence, much less positive holiness (Genesis 8:21, Genesis 9:20-21). The futility of the ground due to the curse on Adam was rescinded after the flood (Genesis 5:29, Genesis 8:21). From the three sons of Noah much of the ancient near east is populated (Genesis 9:18-19, Genesis 10, Genesis 11), either directly or indirectly related to this family.

And now we, and the original audience of the Israelites waiting to cross into Canaan, understand the background, genealogies, and purposes of God in calling Abram out of the city of Ur in Mesopotamia to make a covenant with him so that he can be blessed, be a blessing to the nations, and how in Abraham all the families of the earth shall be blessed (Genesis 11:31-32, Genesis 12, Genesis 15, Genesis 17). The promise that the seed of the woman will crush the serpent's head is not dead. God in His mercy and grace will still use the descendants of Adam to bless all the families of the earth. And to one particular seed, Jesus Christ, this promise will be finally fulfilled (Galatians 3:13-14, Galatians 3:16).

As we have hopefully started to see in this book, I believe a faithful interpretation of the Bible allows for other human beings that were already on the earth when God created Adam and Eve. But this raises a number of theological issues to maintain an orthodox Christianity. And it is to those that we turn in the next section.

Part 3

Implications for Christian theology

Chapter 12

The Image of God

One of the foundational doctrinal truths of Christianity is that mankind is made in the image of God. Although other religions and secular philosophies try to search for justifications and grounds for the dignity of mankind to produce laws, ethics, and human rights, I would argue that they really have no warrant for doing so. I don't mean they don't believe in human worth and dignity, but I would argue they can't justify those values based on their beliefs. That is not to say that they haven't tried. But when you start with the assumption that a human being is just another kind of animal and devoid of any immaterial parts, like a soul, it is hard to reason convincingly that people have inherent dignity and worth.

Many scientists, especially biologists and psychologists, have tried to create theories of morality and values based on evolutionary psychology. They argue that our moral and social emotions have evolved through time and produced motivations that make us feel empathy and kindness by disliking the pain of others and by having a theory of mind of what the other person must be feeling. Of having a sense of justice and fairness through punishing cheaters and those who reap benefits without doing any work. Of loyalty and social cohesion through selection pressures for those mental traits that make us work together as a social species. And of purity, sanctity, and cleanliness through the motivation of avoiding disease.

Social psychologist Jonathan Haidt, psychologist Jesse Graham, and colleagues have developed a theory they call Moral Foundations Theory. Which tries to explain morality based on evolutionary psychology by proposing moral modules that are universally shared by people across time and culture but produce different moral cultures based on which modules are emphasized. Therefore, they claim to explain why cultures can vary morally so much from one another, while sharing the same basic moral intuitions and modules rooted in evolutionary psychology. (Jonathan Haidt, 2023) This was done building on the work of previous biologists, psychologists, and sociologists synthesizing biology with human behavior and culture.

While it provides an explanation that many find satisfying, they are still left with the nagging idea from biology that the body, including the brain, are simply vehicles for the germline cells. The cells that produce gametes like sperm and eggs to further reproduction and the continuation of the species. In other words, if morality and the motivations to practice it have their ultimate utility and ends in sex and reproduction, then there is no objective basis for that morality. So, in my view, they are left with morality as simply utility or a by-product of a brain that helps a social species like us produce more offspring. But to be fair to Dr. Haidt and colleagues, they admit that their theory is only a descriptive theory trying to explain how people and cultures actually behave, not a normative theory about what is morally good or bad. I believe that in the final analysis, the evolutionary psychologists must admit that there is no objective "right" thing to do, only emotions and motivations shaped by evolution.

Others like historian Tom Holland have argued that the western ethical framework is largely a byproduct of Christian influence. That many cultures in the west prior to Christian dominance, including the Greeks and early Roman Empire, did not see anything immoral about slavery, brutal conquest, imperialism, denying the sexual rights of women and slaves, and seemed unashamed about violence, genocide, and conquest. That our western culture is so steeped in Christian assumptions and values that have shaped our moral and social norms that they are like the air we breathe and are by no means universal or self-evident. (Holland, 2019) And I think Tom Holland is insightful and largely correct. If we look at different cultures throughout time, including those of the ancient near east we have looked at in this book, most of us in the modern west would be incensed at their beliefs and practices. Some historians try to assure us that these cultures had people with the same human nature as us, including the same moral framework. But I don't think the evidence bears that out. Both the Egyptians and Mesopotamians wrote a lot about people outside of their cultures as uncivilized barbarians and more akin to animals. And they both seemed to be fine with waging regular military campaigns into neighboring regions for the sake of territory and gain. But on the other hand, the law codes, inscriptions, and yearly religious festivals from Mesopotamia included the kings expressing (or at least professing) that they did not oppress the weak or the widow or take advantage of people. And in Egypt it was thought that the heart of the deceased was weighed on the scales against a feather to see if that person was morally worthy of the afterlife or if they were to be destroyed. But the Akkadian, Assyrian, and Babylonian empires were some of the most ruthless and savage of imperialists. And in Egypt during the Intermediate periods, people lamented that Maat (justice, order, and goodness) had fallen apart. And the evidence was that poor people, slaves, women, and others were taking positions and roles that were opposed to Maat. So, there was definitely the idea (at least in the literate class's mind) of a social caste system. Which is to say they were inconsistent and hypocrites like the rest of us. But to say that they believed like the Christians that all people are equal and made in the image of God, that the meek will inherit the earth, that the first will be last and the last first, and that humility and lowliness are virtues doesn't seem to be supported by the evidence. But this nurture over nature, or environment over genes, argument still doesn't give us any real justification for objective morality or human dignity.

Other secular people have tried to go the way of rationalism to create an objective morality. By saying that we can create rules that apply equally to everyone and that it is irrational to go against them. For instance, applying a version of the golden rule that you can't do something to others that you would not want them to do to you. Apart from the fact that this is less loving and restrictive than the commands of Jesus (Matthew 7:12, Luke 6:27-36), it is still a value judgement. If someone says, "I don't care about your rules for fair behavior. I think I'm stronger and better, therefore I should dominate the weak." They are not being irrational, but they are being immoral. We would say, "Well you *ought* to care about fair behavior!"

I think people have given logic and rationality a place that is unwarranted in moral philosophy. A computer operates on the basis that its hardware is built from logic gates. So, it is rational in the sense that it avoids contradictions in applying rules. But a computer cannot tell you the morally right thing to do without humans giving it rules of morality. Therefore, in my view, it just begs the question of what is the morally right thing to do and what gives people dignity and worth? I think the 18th century Scottish philosopher David Hume understood this limitation to reason. He argued that there is nothing contrary to reason in preferring the destruction of the world to getting your finger scratched. And many studies in abnormal human psychology have shown that people without proper moral intuitions and emotions could ace a test on human morality based on what their peers think is the right thing to do in various situations. But they simply don't care and therefore behave immorally.

So, the secular person is left with a dilemma. Both the evolutionary psychologist and the moral rationalist deny moral realism. The idea that morality is objective, and a standard exists independently of our human minds. Although this is a broad generalization and not fair to many in moral philosophy, I think it encapsulates the majority view of the above camps. They are often left with the more honest nihilism of the old atheists, who understood that objective morality could not be grounded without God.

As Christians, we believe that morality is objective and rooted in the very nature of God Himself. And I would argue that virtually all human beings, religious and non-religious, live as if morality is objective and true. People every day talk about justice and injustice, fairness, human rights, human dignity, care, kindness, loyalty, truthfulness, trust, love, respect, and the like. So much of modern life in the west, with our ideals of social progress, is predicated on the objective dignity and worth of humanity. We often hear people say that they don't need Christianity or religion to be a good person, but I believe that they just haven't thought about it long enough to make such a statement. Nor do they appreciate the history of western civilization and the assumptions and foundations that Christianity provided for their thinking. If this universe and planet is just a cosmic accident, and biological life, including humans, are just the product of random mutations producing genes and proteins that makes some people more successful at reproduction, then there is no justification for real morality and human dignity. But I believe that even non-Christians cannot escape the moral law and its implications. Most of us live as if morality were true and objective. And we Christians believe it is.

Right from the opening chapter of the Bible, and the rest of God's word further illuminating it, is this doctrine of the dignity and worth of mankind based on us being created in the image of God. God's image is the basis for our dignity and worth. It is something unique to man and does not apply to the other forms of life on Earth. I will use the male noun man here, but the term man I am using is just generic for mankind. I believe that no other religion or philosophy gives men and women the same indisputable dignity as Christianity. Not that we are star dust, inhabitants of the pale blue dot, or part of the universal mind.

The way we view other people is a direct result of the view of humanity that we hold, and Christians are called to view other human beings as bearers of God's image. People with the capacity to reflect His image to the world and bring Him glory. This should bring great humility to the believer and should spring us to action and love for mankind. The image of God is unconcerned about race, ethnicity, genetics, social class, caste, heritage, and the like. It is a fundamental constituent part of what it means to be human. If fact, the more that we reflect the image of God, the more human we are. Our humanity is found in accurately reflecting God.

But what is it? What does God's word teach about the image of God? And how does it relate to those humans living on the planet outside of the garden? We have already argued that God making Adam in His image and placing him in the garden was a correction and polemic against the pagan view of mankind and creating idols to reside in their temples. Whereas the Mesopotamians, Egyptians, and other polytheistic cultures in the ancient near east made images (idols) and placed them into their temples as incarnations of the gods, the Lord God made an image of Himself and placed him into the temple garden of Eden. Whereas the Egyptians and Mesopotamians had their rituals for animating and bringing to life the image of the god, the Lord God animated Adam and placed him in the temple garden. (Richter, 2018) In contrast to the pagan mythologies, where the gods were annoyed by the noise of humans and tried various schemes to limit their population, the Lord God blessed them and told them to be fruitful and multiply. But that doesn't answer the question of what constitutes the image and likeness of God. What is that makes human beings made in the image of God so special. And how does it relate to our broader thesis in this book? Before we do that, we need to look at the related concept of the soul. I don't believe we can talk about the image of God in man without defining what the soul is, what constitutes the nature of man, and makes them what they are.

The Bible teaches that man is a unity of being, but that he is made of both material and immaterial parts. He is body and soul. It is more biblical to say that man *is* body and soul, than to say that man *has* a body and a soul. But the Bible does not teach some sort of Greek dualism where the body in intrinsically evil and is the prison house of the soul. The body and the material constituents of it are not evil in and of themselves. It could be argued that the Old Testament is somewhat ambiguous about the dual nature of man. Whether the spirit, or breath of life, from God is constituting an immaterial soul or whether it refers to the animating energy of life from God to the body. I do not agree with some who say that the Old Testament saints were unaware of a soul or the spiritual existence of man in the presence of God absent from their bodies. If you read the Psalms, it is very hard to see the praise and exultation of the authors in the promises of God relating only to this life. The Old Testament Jews must have had developed theologies based on Enoch and Elijah being translated into heaven, neither seeing death. And mankind by our own experience is conscious of the fact that we are more than our bodies.

What is clearer is that most of those sitting under the teaching of the Old Covenant in the 1st century AD believed in a resurrection of the body and life eternal. The new heavens and the new earth in Isaiah and passages from Daniel and Job are hard to paint in any other terms. The Sadducees did not believe in the immortal soul and final resurrection. The Essenes were a monastic Jewish sect distinct from the Pharisees and Sadducees who practiced communal living, lack of private property, piety, good works, and strict observance of the Old Covenant, except for temple worship and sacrifice. Thanks to them we have the Dead Sea Scrolls, which most scholars believe were written by them. They believed in the immortal soul but not the resurrection of the body. But the majority theological position of the Jews during the time of Christ was that of the Pharisees. The Pharisees believed in both the resurrection of the dead and the immortality of the soul.

I believe that as special revelation progressed and became more transparent, the New Testament gives us a much richer understanding of the dual nature of body and soul. But the whole man is still viewed as a unity composed of both. The body is referred to as our tent that we inhabit, earthen vessels, the temple of the Holy Spirit, and the outer man. The immaterial part of man is that which men cannot destroy, so we should fear God alone, who can destroy both body and soul in hell. To be absent from the body is to be present with the Lord. Jesus tells the thief on the cross that he will be with Him today in Paradise. The inner man is the ego or personality which can survive separated from the body and maintains the continuity of consciousness after physical death. We know that the outer man is corruptible and is undergoing decay, but the inner man is being renewed day by day by the power of the Holy Spirit.

I believe that the word of God from Old to New Testament presents man as a unity and does not make a distinction about when the soul comes into being. In fact, I would argue that the soul in the Bible is fundamental to the constituent nature of man, so asking when it comes into being is like asking when a person comes into being. It is simple and not composed of parts, it cannot be divided. It is spiritual and not to be found in genetics or biochemistry. It is part and parcel to being human. But the soul is also synonymous with the animating principle of life. This principle is shared with both humans and the other creatures. God gives and takes it away for all His creatures. I will argue that what makes the soul spiritual, eternal, and distinguishes man from beast is the image of God imprinted on it. But first, what is the soul?

Genesis 2:7

7 Then the Lord God formed man of dust from the ground, and breathed into his nostrils the breath of life; and man became a living being.

Lest the readers should think that man was not a creature and somehow divine or heavenly, this passage brings the knowledge of Adam's origins back down to earth, so to speak. Adam was formed from the dust of the ground and God animated him. These passages seem to me to make it hard to maintain that Adam was a hominin or some member of a hominin species prior to Homo Sapiens. He appears as a fully matured male human endowed with reason, intellect, morality, language, and the propensity for culture. The cultural context of the rest of Genesis 2-4 makes it hard to maintain a date of hundreds of thousands or millions of years ago. It also makes the idea of man already existing in a population and then infused with a soul contrary to the Bible's teaching on the soul and antithetical to the text. The passage describes God animating Adam from inanimate dust as a unity of being, opposed to just giving him a soul. God took inanimate dust from the ground, fashioned it, and animated him. He gave life to non-life. This is the God we serve. In His word, He repeatedly takes that which is dead and brings it to life. In this case, Adam to be placed in the garden temple.

This Hebrew construction of the breath of life in Genesis 2:7 is not unique to humans though. It is the same phrase used to describe the animals who are also animated and have the breath of life in them.

Genesis 6:17

Behold, I, even I am bringing the flood of water upon the earth, to destroy all flesh in which is the breath of life, from under heaven; everything that is on the earth shall perish.

Genesis 7:15

15 So they went into the ark to Noah, by twos of all flesh in which was the breath of life.

All animals have the breath of life in them. God's cares for His creation. He is concerned about the animals and their coming and going. What is unique in the account is the fact that God directly breathes the breath of life into the nostrils of man. There is an intimacy in the act of animating Adam into a living being. But God gives all creatures, including mankind, the breath of life (Acts 17:25).

Job 27:3

3 For as long as life is in me,

And the breath of God is in my nostrils,

Job 33:4

4 "The Spirit of God has made me,

And the breath of the Almighty gives me life.

Revelation 11:11

11 But after the three and a half days, the breath of life from God came into them, and they stood on their feet; and great fear fell upon those who were watching them.

So, we see that both man and the animals are animated with the breath of life from God and that this is synonymous with the soul. And we also observe that Adam was different from the animals in the sense that the animating of his body was a direct, intimate cause.

Mankind no doubt occupies a unique place in God's creation. No animal stares up at the night sky and wonders at the creation. No animal experiences the guilt and shame of violating God's moral law written on their hearts. None besides man feels the tug of "I ought" to do something. For instance, I ought to help someone in need or danger, even if that person is not in my family and doesn't carry my genes. I ought to care about fair behavior, even if I am stronger or of a higher social status. I ought to love God with all my heart, soul, mind, and strength, and my neighbor as myself.

No animal has beliefs about their beliefs and can resolve to change them. Humans have language with syntax and grammar and can form novel ideas and express them in new ways. We think about our mortality and plan our futures accordingly. We have a sense of existence beyond the physical. We intuitively feel that we are more than our bodies. We are moral creatures. We have the sense that certain things have objective value and beauty. We are indebted to a God who intimately breathes into the first man with whom He wants to have a covenant relationship with, and by extension of that covenant head, the rest of humanity. What gives man these unique properties not shared with the rest of the creatures is not the soul by itself, as we have seen it is common to all living beings. It is the soul being imprinted with the image of God, which I will explain shortly after we have defined what the image of God is.

But Christianity does ***not*** teach that the soul is eternal. It did not pre-exist and is said repeatedly in the Bible as coming from God who gives it. We will get to the transmission of the soul, the image of God, and original sin at the end of next chapter, but suffice it to say here that the creation of the soul is more biblically faithful than the idea that it is transmitted from parents. But I think an even better description is that the soul is simply a property of human nature. It is inherent within the body and is inseparable from the physical body outside of death. Each soul is unique to the person and forms the wholeness of each human being. I think it is warranted from the biblical texts to assume that a person at conception is a person composed of both body and soul.

Some might say at this point, "Science has never shown us that there is an immaterial part to humans." Apart from the fact that the word of God is our ultimate authority, many would still disagree. Based on data from the literature of after death experiences, people have described things around them after they are clinically dead that they should have no knowledge of. Many philosophers, and some scientists, have argued for the existence of a soul, or an immaterial part of humans, based on consciousness, subjective experience, or that minds and brains are different things. Try this as an analogy to see if it helps: Astrophysicists think that there are such things as dark matter and dark energy. Either that or our theory of gravity is wrong or incomplete. They say we believe they exist because of the way galaxies behave, the expansion of the universe, and the total mass of the universe. As of the time of this writing, we haven't yet gathered any direct information or evidence about dark energy or dark matter from our current instruments. But they have strong confidence that these forms of energy and matter exist based on the effects they observe from the energy and matter they can observe. Will they observe and describe these forms of dark matter and energy? Or will they change the theories that describe the behavior of galaxies and the expansion of the universe? I don't think anyone knows yet, but it is an apt analogy for inferring existence through effects, even though we can't directly measure them. If science does not have the knowledge and technology to describe what dark matter and energy is, why should it surprise us that we can't directly measure the soul? Just like no matter how far into space we could travel, even to the edge of the observable universe, we would not expect to find heaven. We would not expect to find an area outside of spacetime by exploring within it.

But I think a lot of confusion and speculation comes from giving the soul functions that are unwarranted by the Bible itself. A lot of secular scientists and philosophers use the pejorative "ghost in the machine" when discussing the idea of a soul. This idea that an immaterial soul is calling the shots and telling the brain what to do. So, psychologists and neuroscientists point out example after example of how the brain works and that damaging a part of the brain impedes the function of the person. Or that certain areas of the brain and clusters of neurons light up when doing various mental tasks. But I think it is an error to leap from that to concluding that the soul does not exist. If we sat down a world-class pianist at a broken piano and what was produced was awful music, we would not conclude that the pianist did not exist. We would understand right away that the instrument to produce the music was broken. Of course, this analogy goes back to the idea of the soul controlling the brain, so it is not ideal. The brain is part of the body and there is no man without the body. Likewise, there is no man without a soul. From the perspective of the Bible and modern science combined, I think it would be accurate to say that the soul interacts with the physical world through biochemistry, the brain, and the body. While it interacts directly in the believer with the Holy Spirit, God, and the spiritual world. Or as theologians would say, it interacts through mediated causes in the body and world, while it interacts through immediate causes in the spiritual world.

From the Bible's perspective it is abnormal for the soul to be separated from the body. Hence, we long for our resurrected bodies after death. When we then view man and his psychology in biblical categories I argue that we see man as a unity. The soul is not acting alone and then the brain and body react. Man's thoughts, desires, appetites, drives, emotions, will, volition, and behaviors come from the whole person, including the brain and body with all its neurons, hormones, and biochemicals. Man ***is*** body ***and*** soul.

The Bible does not present us with a view of natural (unregenerate) man who has a lower base nature from which sin springs forth and is irrational, and a higher spiritual nature which is rational and from which goodness and piety come from. The biblical anthropology is one that describes man as a unity of body and soul that is either infiltrated in all its parts by sin or one that is released from the bondage of sin and is being renewed by the Holy Spirit. The Bible presents us with man who is body and soul, material and immaterial. No dichotomy of one taking on a sinful nature and the other taking on a righteous one.

But you may be thinking, "Doesn't Paul talk about not obeying the lusts of the flesh and that if you walk by the flesh you will die?" The words translated flesh in the New Testament have two main meanings. One is with reference to the physical body; the other is with reference to the sinful nature. Paul is addressing the sinful nature here and in many of his references to the flesh. So, we must either look at the Greek, determine it by the context, or both. But the body in and of itself is not sinful. God created it. Jesus had one. But sin permeates all parts of body and soul in fallen man.

Let's go back to God's word to start putting together what the image and likeness of God is.

Genesis 1:26-27

26 Then God said, "Let Us make man in Our image, according to Our likeness; and let them rule over the fish of the sea and over the birds of the sky and over the cattle and over all the earth, and over every creeping thing that creeps on the earth." 27 God created man in His own image, in the image of God He created him; male and female He created them.

Up to this point of the creation account, all the plants and animals are described as originating from God's spoken command, being created after their kinds. God spoke and the earth brought forth plants and fruit trees after their kinds. God created the sea creatures and the birds after their kinds. God spoke and the earth brought forth the land animals after their kinds. Then we have a decidedly unique break in the common refrains of the chapter. We do not have what we would expect if the account continued in the same fashion: Then God said let the earth bring forth man after his kind, and it was so. No! The account shifts gears and we have what many refer to as a Divine counsel. As I have argued previously, I believe the best interpretation is that God is conferring with Himself about creating a special kind of being. One that will be given dominion as vicegerent over His creation and will have the unique capacity to reflect their Creator back to the world for His glory. Moses is making clear that the male and female are special creatures, and they hold a special place within creation. They are not made according to their kinds. They are intentional in the way that they were created and in the purpose that they had on the Earth.

As we have seen, there are different interpretations about what the plural Us and Our mean in this passage. Some skeptics think this is an adaptation of an earlier polytheistic creation account, and the redactors or editors of Genesis were not clever enough to remember to remove the plural forms of the Hebrew for their monotheistic religion. Or that the pull of tradition was so strong they did not want to rock the boat by changing it. They make a similar argument for the so-called two creation accounts of Genesis 1 and 2. We have avoided all of this by pointing out that these plurals refer to the Trinity and by recognizing the different types of literature for the first two chapters of Genesis. Many biblical scholars hold the view that God was conferring with the angels in heaven. But I don't see any evidence from Scripture that we are made in the image of angels. Others say that it is a form of the royal "we". But that would seem to me to be out of context and not known in the Hebrew. I again think the best interpretation is that this is a conference between the Persons of the Trinity. We have already been introduced to God the Holy Spirit in Genesis 1 hovering over the deep and soon we get the promise of the Messiah in the woman's seed as God the Son in Genesis 3. In the gospel of John, we are given further revelation that God the Son was the Father's agent in creation as the Word, so all three Persons of the Trinity were present in Genesis 1. From the immediate context, we have the plurals Us, Our, and Our in verse 26 and the singular "His own image" in verse 27. We have plurality and singularity referring both to God within two verses. One in substance, nature, or being and a plurality of Persons, right here in the immediate context.

So, we begin with the triune God creating man in His image. What are the meanings of the words image and likeness here in the Hebrew? Image is a rendering of the Hebrew (tselem), coming from a root word meaning shadow. A shadow behaves just like the person casting it. It can be translated image, resemblance, shadow, or likeness. The same word is applied to Seth as being created in the image (tselem) of Adam (Genesis 5:3). The word likeness is a rendering of the Hebrew (demut), coming from a root meaning blood. One who begets another is often the same in likeness, appearance, and behavior. It can be rendered likeness, similitude, alike. This word is also applied to Seth in the likeness of Adam in Genesis 5:3. Hebrew has other words for image/god/idols. One common one is the Hebrew (pesel), meaning to carve or fashion. It generally means to carve an image/idol, to fashion something carved and shaped by human hands to be worshipped. It is applied in the second commandment in Exodus 20 prohibiting the creation of idols to be worshipped. The word translated likeness in this commandment is the Hebrew (temunah) which can be rendered form, likeness, image, representation.

So, the triune God makes mankind in His tselem and demut. Adam and Eve are representations and likenesses of God Himself. This is the reason that Israel was prohibited against making idols or graven images to bow down to. God had already made man in His image and likeness. Instead of the pagan polytheists who made images of the gods, the invisible Lord God without form created mankind in His image.

In this period of history in the ancient near east when Genesis was written (c. 1250 BC), all the cultures contemporary with the Israelites were polytheistic cultures with many gods and many temples to house these gods. The places where worshippers could offer material things that the gods needed to survive and flourish. Things like food, clothing, gifts, cosmetics, money, and the like. Not to mention taking the gods out to socialize with each other every so often, especially during yearly celebrations and rituals. Each deity generally had its own temple, and in each temple there would be an image of the god that it was supposed to embody. In other words, the polytheists would carve an image (tselem) of their god (elohim) and then animate the image through ritual. Many of the prophets mock these pagans for their idolatry and deride them for creating not images (tselem), but graven images (pesel), by human hands using wood, metal, or stone and then bowing down to the works of their hands. When worshippers came to the temple of their gods and tried to propitiate (appease, make favorable) the gods with their prayers and gifts, they would address the image assuming full well that the image represented the reality and manifestation of that god or goddess. To put it another way, when people were speaking or making offerings to the image, they believed they were addressing the god himself or herself. The image or idol was an incarnation of the god, "animated" through various rituals.

Also present in these cultures was the idea that the rulers, kings, or pharaohs were thought to be descendants, manifestations, or appointed directly by the gods, thereby legitimizing their authority to rule. They were also sometimes said to be the image of the gods. You could not question the authority of the king because the gods had sanctioned his rule. This type of psychological manipulation may have been very effective in keeping down the masses and suppressing free thought, but on the plus side it made for very long-lived kingdoms like Egypt. This concept of the images of gods being worshipped applied to the Roman pantheon in the 1st century AD as well.

I should note here that I believe the best exegetical case is that the words translated image and likeness in our English Bibles from Genesis 1 and 5 have the same meaning. Theologians over the years have tried to distinguish what the image entails versus the likeness, but I think they are used interchangeably. Both image and likeness are used in verse 26, while only image is used in verse 27. They both have similar meanings in Hebrew and are employed interchangeably.

We then have the triune God creating mankind in His image and likeness to be like Him in capacity or behavior in some sense. Theologians call these the communicable attributes. There is still the Creator/creature distinction, but we also have knowledge, love, goodness, kindness, righteousness, dominion, and relationships. Even Adam did not have these in infinite perfection like God does, but by way of analogy we can show what God is like. And Jesus Christ was the true image of God and His exact representation. Jesus shows us the Father. These are the things that we share with God that can reflect the character of God to the world.

We are given two aspects of the image in this passage in Genesis 1. One is our dominion over the other creatures on earth. We are called to rule and subdue the Earth as wise and righteous vicegerents, like God is the King and Ruler over the heavens and the earth. The other is that the image consists of male and female. There is a relational aspect to the image, like God is in relationship between the three persons of the Trinity from all eternity. As well as to His creatures since creation. So, from this first instance of the image of God in scripture we already know a lot about it. We will look at the testimony of scripture in other places to further unpack the image and likeness, but we begin to see why creating graven images and worshipping them is so serious and sinful.

Exodus 20:4-6

4 "You shall not make for yourself an idol, or any likeness of what is in heaven above or on the earth beneath or in the water under the earth. 5 You shall not worship them or serve them; for I, the Lord your God, am a jealous God, visiting the iniquity of the fathers on the children, on the third and the fourth generations of those who hate Me, 6 but showing lovingkindness to thousands, to those who love Me and keep My commandments.

Leviticus 19:4

Do not turn to idols or make for yourselves molten gods; I am the Lord your God.

Leviticus 26:1

'You shall not make for yourselves idols, nor shall you set up for yourselves an image or a sacred pillar, nor shall you place a figured stone in your land to bow down to it; for I am the Lord your God.

Leviticus 26:30

I then will destroy your high places, and cut down your incense altars, and heap your remains on the remains of your idols, for My soul shall abhor you.

Deuteronomy 29:17

moreover, you have seen their abominations and their idols of wood, stone, silver, and gold, which they had with them);

Deuteronomy 32:21

'They have made Me jealous with what is not God; They have provoked Me to anger with their idols. So I will make them jealous with those who are not a people; I will provoke them to anger with a foolish nation,

2 Kings 17:12

They served idols, concerning which the Lord had said to them, "You shall not do this thing."

1 Chronicles 16:26

For all the gods of the peoples are idols, But the Lord made the heavens.

Psalm 135:15

The idols of the nations are but silver and gold, The work of man's hands.

Isaiah 2:8

Their land has also been filled with idols; They worship the work of their hands, That which their fingers have made.

Isaiah 10:11

Shall I not do to Jerusalem and her images Just as I have done to Samaria and her idols?"

Isaiah 31:7

For in that day every man will cast away his silver idols and his gold idols, which your sinful hands have made for you as a sin.

Isaiah 42:17

They will be turned back and be utterly put to shame, Who trust in idols, Who say to molten images, "You are our gods."

Jeremiah 8:19

Behold, listen! The cry of the daughter of my people from a distant land: "Is the Lord not in Zion? Is her King not within her?" "Why have they provoked Me with their graven images, with foreign idols?"

Jeremiah 10:14

Every man is stupid, devoid of knowledge; Every goldsmith is put to shame by his idols; For his molten images are deceitful, And there is no breath in them.

Acts 19:35

35 After quieting the crowd, the town clerk *said, "Men of Ephesus, what man is there after all who does not know that the city of the Ephesians is guardian of the temple of the great Artemis and of the image which fell down from heaven?

Romans 1:23

23 and exchanged the glory of the incorruptible God for an image in the form of corruptible man and of birds and four-footed animals and crawling creatures.

1 Corinthians 10:19-21

19 What do I mean then? That a thing sacrificed to idols is anything, or that an idol is anything? 20 No, but I say that the things which the Gentiles sacrifice, they sacrifice to demons and not to God; and I do not want you to

become sharers in demons. 21 You cannot drink the cup of the Lord and the cup of demons; you cannot partake of the table of the Lord and the table of demons.

Idolatry has plagued humanity since the image of God was manifested in Adam. They have suppressed that knowledge in unrighteousness and made images for themselves to worship.

The ancient people reading Genesis would immediately be shocked by the language used here by Moses. I believe this is another instance of the polemical nature of Genesis 1 against the pagan religions that we talked about earlier. Moses is saying that the images of the gods in the temples are not to be worshipped. The king or pharaoh is not the image of his gods on the throne. Instead, the Creator God made human beings in His image to be a reflection of Himself in His creation. God can make an image in His likeness, but man cannot make images in the likeness of their gods. There is no one special, favored class of people that the image of God is seen through. All human beings are made in the image of God. These human images (male and female) are not meant to be worshipped like the polytheists. They are wholly separate from the Creator to reflect something of Himself. This would certainly be a scandal in the ancient near east.

What else does this image consist of? Is it about what man is physically or mentally? Certainly, it is not a reflection of what God looks like bodily, as God is spirit. Is it in man's capacities for reason, intelligence, spirituality, creativity, or moral choices? Does it have to do with more than his role as vicegerent or his relationship with other humans and with God Himself? Picking up with Genesis 2.

God creates the garden, causes the trees to grow that are pleasant to the sight and good for food. He places Adam there where he was meant to cultivate and keep it. God creates the animals and brings them to Adam, and he names them as an exercise of his dominion that he has been given by God. Names played a much more significant role in Israelite and other ancient near eastern cultures than they do today for us in the west. Your name was to represent your role and purpose. It said something about who you were or expected to be. We can see this from these early chapters of Genesis where the names of the people and places reflect what their roles were. Adam is naming the animals and what their respective roles are. And the one doing the naming was exercising authority. Adam is exercising dominion over the earth as commanded. As God is sovereign over all the universe, He is giving man a role of vicegerent to exercise authority over the ordered world which God has made for his habitation. Man is uniquely qualified in his abilities for this task. Whether he has been effective or wise since this mandate is another question. But we can see Adam and Eve exercising one of the characteristics of the image, that of dominion.

A suitable helper is not found for Adam, so God creates Eve from his side. We are told that both male and female are made in the image of God in Genesis 1, so there can be little doubt at this stage in the account that both Adam and Eve possess the image. Adam's relationship with Eve is also part of the image. One attribute of the image consists in our relationship with God, with our spouse, and with other human beings. These relationships reflect the relational aspect of God within the Trinity between the Father, Son, and Holy Spirit from all eternity. When we are in right, loving relationship with one another, we are displaying the image of God to the world. We then see Adam and Eve in their relationship displaying an aspect of the image.

Man was not meant to be alone, and it is difficult to show love to the world without there being an object of that love. Jesus put our love in the proper ordering when telling us that if we don't love God more than anyone or anything else, we are not worthy to be called His disciples. Our first love is always to be towards God. But loving God will always allow us to love others more fully. If we are secure in our love for God and in His love for us, we are freed to love others more fully and not to be people pleasers, always rising and falling on the acceptance of others. And once we see all humans as made in the image of God, these two concepts have dramatic implications for what we will say and do to fellow image-bearers.

Adam and Eve were both naked and not ashamed. This gets us to the point of asking what was the state of Adam and Eve before the Fall? And how does this relate to the image? We know that Adam and Eve had direct communion with God in the garden, and they were not afraid or ashamed. We know from Scripture that God is holy and cannot tolerate the presence of sin. Whenever someone after the Fall comes into the presence of angels, much less God Himself, there is a sense of dread and uncleanness. We are undone in the presence of a holy God. We can feel good about ourselves when we are comparing ourselves to others, but this false pretense vanishes when we meet the holy Other and His law. Sinful people need a mediator between themselves and God. But the fact that Adam and Eve have this intimacy with God in the garden should be our first clue that they are in a state of positive holiness and undefiled by sin.

Ecclesiastes 7:29

29 Behold, I have found only this, that God made men upright, but they have sought out many devices."

If the Bible ended in Genesis 2, we would be left wondering what is this image of God and how does it relate to man today? Did Adam lose the image after the Fall? Did the effects of the Fall change this image in some way? Before we see how the Fall affected the image, if at all, we need to examine the rest of the Bible to illuminate what else might be the constituent parts of the image.

Colossians 3:10

10 and have put on the new self who is being renewed to a true knowledge according to the image of the One who created him—

Paul is talking about the born-again Christian, in thinking about the reality of his union with Christ, to put off the old self and put on the new self. This is another verification of Paul's teaching that we still have the old nature when we are born again. We are to kill the deeds of the flesh (the old man), not give the flesh opportunity for sin, and instead walk by the Spirit. To be renewed in our inner man. Here I believe we have one more aspect of what makes up the image of God in Adam: true knowledge.

Our knowledge is being renewed according to the image of God. And if it is being renewed, it must have been marred or defaced in some sense with the Fall. I believe this knowledge consists in thinking properly about God and towards God. Consider how Adam would have thought about God before the Fall. He knew that He was the sovereign Creator. He knew that he owed every praise and gratitude to God for giving him life, a wife and helper, and a garden that contained all the bounty that he required. Adam knew that God cared about him and had his best interests in mind. Adam knew that true joy, delight, and satisfaction lie in obeying

God and reflecting Him to the world. Adam's knowledge would not have been infinite. Adam was not omniscient. His knowledge was true and accurate but not exhaustive. But the irony was that eating from the tree of the knowledge of good and evil defaced and ruined the true knowledge of Adam and Eve. They could once clearly apprehend and know God, now their knowledge was cloudy, and their foolish hearts were darkened.

We now have a very dim knowledge of God. But Christians are being renewed day by day, glory to glory, in gaining back an accurate knowledge of God and the joy that comes with following Him. We struggle to place God in His rightful place in our thinking. This world and our temporal life fight for the preeminence of our thoughts. The unbeliever's knowledge of God is defaced and inaccurate. But even so, they know something of God. They can read God's word and even know it very well, but it does not affect their hearts. They don't treasure it or hear God speaking to them through it. They look to the heavens and the earth and the picture that they see is not clear.

We know from a study of history (and even today) that many cultures have seen gods behind natural phenomena. A god for each aspect of nature. Others have thought that god is everything. That god or gods are not separate from the creation but are part of it in substance and essence. Others have a pantheon of various gods, each of which has a certain role and dictates the outcomes and destinies of various natural processes. And many in our modern day of science and materialism see no God at all when they look to the study of the heavens, earth, and man. They see deterministic laws, molecules begetting molecules, atoms colliding in chains of causation from which there is no escape from the inviolable laws of nature. They see man as a grouping of biomolecules with the unfortunate remnants of faux spirituality and morality as artifacts of natural and sexual selection.

Our knowledge of God is very dim indeed. I believe the first few chapters of Romans teach us that people have a knowledge of God. But they cannot appropriate that knowledge without grace. They exchange the knowledge of God for futile speculations and God gives them over to a mind of stupor and immoral behavior. Once the Holy Spirit brings us to life, our knowledge is cleared, and we can see God and more accurately sense His wisdom and power through nature.

When we read His word, it is a living book and sharper than any two-edged sword.

Romans 12:2

2 And do not be conformed to this world, but be transformed by the renewing of your mind, so that you may prove what the will of God is, that which is good and acceptable and perfect.

1 Corinthians 2:11

11 For who among men knows the thoughts of a man except the spirit of the man which is in him? Even so the thoughts of God no one knows except the Spirit of God.

Therefore, many theologians see true knowledge as an aspect of the image of God. What else can we say makes up the constituent parts of the image?

Ephesians 4:23-24

23 and that you be renewed in the spirit of your mind, 24 and put on the new self, which in the likeness of God has been created in righteousness and holiness of the truth.

In the book of Ephesians Paul is again exhorting the believers in Ephesus to walk in a manner that is consistent with the objective reality of their union with Christ. They are to put off the old self and its desires which are corrupt and to walk in the newness of life. Paul gives us two more aspects of what theologians consider the original image consisted of: righteousness and holiness. These are also being renewed in sanctification (progressive conformity to the image of God) and were therefore corrupted in the Fall.

Righteousness is the quality of describing someone who does what they ought to do. It described someone who fulfills moral obligations. It relates to the idea of conforming to a standard. To be righteous means to obey God. Not only in action, but also in faith, emotion, intention, and love. We know that Enoch, Noah, Job, Abraham, Moses, and others were found to be righteous men. They walked with God and found joy in trusting His promises through faith and being obedient. They could be said to be righteous in the sense that the overall picture of their lives was one of obeying as they should. But Adam was righteous in an additional sense as well. He did not have imputed guilt nor moral corruption of his nature. He would have found the commands of God to be right, just, and true. He would have enjoyed obeying God and having intimate fellowship with Him in the garden. He and Eve would have had righteous children, the whole family enjoying God's presence and delighting in His presence. The work given by God to cultivate, guard, serve, and keep the garden would have been enjoyable and satisfying. They would have cherished the commission to expand the kingdom of God outside of the garden until it filled the entire Earth. Adam and Eve were therefore not just innocent, but positively righteous.

Now, we can observe that we are not righteous. There is none righteous, no not one. The actions of man are desecrated with selfishness, greed, lies, idolatry, and pride. We have a hard time not thinking about ourselves. We have a difficult time being happy for other people when things in our own lives are going badly. When we do acts that conform to God's law and would otherwise be commendable, they often have no regard for His glory. We may do things to make ourselves feel good about our ethics and morality, but they don't have their foundations in pleasing God. The problem is that we are not satisfied with God.

We want to find merit in our own deeds and want to be judged righteous based on what we do. We all have the temptation to say, "I may not be perfect, but at least I'm not like so and so." We lean towards self-justifying and impatient attitudes. We want to assign the bad behavior of others to their character, while excusing our own due to circumstances. We want to make the law of God more palatable and something that we can measure up to. We like to think that God cannot really get that bent out of shape over sin. It doesn't seem so bad to us. But God is not like us. He is holy and pure. Only by the blood and work of Christ can we be forgiven and brought into a right relationship with God. We have fallen a long way indeed.

Holiness means to be set apart and sanctified. To be dedicated or consecrated. It means keeping oneself free from sin and temptation. God requires holiness, for He tells us to be holy as He is holy. God needed to remind the Israelites again and again that they were to set up their culture in a way that made them visibly distinct from their contemporaries. They were to look different than the other nations and, as a result, to fulfill their mandate to be a blessing and draw the other nations to worship the one true living God. They were to have their hearts circumcised and walk by faith. But repeatedly they found themselves hardened in their hearts and practicing idolatry with their neighbors. They repeatedly became a derision and curse as opposed to a blessing. Lest we think we would have done better in that covenantal relationship; we can be assured that our sinful hearts would have led us astray as well.

Adam would not have known an experiential knowledge of sin and the misery that it brings. This concept would have remained fully alien to him as he was holy and separate from sin and its effects. Adam and Eve would have enjoyed perfect fellowship in their marriage, family, and with God Himself. They would have been a light to the profane world outside of the garden and influenced people for God's glory. They were initially walking in positive holiness.

God has given Christians His Spirit so that we again can be holy. We are to walk in the good works that God has prepared for us. As Christians, we are to be set apart and let our light shine in the world so people will know that we reflect His image and belong to Christ. We are not to run in the dissipations of our former life. And we should not be surprised when our friends marvel that we don't want to participate with them in these things. God does not call us out of the world, tell us to run to the nearest monastery, or set up our own insulated counterculture. But instead to be in the world but not of the world. We should be a beacon to the culture around us, but people should know that we are different. We are being renewed in holiness. Christians should be the humblest of all people, remembering what we have been forgiven of, and that sin still lurks and wants to rule over us. But at the same time unflinching for the truth and different from the world. Holiness is seen as awkward to modern eyes, as it is the opposite of worldliness. It is seen as oppressive and narrow-minded. But without holiness, we will not see God.

Romans 3:23

23 for all have sinned and fall short of the glory of God

In sinning, all fall short of God's glory. Or to rephrase it, in sinning we fall short of accurately reflecting the image of God. Adam and Eve, prior to the Fall and without sin, could be said to encompass the image of God in the fullest sense for men and women. They had the positive moral characteristics of true knowledge, righteousness, and holiness. They also had the capacities and faculties for exercising dominion and living in right relationships. This required intelligence, reason, moral conscience, language, free will, religious affections, social emotions and intuitions, foresight, technology, and culture building. In other words, Adam and Eve had rational souls and positive righteousness.

So, do we still have the image of God after the Fall? I have tried to show that if we do, it is not in the same sense as it was before the Fall. After the Fall and the curses in Genesis chapter 4, we are given the account of Cain and Abel, Cain descendants, and Seth's birth. Starting in chapter 5, we are given the line of Seth through Adam.

Genesis 5:1-3

5 This is the book of the generations of Adam. In the day when God created man, He made him in the likeness of God. 2 He created them male and female, and He blessed them and named them Man in the day when they were created. 3 When Adam had lived one hundred and thirty years, he became the father of a son in his own likeness, according to his image, and named him Seth.

We are again given the affirmation of Genesis 1 that Adam and Eve were made in the image of God, this time just using the word likeness. We are told that Adam had a son he named Seth in his own likeness, according to his image. I believe that the Holy Spirit put this in the passage to confirm to us that being made in the image of God was still present even after the murder of Abel and all the sin since the Fall. Although, it was in a different sense than the perfect image before the Fall. But the purpose of the image was to continue through the line of Seth.

After Noah's flood and the declaration by God that the intent of man's heart was evil from his youth, God gives us these statements.

Genesis 9:6

6 "Whoever sheds man's blood,

By man his blood shall be shed,

For in the image of God

He made man.

After the all-encompassing declarations that man is sinful, we are told he is yet in the image of God. Lest we should think that maybe only Seth's line continued to have the image of God; this decree is given regarding humanity in general. Some may suggest that maybe Adam was made in the image of God but the prohibition against murder is pointing back to the fact that man was ***initially*** made in the image of God. Maybe he no longer retains any of that image?

1 Corinthians 11:7

7 For a man ought not to have his head covered, since he is the image and glory of God; but the woman is the glory of man.

This passage confirms that man is still in the image of God. It does not mean that women are no longer in the image of God, as that would clearly contradict the rest of scripture on this matter. But it could still be argued that maybe only those who are saved have the image, as the context of this verse is for believers in the church.

James 3:8-9

8 But no one can tame the tongue; it is a restless evil and full of deadly poison. 9 With it we bless our Lord and Father, and with it we curse men, who have been made in the likeness of God;

This is yet another proof text of the fact that man is made in the image and likeness of God. We should not curse men who are made in the likeness of God. Cursing one who is capable of reflecting God's image to the world is sin. The Greek verb tense used in the phrase "have been made" refer to something of a present reality. Not pointing to something that once happened in the past. So, I think we are left with the clear conclusion that cursing men (generic) is a sin because humankind, male and female, are still in the image of God.

So far, we have determined that Adam and Eve were miraculously created by God and made in His likeness. We have shown that in the original image they walked in positive knowledge, righteousness, and holiness. And the New Testament confirms that these moral qualities need to be renewed according to the original image. But how is this possible?

Romans 8:28-29

28 And we know that God causes all things to work together for good to those who love God, to those who are called according to His purpose. 29 For those whom He foreknew, He also predestined to become conformed to the image of His Son, so that He would be the firstborn among many brethren;

The answer is Jesus Christ. Christians are predestined to be conformed to the image of Jesus Christ, who is the true image of God.

2 Corinthians 4:4

4 in whose case the god of this world has blinded the minds of the unbelieving so that they might not see the light of the gospel of the glory of Christ, who is the image of God.

Colossians 1:15

15 He is the image of the invisible God, the firstborn of all creation.

Hebrews 1:3

3 And He is the radiance of His glory and the exact representation of His nature, and upholds all things by the word of His power. When He had made purification of sins, He sat down at the right hand of the Majesty on high,

Jesus is the image of God in the fullest sense. When Jesus talks, God talks. We are given revelation when Jesus speaks. When He acts, he acts according to the will of the Father. He retains all the attributes of God in His divine nature. If you want to see what God is like, look to Jesus.

John 14:9

9 Jesus *said to him, "Have I been so long with you, and yet you have not come to know Me, Philip? He who has seen Me has seen the Father; how can you say, 'Show us the Father'?

Jesus Christ is the true image of God, both in His divine nature and His human nature. He is the last Adam. Christians are being renewed back into the original image of God when we follow Jesus and become more like Him. He shows us what the original image of God looked like and what it can be. Adam failed to maintain the original image of God in its fullness and thereby failed in his mission to be a light to the profane world outside of the temple garden. Israel failed to reflect the image of God to the Gentile nations and to be a blessing to all the families of the earth. Jesus Christ did not fail as He walked in true knowledge, righteousness, and holiness. And now He sends out His disciples to go out into all the world, set apart as a light on a hill, holy and sanctified, to be progressively conformed to His image and complete the mission that Adam and Israel failed to accomplish. May we seek His help in doing this.

I then conclude that the best exegetical case can be made for the fact that the original image of God in Adam and Eve consisted of true knowledge, righteousness, and holiness. It also consisted of having a rational soul capable of exercising dominion over the earth and being righteous vicegerents. Lastly, it entailed living in loving and godly relationships with God Himself and other fellow image-bearers. When Adam and Eve were created, they had hearts of flesh. Which is to say, they had an undefiled soul, mind, will, and emotions. They had the unclouded reason and intellect to exercise a true knowledge of God, and they possessed a true love of God. They longed to obey and live in righteousness and had the ability to conform to that longing. They were able and willing to obey the command, you shall be holy for I am holy. (Leviticus 11:45, 1 Peter 1:15-16) They possessed the ability to keep themselves separate from sin and defilement.

What then was lost in the Fall? After the Fall, Adam and Eve lost the positive moral qualities of true knowledge, righteousness, and holiness. These now needed to be renewed. But they maintained a rational soul and a moral conscience. They still possessed the potential capacities to reflect the image of God but not without God bringing them back to life through grace. And even in that sense, they and the rest of humanity could never attain the true image given to Adam and Eve before the Fall until they were glorified in heaven. Adam and Eve's hearts of flesh had been replaced with hearts of stone, they were dead in trespasses and sins, and their minds were corrupted and defiled. They no longer had a true, clear knowledge of God. Their actions were no longer guided by glorifying God, and they found that their slavery to sin enticed them to flirt with temptation and sin versus being separated from it.

The image of God after the Fall then is the constituent nature of man reflected in his rational soul, moral conscience, and eternal existence after creation. It is what separates him from the animals. The soul has been stamped with image of God. As God is a rational and volitional Being, so then is man. As God is a moral Being, so then is man. As God exists eternally, so then man exists eternally after his creation. As God is King over the universe and exercises dominion and rule, so then does man over the Earth. As God is a relational Being within the Godhead, so man is relational. We can then rightly say that man is still in the image and likeness of God, a creature exalted above the others. But one who has lost positive righteousness and needs to be renewed. This image is what makes him man. It is what gives him dignity and worth. It is why he is the crown of creation.

But the only way to begin the renewal of the image of God in the moral sense given to Adam and Eve is if God regenerates us by the Holy Spirit. The Spirit brings us back to life from spiritual death, gives us a back a heart of flesh, and makes us love God again. He unclouds our vision so that we may see God through His creation more clearly, unites our souls with Christ, and gives us a seed of the incorruptible in our souls. We are again given the ability to attain true knowledge, righteousness, and holiness. Albeit in an incomplete and imperfect way until heaven. The image of God that was defaced by the Fall begins to be renewed back to the true image of the One who created Adam and to the image of the last Adam.

Let's then summarize what we've learned about the image of God. God created Adam in His image and likeness to reflect Himself to the broader world. He fashions this image-bearer from the dust of the earth, animates him with His breath, and places His image in the temple garden of Eden. Adam exercises dominion over the creatures by naming them but no helper is found suitable for Adam. God creates Eve from Adam's side, and the two image-bearer priests are in the sacred temple space of the garden to guard and keep it. They are without sin or defilement of their nature, serving God in true knowledge, righteousness, and holiness. They have the capacity ***and*** ability to live in positive righteousness and rightly exercise dominion over the garden space and extend the boundaries of the garden to influence the world and spread the kingdom of God throughout the Earth. They are in right relationship with God, with each other, and with the creatures and environment. They are accurately reflecting the image of God. Were they to carry out their mandate and mission outside of the garden, they would have looked a lot like Jesus Christ in His earthly ministry, minus the divine qualities of the Savior. They would have been in constant communion with God, loving people, upholding justice, being resolute about the truth, and being a light to the profane world.

But they disobeyed, fell from grace, and failed in their mission. The image of God was corrupted and defaced. They no longer walked in true knowledge, righteousness, and holiness. They were cast out from the temple garden, were frustrated in exercising dominion, and had strife and enmity in their relationships with God, each other, and those around them. They still had the image of God, but it was marred and defaced and needed to be renewed.

Subsequent generations made images and idols instead of appreciating the image of God in man and their foolish hearts were darkened. They exchanged true knowledge for lies and bowed down and worshipped images made of stone, wood, and metal, the works of their own hands. Because of this God gave them over to a depraved mind and sinful passions and chose for Himself a people from the line of Shem through Abram to once again create a people set apart and holy. He made a covenant with Abraham based on promises that in him all the nations and families of the Earth would be blessed. 430 years later he made another covenant with the descendants of Abraham through Jacob's twelve sons at Mount Sinai. But the law of Moses was powerless in motivating obedience and Israel failed to renew the image of God and to be a light to the nations.

The promises of God to Abraham were finally fulfilled in Jesus Christ, the last and final Adam, who was the true image of God. And now Jesus sends out His disciples who are being progressively renewed into His image by the power of the Holy Spirit to be a light and witness to all the families of the earth.

What are we to say then about all the people living on the planet when Adam and Eve were created? Did they already have the image of God prior to Adam and Eve? Were they given the image around the same time? Did they not have the image and are therefore less human? Did they have souls? How was the soul and image transmitted? We will address these issues, but we first need to examine the doctrines of the Fall and Original Sin.

Chapter 13

The Fall and Original Sin

The doctrines of the Fall, original guilt (referring to Adam's first sin and our condemnation), and original sin (the sinful nature that we are born with) are of tantamount importance to our discussion in this book and any treatment of the biblical anthropology of man. This is often the place where liberalism sneaks in and orthodox Christian doctrines are abandoned. The Christians who believe in a young Earth often claim the doctrine of the Fall is minimized by the "other" side. And those who believe in an old Earth and evolution assert that the "other" side is too stuck in an Augustinian view of the Fall and makes too much of it. It is often cautioned at this point that if one abandons the doctrine of the whole human race biologically descending from Adam and Eve, then our whole systematic theology falls apart and we need to employ increasingly creative hermeneutics to interpret the Bible. Especially regarding the doctrines of the Fall, original sin, and the image of God.

Some go as far as saying that the ancient biblical authors really can't be faulted for thinking in certain ancient ways conditioned by their culture. Here modernist assumptions sometimes come in to undermine the inspiration of the Bible. Or the work of the Holy Spirit in leading men to write down the words of God is subtly abandoned while couching it in confusing language that makes it hard to determine what the author really believes or is saying. Let God help me not be guilty of such confusion regarding my thesis.

We have already laid a lot of the groundwork in previous chapters for our view of the Fall. We began by discussing the type of literature in Genesis 1 and concluded that it is not historical narrative, but a type of highly stylized prose that does not lend itself to a literal historical interpretation in the usual sense. And then argued that the original audience would not have understood it as historical narrative. This interpretation affirms God creating out of nothing and being the direct Agent responsible for creation but makes no claims about the age of the universe and Earth or the mechanisms and timing of how the spheres of creation were ordered and filled. We saw that cosmology and astrophysics is entirely consistent with God creating the universe from nothing, and we argued that it takes more of a blind leap of faith to believe that this finely tuned universe has a materialistic explanation. And I tried to hammer home the idea that God doing something in His creation and describing natural laws and process were not mutually exclusive from a proper biblical theology.

We exposited the early chapters of Genesis and saw that there was still darkness and the sea in the original ordered creation in Genesis 1, and that this contrasts with the new heavens and earth, and new Jerusalem in Revelation. This lends credence to the interpretation that the original finished creation was a place that already had forces hostile to life, although bounded and controlled by the sovereign God. (Waltke, 2001) We further bolstered this theology by noticing that God's pronouncements of the creation being good and very good were not warranted by the text as referring to moral goodness, but as good in the sense of being functional and ordered for His image-bearers and other creatures. But in order to explain the consequences of the Fall and original sin, I think we need to go over the biblical exegesis once again for the Fall and then tie it into the doctrine of original sin. We can then appreciate its implications for our argument in this book and then apply it to history and the people who were alive when Adam and Eve were created.

If we are going to talk about sin and its consequences, we should start by defining what sin is. Many people talk about sin as anything that produces frustration, suffering, and death. Some equate natural "evil", like drought, disease, and natural disasters with sin. But sin in the Bible always has a moral character to it. It exists in the ethical realm. And it is always with reference to either man or angels. Animals cannot sin. They lack the intellectual and moral capabilities to understand God's law and respond, and they are not made in the image of God. Therefore, they are not held liable for not loving God or loving others as themselves.

But having the ability to do something does not always precede culpability. Unregenerate man in the Bible is said to a be slave to sin. And the more they sin, the harder their hearts become, and the more God gives them over to sinning. But God still holds them responsible. And although sin involves the will of man choosing to pursue a lack of conformity to or transgression of God's will, the will of man itself is subject to the inner disposition of the man. No normal, functioning adult chooses to do something without regard to what they want, their end goals, certain moral considerations, and what they believe to be true. And this is the Bible's teaching. That out of the heart (man's internal nature) springs forth what man does. Man's behavior is determined by man's nature.

Then what is sin? It is the transgression of, or lack of conformity to, the law of God. Sin is lawlessness (1 John 3:4). And this moral law of God is either written on the heart, given in special revelation, or both. The innate law written on the heart is felt and experienced by the man without a hardened conscience through the guilt or acquittal of the moral conscience. The written law given in special revelation corresponds in moral terms with the innate law of the heart, but can be read and understood by any man, regardless of the hardness of his heart. And the penalties for transgressing the law are determined by the Lawgiver. With that definition of sin, let's move on.

I argued that the Fall of Adam and Eve did ***not*** introduce physical death, disease, and natural disasters to the world. I will again support the following assertions from the biblical text in a bit, but let's start by first simply asserting them. The Fall introduced spiritual death to the whole of the human race who had the image of God when the Fall happened, and to all those subsequent to this period in history. This is because God chose Adam to represent us all. God inaugurated the covenant of grace in history after Adam fell and provided the means of faith to trust in His promises for redemption, to be completely paid for in Christ in the fullness of time.

We have already argued that Genesis 1 is a unique type of highly stylized genre of biblical literature and that it is much more a polemic against the pagan cosmogonies and idolatries of the surrounding nations. And how God set up the creation to function for the benefit of mankind and the creatures, to exercise the mandates of being fruitful and multiplying, and to exercise his dominion over the created order as vicegerents for God. It is less an account of material processes to be read as historical narrative than it is an account of God's spoken fiats to create the universe and produce an orderly world full of creatures and fecundity in which humankind is to glorify their Maker and spread His glory and kingdom to the ends of the Earth. I have also tried to show that Genesis 2 changes genre to historical narrative and the propositional truth given in that chapter and the rest of Genesis is to be taken in the normal sense for biblical narrative literature. But Genesis 1 gives us theological truths and is a vital part of the entire storyline from creation to consummation. So, let's start by examining what many confessions and creeds have done by extrapolating doctrines that I believe are not warranted by the text itself.

Let's begin with the first statement of historical orthodoxy that I think is a stretch of exegesis from the Bible. The doctrine that when God was done creating the universe in Genesis 1, that His pronouncement of things being good on all the days (except day two), and His pronouncement of all things being very good at the end of day six, implies that the universe was perfect. Without death, disease, predation, natural disaster, and the like.

Is the idea of the universe as a place of perfection to be read out of the text? Are we being influenced too much by an Aristotelian view of the perfect cosmos? With perfect concentric circles for the orbits of the planets and perfect spheres for the heavenly bodies? Does not the context of Genesis 2 give us an idea that the garden of Eden was a much smaller sacred place hedged in as a paradise and temple where Adam and Eve had intimate communion with God? Thereby implying that the surrounding environment was different? Can we form a whole systematic natural theology based on this chapter and a couple verses from Romans 8 (which we will address)? Can we use those verses about creation groaning and being subjected to futility, waiting for its redemption as proof texts for an idyllic initial creation? Do the new heavens and new earth, along with the new Jerusalem, described in Isaiah and Revelation give us a picture of a global paradise regained that was lost? Or does it indicate something much better than the original creation and more akin to the conditions in the garden? Phrased a different way, are the new heavens and new earth going to look like the original heavens and earth before the Fall?

We'll start trying to answer those questions in a bit, but first we need to ask if a creation created in futility and having death impugns the character of God? It is paramount for Christians to affirm the biblical doctrines that God is both good and that He is not the author of sin. But what would be considered sin in a world without culpable moral beings? We have already seen that plants and animals cannot sin nor be held morally responsible. We will dive down into the implications for Adam and those he represents later, but I would argue that a planet with plants, animals, death, disease, and natural disasters prior to the entrance of sin into the world through Adam does not meet any biblical definition of sin, that they were not a penalty nor consequence of sin prior to Adam and Eve, nor does it impugn the goodness of God. I'll make my case for that as we work our way through this chapter.

Going back to what the original creation looked like and if it differs from the new heavens and earth, our first clue from the text comes from Genesis 1 and how it describes the creation. God calls forth light and separates the light from the darkness. There is both day and night in the original creation. God has brought forth the light and separated it from the darkness, but the darkness is still present. John uses this imagery in the beginning of his gospel by employing language from Genesis 1. John tells us that Jesus was the Light of the world, that the Light shines in the darkness, and the darkness did not comprehend it. (John 1:1-5) John uses the motif of darkness throughout his gospel as a metaphor for the realm of spiritual death. We assume that day and night operated in the whole of the original creation just like today. What information are we given by this same apostle John in the book of Revelation about the new Jerusalem regarding darkness?

Revelation 21:22-23

22 I saw no temple in it, for the Lord God the Almighty and the Lamb are its temple. 23 And the city has no need of the sun or of the moon to shine on it, for the glory of God has illumined it, and its lamp is the Lamb.

Revelation 21:25

25 In the daytime (for there will be no night there) its gates will never be closed

Revelation 22:5

5 And there will no longer be any night; and they will not have need of the light of a lamp nor the light of the sun, because the Lord God will illumine them; and they will reign forever and ever.

If we are to take the new Jerusalem as a recapitulation of the original creation that was perfect, why was there still darkness before the Fall? Why was there a need for a garden temple but there is no need in the new Jerusalem? Any time we are trying to interpret apocalyptic literature like Revelation, we should be wary of wooden, literal interpretations. We realize that Revelation is apocalyptic literature, and that type of literature employs symbolism and metaphor. Another point of note is that the only way to explain something to somebody they are unfamiliar with is by using things that are common to their experience. In explaining how a nuclear reactor works to a remote tribesman who has never been educated or had the benefit of electricity, you would have to employ things he does understand to try to make analogies to give him understanding. I think the same can be said of God giving us glimpses of heaven through apocalyptic visions.

Therefore, I personally don't think we can dogmatically say whether there will still be the sun, moon, and stars in the new heavens, if they are just absent in the new Jerusalem, or if it is symbolic language for things our minds cannot grasp. I personally don't think we can be dogmatic based on the apocalyptic literature of Revelation. But I am sure that God has conceived of things for His people that will be nothing short of perfection, and I will be filled with awe and worship if our good and benevolent Lord has us walking on streets of gold and there is no need for celestial bodies.

But the main point here is that there is no need for these celestial lights in the new Jerusalem, as God is illuminating the holy temple city. And the relevance for my thesis is that there ***was*** a need for these illuminating bodies and a temple in the original creation, but they are no longer needed in the new creation or holy city. This leads us into the second point of difference between the original finished creation in Genesis 1 and the new heavens and new earth of Revelation.

It is obvious from the text of Genesis 1 that the seas were a part of the original creation. Approximately 70% of Earth's surface is covered by the sea. The way the Old Testament talks about the sea is considerably variable. Israel was not a sea-going people and had little experience with sea travel and navigation. In one sense, the Old Testament mainly talks about the sea as something created by God, filled with His creatures (even great sea monsters), bounded by His will, given a boundary that it cannot pass, as God dividing the sea in the Exodus and using it to judge Pharaoh's army, and in a manner of fact way with reference to geography. In another sense, the sea is used as a symbol for judgement, for God whipping up the waves in anger, God drying up the sea in His rebuke, in God delivering people by calming the sea, and as a metaphor for defeating Israel's enemies. And then there are those passages that critical scholars and teachers of comparative religion like to point to as supposedly coming from Canaanite religion and myth. Let's take a little excursus from our main argument to deal with those verses, starting with Leviathan first.

Psalm 74:13-14

13 You divided the sea by Your strength;

You broke the heads of the sea monsters in the waters.

14 You crushed the heads of Leviathan;

You gave him as food for the creatures of the wilderness.

Psalm 104:25-26

25 There is the sea, great and broad,

In which are swarms without number,

Animals both small and great.

26 There the ships move along,

And Leviathan, which You have formed to sport in it.

Isaiah 27:1

27 In that day the Lord will punish Leviathan the fleeing serpent,

With His fierce and great and mighty sword,

Even Leviathan the twisted serpent;

And He will kill the dragon who lives in the sea.

And with poetic verse the Lord responds to Job in chapters 38-41, with chapter 41 specifically talking about Leviathan.

And then these verses about Rahab.

Psalm 89:9-10

9 You rule the swelling of the sea;

When its waves rise, You still them.

10 You Yourself crushed Rahab like one who is slain;

You scattered Your enemies with Your mighty arm.

Isaiah 51:9

9 Awake, awake, put on strength, O arm of the Lord;

Awake as in the days of old, the generations of long ago.

Was it not You who cut Rahab in pieces,

Who pierced the dragon?

Job 26:11-13

11 "The pillars of heaven tremble

And are amazed at His rebuke.

12 "He quieted the sea with His power,

And by His understanding He shattered Rahab.

13 "By His breath the heavens are cleared;

His hand has pierced the fleeing serpent.

Psalm 87:4

4 "I shall mention Rahab and Babylon among those who know Me;

Behold, Philistia and Tyre with Ethiopia:

'This one was born there.'"

Isaiah 30:7

7 Even Egypt, whose help is vain and empty.

Therefore, I have called her

"Rahab who has been exterminated."

First, the context. The context of Psalm 74 is one where the author is lamenting the destruction of Jerusalem and the temple and wondering why God is letting them spurn His holy name. And then the author recalls all that God has done in the past, including mighty deeds of deliverance, then he mentions that God divided the sea (verse 13), which is followed immediately by the references to the sea monsters and Leviathan. The context of Psalm 104 is one of praise for God's creation and His providence over that creation. The use of Leviathan here as a sea monster is clearly a reference to Genesis 1:21 as one of the sea monsters created on the fifth day. The context of Isaiah 27 is in the middle of a song about the future deliverance and restoration of Israel, and part of a larger poetic section on God's justice.

While Job chapters 38-41 are in poetic verse describing God's sovereignty over all things to Job. In chapter 40, God describes Behemoth and in chapter 41 Leviathan. Many commentators think Behemoth is a description of the hippopotamus and Leviathan the crocodile, with high poetic language to emphasize the strength, fierceness, and aggression of these animals. And the theological point to Job is that these creatures who are made by God and are far more powerful than man and are untamed by him, are nothing compared to their Maker. While this interpretation is definitely feasible in the context, it doesn't really address the twisting serpent or multiple headed description of Leviathan from other texts. But God does call him the **"king over all the sons of pride"** (Job 41:34).

Psalm 89 is about God's sovereignty and covenant faithfulness. How God is exalted in His creation and is in control of all things. How God stills the sea, crushed Rahab, and scatters His enemies. This is most likely a symbol for Egypt in this context. In Isaiah 51 the context is for those in Israel who pursue righteousness to find solace in the God who will vindicate the righteous and judge the wicked. The God who will protect those who keep the covenant and to not be afraid of those outside of God's covenant blessings who will parish. And then the author recalls the days of old when God cut Rahab in pieces and pierced the dragon. And then the following verses talk about the Red Sea being parted and the redeemed walking through on dry land. So, many commentators see Rahab as a clear symbol for Egypt. While in Job 26, Job is responding to his friends with the greatness and sovereignty of God in poetic language. In Psalm 87, Rahab is clearly a reference to Egypt, as it is grouped together with other nations. And finally in Isaiah 30, Judah is warned about seeking safety and an alliance with Egypt and God calls Egypt Rahab (or arrogance).

So, Leviathan is associated with pride and Rahab with arrogance. Leviathan can be a reference to the sea monsters that God created among the other creatures to fill the sea in Genesis chapter 1, an animal that is prideful and ferocious, or one that has characteristics like a twisting serpent with multiple heads. Rahab can be a reference to Egypt and referred to as a dragon or serpent. And both are used as symbols for Israel's enemies.

And this brings us back to the idea of interpreting scripture according to the type of literature we are reading. When looking at these verses, it is important to recognize that we are dealing with poetry and therefore interpreting all of it literally will probably lead us astray. Just like we would be off base by interpreting Job 38:31-32 to mean that God literally binds and loosens the constellations of stars with chains and cords and leads them around the heavens. So too here we are mainly dealing with poetry, symbolism, and metaphor. This is important to keep in mind.

Leviathan and sea monsters are both referred to in Psalm 74 above. The same Hebrew word for sea monsters is used in Genesis 1:21 when God created the great sea monsters on day five. But there is no sense of chaos or conflict there on day five, and God saw that it was good. So, sea monsters can simply refer to large aquatic animals, like whales, orcas, and sharks (Genesis 1:21, Psalms 148:7). Although the Hebrew for sea monsters and Leviathan are different, they seem to have a similar meaning. The word rendered sea monsters (Tanniyn) in Psalm 74 and elaborated as Leviathan in a second verse common to Hebrew poetry, is possibly an allusion to a multi-headed sea monster called Tunnanu from the Baal Cycle in Ugaritic mythology (in modern western Syria). Who appears to be a monster on the side of the sea god Yamm when Baal and Yamm do battle. But it appears to be Baal's sister Anat who kills the creature who is described as a twisting serpent with seven heads. Tanniyn and Tunnanu are most likely related words. But Hebrew and all the other languages and dialects around Canaan are very closely related, so similar words describing seas, serpents, or dragons is not that significant.

So, the references to Leviathan, Rahab, and the sea monsters then can used symbolically for God's dominion of the sea, as a symbol or metaphor for the defeat of Israel's enemies, associated with pride and arrogance, and probably also simultaneously a polemic against the worshippers of the Ugaritic pantheon. Which in the Baal Cycle is the story now familiar to us of a multitude of gods acting like people, generations of gods sexually producing younger generations from god and goddess couples, conflict, scheming, and the younger gods trying to displace older gods. As I mentioned in chapter 4, some scholars think that the Baal Cycle (or a similar myth) with its storm god equipped with weapons and defeating a female sea goddess, in this case Yamm, was prior to and the inspiration for the Enuma Elish, in which Marduk defeats the sea goddess Tiamat with weapons of storms and winds from Anu. But unlike Enuma Elish, the Baal Cycle is not a story about creation. At least not from the battle of Baal and Yamm. The creation of the gods and the world are either absent from the myth or may have been initially on the missing and damaged tablets from Ugarit.

I don't think the point is to affirm the truth of Ugaritic mythology, or even the presence of such creatures, but to symbolically describe God's sovereign rule, even over the enemies of Israel by using the symbol of the serpent in the sea. We have a seven headed dragon in Revelation 12:3, but almost no one is suggesting that this dragon is to be interpreted literally. The symbolism is even interpreted for us in the text as the devil and Satan, the one behind the God opposing nations of the world and its rulers in the symbol of the beast, employing apocalyptic imagery from books like Daniel. The God of the Bible is not defeating a sea goddess from which to make the world, nor is He slaying a sea monster so He can ascend to rule the gods. In Genesis chapter 1 He is hovering over the waters, calling things into existence, and forming and filling the spheres of creation for His creatures and His glory. And in these poetic verses above they are a metaphor or symbol for His sovereignty and deliverance.

The important thing is that all these references occur in poetic literature, which stands out clearly when looking at the context. The Israelites would have been very familiar with these Ugaritic religious texts, their own experiences and history of idolatry with the god Baal and the goddess Asherah, and could therefore understand the symbolism being applied.

I then conclude that the sea in the Old Testament is presented as something that was created by God, filled with aquatic creatures, including great sea monsters, and God saw that it was good (Genesis 1:21). But it is often used in poetic literature as a symbol and a metaphor for judgement, for God whipping up the waves in anger, God drying up the sea in His rebuke, in God delivering people by calming the sea, and as a metaphor for Israel's enemies being defeated. Even sometimes referencing pagan mythology as a metaphor for God's greatness and superiority over the surrounding nations, and to symbolize deliverances that happened in real history. But it is always something God controls and is bound by His will. Jesus shows us again that the storm and raging sea obey His commands (Matthew 8:23-27).

But like I have tried to argue many times already, we can't expect to make literal interpretations from texts that were not meant to be read that way. As another example, take 2 Samuel 22 where David sings a song to the Lord for delivering him from his enemies and Saul. Here the language is most certainly poetic, symbolic, and metaphorical for the historical events that have just unfolded in chapter 21. It doesn't mean that God is not really working and involved in the outcomes, only that the inspired authors employ majestic language in poetry and song to praise the One who controls all things, even if the spiritual realm is only revealed by the Spirit. And if we don't understand this, then every time we read about the sun being darkened, the moon turning to blood, the stars falling to the earth, the pillars of the heavens being shaken, smoke coming from the nostrils and fire from the mouth of God, and other judgement language like this we will be apt to misinterpret the text.

But despite the different characterizations of the sea in the Old Testament, I think we can safely say that it is often an image and symbol for things that prevent peace and harmony. Something that can symbolize enemy forces and judgement, which are both contrary to calm and security. So, how does this characterization of the sea apply to our thesis?

On day three, God separates the land from the sea and calls it good. We have argued that the reason that they were pronounced good is not some intrinsically moral aspect of them, but that the land is where human beings dwell. We are creatures made for the land. What then does Revelation have to say about the new heavens and new earth regarding the sea?

Revelation 21:1

21 Then I saw a new heaven and a new earth; for the first heaven and the first earth passed away, and there is no longer any sea.

Again, we are not so concerned about the hydrology of the new heavens and earth. Nor whether the water cycle operates in the same fashion as it does on Earth today. I personally don't think we can be dogmatic based on the type of literature of Revelation. But the first conclusion we can draw about the nature of the original ordered creation before the Fall is that it was qualitatively different than the new heavens, new earth, and new Jerusalem are going to be. There are no elements opposed to life or things that prevent peace and security in the new heavens, new earth, and new Jerusalem.

The next thing we know from our study is that the garden of Eden was a type of temple. We have shown that this makes sense based on how God was installing his image (Adam) into the temple garden after animating him. (Richter, 2018) It also makes sense based on the language given to Adam and Eve to serve and guard the garden as priests. (Beale, 2018) And serving and guarding implies that there are elements of the common and profane that need to be guarded against. And the later decorations and iconography that is used in the tabernacle and temple for trees, open flowers, and cherubim harken back to the garden. If the garden of Eden is a temple with Adam and Eve as priests, then their exercising dominion and subduing implies a common and profane world outside of the boundaries of the garden. Additional evidence for this argument is the fact that Adam and Eve are expelled from the garden after the Fall. Combining all of this, we can infer that the garden is a special sacred space and outside of it is the common and profane. What about the pronouncement that all things were very good?

When God declares at the end of the 6th day that all He made was very good, we can see that He has separated the spheres of creation and filled them.

They are no longer formless and void. The creation is now set up in a way that God's vicegerents can exercise wise dominion and begin the process of subduing the earth for God's glory and our joy. The creation was now ordered according to His plan, and He was pleased with the outcome. He steps back as it were to admire the work. We strengthened this argument by noticing the functional names and purposes given by God for the material elements He created in Genesis 1. And also by noticing that "it was good" is missing on the 2nd day, providing evidence that the evaluation of goodness was how it functioned in relation to mankind and God's purposes for the world. But many of us want to take this declaration of being very good and make a lot of moral assumptions and try to force into it a whole host of ideas about what the creation before the Fall looked like and how it operated. But where in the text can we find such ideas?

One thing many people point to is Genesis 1:29-30 where God tells the male and female created in His image that He has given them every plant yielding seed and fruit trees yielding seed for their food, and that He has given the green plants as food for the animals. And then they point to Genesis 9:2-3 where God now seems to allow them to eat animals for meat. They argue that this shows both man and beast were initially vegetarian, and it is only after the Fall and the flood that God allowed predation. And I think there is some truth to this. But Derek Kidner pointed out that forcing these verses to mean that all people and animals on the planet were herbivores before the Fall and the flood is as unlikely as pressing the meaning of the text to say that all plants on the earth were equally as edible. (Kidner, 1967) But I do think that Adam and Eve are portrayed as vegetarian from the text and that they did not eat meat. Because when the curses are pronounced on Adam in Genesis 3:17-19, Adam seems to be devastatingly frustrated because the plants that he is growing to become bread are now going to be hard won.

But before we go on, let's remember the type of literature in Genesis 1 and the scope of who is in view. It is a strong polemic against the creation myths of the surrounding nations and one of the reasons for the creation of mankind in the pagan myths is to produce food for the gods. Here in contrast is God providing food for mankind and animals. Once again, God the Holy Spirit is giving us the correct theology through the writing of Moses that the Lord God does not need anything from man, He provides life and breath and everything for us. And although we have argued that Adam and Eve are the primary male and female in view in Genesis 1, Adam also represents mankind in general. In other words, Adam and Eve being set apart and sanctified in the pristine image of God within the temple garden does not tell us about the conditions or the people and animals outside of it.

I think we can assume that Adam and Eve were initially created vegetarian and that the environment after the flood with all its sin, depravity, capital punishment, atonement, and sacrifice is a new reality with different rules for the second Adam, Noah. But to say that all animals outside of the garden before the Fall were herbivores is I think reading too much back into the text. The green plants, grasses, etc. make up the base of the food chain for all the animal kingdom and God provides for them all. (Kidner, 1967)

1 Timothy 4:3-4

3 men who forbid marriage and advocate abstaining from foods which God has created to be gratefully shared in by those who believe and know the truth. 4 For everything created by God is good, and nothing is to be rejected if it is received with gratitude;

We know from the Bible, especially in the Psalms and God's response to Job, that the Lord provides the prey for the lions and other animals, and that He takes away the spirit of man and animals at death. Is the physical death of animals a judgement for Adam and Eve's sin? I don't think that the Bible gives us any warrant to say that God providing for the predators is a judgement on them for man's disobedience. That would be odd for God to provide prey if animal death is in any way connected to sin. Neither do I see biblical evidence that the death of plants and animals was a judgement from God. But what about the imagery from Isaiah describing the new heavens and earth? I believe that the predators and prey living in harmony in these verses are metaphorical for peace. The main reason I hold this interpretation is that it is clearly in the context of the new heavens and new earth, but that people will still die, albeit at an old age (Isaiah 65:17-25). So, I take the language as metaphorical for peace, harmony, and fulness. There will be no weeping or crying and God's people will live in harmony with Him and their labor with be blessed and not frustrated.

Man's sin does have consequences for the animals, as it does for the earth itself. And they often share in the judgement for man's sin, like in the flood. We are told repeatedly in the Old Testament that the sins of Israel and the other nations pollute the land. And the judgement of God often comes in drought, famine, pestilence, disease, and wild beasts.

What conclusions can we then draw about the state of the world prior to the Fall? I think we can say that the good Lord God has separated and filled it according to His will and for our flourishing, He is pleased with the results, and that He has given all good things including food to man and beast. But there was a need for a garden temple in this ordered creation, implying a profane world outside of it to be subdued. And although there were forces hostile to life as part of this world, they were bounded and controlled by the sovereign Lord.

What then happened with the temptation, sin, and Fall? And how did this affect the broader world? What effects did the Fall produce and are they as far reaching on the physical Earth as many biblical commentators want to subscribe to this historic tragedy? Is our interpretation of the creation being subjected to futility in Romans 8 then reading far too much into the text and how this futility relates to the Fall?

When we come to the account of the Fall in Genesis 3, we are introduced to the serpent, one of the creatures that God had created. We are given more information later in Scripture that this indeed was the devil or Satan, either transformed into the likeness of a snake or more likely using the animal as his mouthpiece for the temptation. This should not surprise us. We have the account of the demons who asked Jesus to send them into the herd of swine after being cast out of the man (Matthew 8:28-34, Mark 5:1-20, Luke 8:26-39). One of the first questions that comes to many peoples' minds when reading the account is: Where did Satan come from? We were not given any prior information about Satan and his fall from heaven. He is just already in the story. Jesus tells us about watching Satan fall from heaven (Luke 10:17-20), although I think the best interpretation there is that Satan is being cast down out of heaven as the accuser of the brethren when the kingdom of God was being manifested during Jesus' ministry and after His ascension (Revelation 12:5-13). Other Old Testament passages give us glimpses of him in 1 Chronicles, Job, and Zechariah. The clearest account of Satan in the Old Testament is from the book of Job where he shows up amongst the angels in heaven to accuse Job before God. We are told that he has been roaming back and forth through the earth, and we know from the New Testament that he is roaming about like a roaring lion seeking whom he may devour. The New Testament gives us a much clearer picture of Satan, his demons, and their schemes, lies, and influence in this world. We know that Satan had already fallen from his heavenly post as an angel (probably high up in the angelic hierarchy), along with the other reprobate angels prior to the time of the tempting in the garden. Otherwise, the chronology doesn't make sense. So clearly Satan and the other damned angels had already sinned against God and been condemned (2 Peter 2:4, Jude 6). From other verses many theologians see the initial transgression of the archangel who was later called Satan as wanting to rise to the level of God and usurp Him. To be like God. The Bible teaches us that the angels are also created beings (Colossians 1:16-17). Which would then make sense of the tactics Satan uses to tempt Eve. To tempt her to be like God.

Satan begins by questioning the word and goodness of God by implying that God is being unfair and restrictive in not allowing Adam and Eve to eat from every tree. Although God has graciously provided all they would ever need from all the other trees and plants. The woman corrects the serpent's lie and repeats the prohibition about the tree of the knowledge of good and evil and adds a separate stipulation about not even touching the tree. We don't know if God had also given them this command, but it seems more likely that the deception of Satan had already begun to work its dirty tendrils into Eve's mind. Then Satan proceeds to directly contradict God and the consequences of surely dying if they eat from the tree. But we know that Satan is the father of lies. Eve is seduced and deceived by the lies of Satan. She sees that the tree is good to make her wise, she takes and eats of the fruit, gives some to Adam, and he eats as well. Their eyes are opened, they know they are naked, and they are ashamed because of it. This is our first glimpse of the effects of moral corruption or pollution. They feel ashamed and their consciences bother them. Satan has employed a half truth about their eyes being opened, but not in the devastating way that actually took place.

They try to cover their nakedness with fig leaves, and they hide themselves from God as He is approaching. This is our first look at sin as guilt. They know they deserve punishment, they now fear God, and so they try to hide from Him. This is a fine illustration of how their communion with God is destroyed by sin. A holy God cannot tolerate sin in His presence. But we will see that God almost immediately provides promises and the means of grace to restore that fellowship, despite the rebellion of His creatures. When Adam is confronted by God about transgressing His law, we can see the relational effects of the Fall amongst the creatures themselves. Adam blames Eve, and by extension God Himself, and Eve blames the serpent. Neither is willing to accept responsibility. The self-justifying actions and deflection of blame for humans transgressing the law of God has begun.

What is the sin that actually took place? Does it have to do with magical fruit? Did the fruit contain some mystical power that made connections in the brain to be able to discern good from evil? The outright act was a transgression of the law of God. He gave them a simple command and they broke this command. But a lot of theologians believe that the real, fundamental sin is one of exerting our autonomy over the rule of God. We don't want God to tell us what to do. We long to be autonomous. Unfortunately to our own peril.

I think it is a combination of things. Adam and Eve had the moral law written on their hearts, but they also had positive holiness. So, they would want to love God with all their hearts, minds, soul, and strength, and to love others as themselves. But God gives them this particular law by special revelation that would prove their obedience. Satan tempts them and they believe his lies. They act in unbelief regarding what God has said, they think God is not good, and that He doesn't have their best interests in mind. They think His command is too restrictive and limits their liberty, they want to determine right and wrong (or good and evil) for themselves, and they desire their autonomy.

We then come to the consequences of the Fall. First to the judgement on the serpent.

Genesis 3:14-15

14 The Lord God said to the serpent,

"Because you have done this,

Cursed are you more than all cattle,

And more than every beast of the field;

On your belly you will go,

And dust you will eat

All the days of your life;

15 And I will put enmity

Between you and the woman,

And between your seed and her seed;

He shall bruise you on the head,

And you shall bruise him on the heel."

What are we to make of this judgement on the serpent? Some people think that up until this point serpents had legs and the curse removed those legs so that snakes now had to crawl. Not likely from the context of the rest of the curse. Many see in this curse the first promise of a coming Redeemer born of the seed of the woman who will overcome the seed of the serpent, the devil, and his children. And we have seen that the New Testament ascribes the final fulfillment of this prophecy to Jesus Christ.

Jesus was talking to the scribes and Pharisees who claimed that they were children of Abraham and God, but Jesus told them that they were of their father the devil.

John 8:42-44

42 Jesus said to them, "If God were your Father, you would love Me, for I proceeded forth and have come from God, for I have not even come on My own initiative, but He sent Me. 43 Why do you not understand what I am saying? It is because you cannot hear My word. 44 You are of your father the devil, and you want to do the desires of your father. He was a murderer from the beginning, and does not stand in the truth because there is no truth in him. Whenever he speaks a lie, he speaks from his own nature, for he is a liar and the father of lies.

We then see within the curses on the serpent the promise of the coming Messiah who will crush the serpent under His feet and break the power of sin and death. These same godless men in John 8 are the ones who had Jesus crucified. But in doing so Jesus crushed the head of the serpent, although the serpent (Satan) bruised his heel (caused him suffering and temporal death). This curse is not about how women are afraid of baby snakes. Neither is it one about snakes losing their legs. It is a curse of humiliation on the serpent, and by extension, on Satan himself. But the glorious promise of the Messiah is included. And then the curses on Eve.

Genesis 3:16

16 To the woman He said,

"I will greatly multiply

Your pain in childbirth,

In pain you will bring forth children;

Yet your desire will be for your husband,

And he will rule over you."

I believe that this curse of childbirth is designated primarily for Eve, as she is never put forth as a covenantal head in the way that Adam is. Eve will now be frustrated in her mandate to be fruitful and multiply. What was a blessing from the Lord will now be mingled with pain. What does it mean for her desire to be for her husband and that he in turn will rule over her? We don't have to read much further to see this same type of language used with respect to Cain and sin.

Genesis 4:6-7

6 Then the Lord said to Cain, "Why are you angry? And why has your countenance fallen? 7 If you do well, will not your countenance be lifted up? And if you do not do well, sin is crouching at the door; and its desire is for you, but you must master it."

In verse 7 we have the same Hebrew construction used for sin. Sin will have a desire to rule over you, but you must master and lord over it. So, the woman will now have a need to rise up and rule the marriage, spurning submission and seeing it as a totalitarian rule, as opposed to working together in their given roles. The man will see this affront to his headship, master the mutiny, and rule over the woman. The relationship is broken and there is now distrust involved where there should have been cooperation. Now on to the curse for the covenant head of humanity, Adam.

Genesis 3:17-19

17 Then to Adam He said, "Because you have listened to the voice of your wife, and have eaten from the tree about which I commanded you, saying, 'You shall not eat from it';

Cursed is the ground because of you;

In toil you will eat of it

All the days of your life.

18 "Both thorns and thistles it shall grow for you;

And you will eat the plants of the field;

19 By the sweat of your face

You will eat bread,

Till you return to the ground,

Because from it you were taken;

For you are dust,

And to dust you shall return."

Here is where I think we must be careful not to read more into the account than it allows, while also not contradicting the teaching of the rest of the Bible. God curses the ground (land or soil, not the planet) because of the sin of Adam and tells him that his manual labor will be difficult. He will come by his food through much hard work and toil. No longer will he have the productive work and fruitful bounty from within the garden to satisfy him and his family. Not to mention the much worse curse of sin, the fear of death, and the judgement to come. As well as being banished and cleansed from the temple garden and the intimate presence of God.

What does it mean that Adam will return to the ground? The Hebrew word for Adam, both as a proper name and generic for mankind, is adam and the word for ground is adamah. In Hebrew the word play is clear showing that Adam is made from the ground. He is the earthy man and those born in the flesh are also earthy. Does it mean that Adam was initially immortal and that one of the curses was mortality and physical death? In one sense, no. I have previously argued (and will again in a moment) that Adam's body was always flesh, blood, and mortal. But in another sense, yes. Adam was banished from the garden, and he could no longer partake of the tree of life. God had warned Adam that in the day that he ate from the tree of the knowledge of good and evil he would surely die. But we know that Adam and Eve did not physically die immediately after they had eaten the fruit. But the repeating refrain of the generations of Adam from Genesis 5 clearly state: And he died, and he died, and he died. Even those in the elect line of Seth who call on the name of the Lord can longer eat from the tree of life.

Implied in the penalty of surely dying for breaking the commandment of eating from the tree of the knowledge of good and evil was the opposite reward for obedience, eternal life. Adam would have been given eternal life in a glorified body as a reward (still of pure grace) for not sinning. Therefore, Adam now had no way to avoid physical death, and with it, the fear of spiritual death apart from God's grace.

Is this what Romans 8 is referring to? Is this the time when the perfect, natural Earth outside of the garden was cursed, and the laws of physics and processes of geology and meteorology changed for the worse? Did earthquakes, hurricanes, tsunamis, and the like begin on this day with this curse? Let us first deal with the issue of the ground being cursed because of Adam.

I have been somewhat surprised by a couple of overlooked passages within a couple chapters of this curse on Adam which seem to get little attention. Although to be fair, some commentators do not think they refer to rescinding the curse of the ground. In Genesis 5 we are given genealogies according to the line of Seth, one of Adam's sons. And then we come to the section with the first mention of Noah.

Genesis 5:28-29

28 Lamech lived one hundred and eighty-two years, and became the father of a son. 29 Now he called his name Noah, saying, "This one will give us rest from our work and from the toil of our hands arising from the ground which the Lord has cursed."

Some think that this oracle of Lamech is talking about the hope of the promise of the seed of the woman from Genesis 3:15 being fulfilled in or through Noah. And I agree that is part of it. Noah will continue the promise of the seed of the woman who will crush the serpent's head. But its near fulfillment seems more likely to be directed at the cursed ground on account of Adam. The ground has been cursed because of Adam's sin, and those subsequent generations have had to labor under this curse, battling with futility from the soil. How is Noah going to give the people rest from the curse of the ground that was a result of Adam's sin? After the flood subsides and Noah sacrifices burnt offerings to the Lord, the Lord makes this declaration.

Genesis 8:21-22

21 The Lord smelled the soothing aroma; and the Lord said to Himself, "I will never again curse the ground on account of man, for the intent of man's heart is evil from his youth; and I will never again destroy every living thing, as I have done.

22 "While the earth remains,

Seedtime and harvest,

And cold and heat,

And summer and winter,

And day and night

Shall not cease."

So, in Genesis 3:17 the ground is cursed, in Genesis 5:29 Lamech prophecies that Noah will give them rest from the ground that is cursed, and finally in Genesis 8:21, the Lord says that He will never again curse the ground on account of man and then gives a promise. The promise being: As the natural order of the world continues, He will ***neither*** curse the ground ***nor*** destroy every living thing from the face of the earth. There are two promises from God here, not one. One refers to the curse of the ground on account of Adam and the other is regarding the flood.

I believe this is the strongest exegetical interpretation from the immediate context, that the curse that God pronounced on the ground because of Adam was rescinded after Noah's flood. And I can't find any other references in the Bible that describe people toiling in futility to produce crops from the ground because of the effect of Adam's sin. As we have previously seen, there are numerous curses of the land not producing associated with breaking the covenant at Sinai. And a lot of blessings connected with keeping the covenant. But a lot of other verses talk about the bounty of the land and the graciousness of God in providing the crops and the rains. But no explicit or implicit mention of needing rest from the cursed ground because of Adam's sin. Let's look at some verses about God blessing the ground and its fruit.

Leviticus 26:4

4 then I shall give you rains in their season, so that the land will yield its produce and the trees of the field will bear their fruit.

Deuteronomy 7:13

13 He will love you and bless you and multiply you; He will also bless the fruit of your womb and the fruit of your ground, your grain and your new wine and your oil, the increase of your herd and the young of your flock, in the land which He swore to your forefathers to give you.

Psalm 65:9-13

9 You visit the earth and cause it to overflow;

You greatly enrich it;

The stream of God is full of water;

You prepare their grain, for thus You prepare the earth.

10 You water its furrows abundantly,

You settle its ridges,

You soften it with showers,

You bless its growth.

11 You have crowned the year with Your bounty,

And Your paths drip with fatness.

12 The pastures of the wilderness drip,

And the hills gird themselves with rejoicing.

13 The meadows are clothed with flocks

And the valleys are covered with grain;

They shout for joy, yes, they sing.

Psalm 104:14

14 He causes the grass to grow for the cattle,

And vegetation for the labor of man,

So that he may bring forth food from the earth,

Psalm 147:14

14 He makes peace in your borders;

He satisfies you with the finest of the wheat.

Proverbs 3:9-10

9 Honor the Lord from your wealth

And from the first of all your produce;

10 So your barns will be filled with plenty

And your vats will overflow with new wine.

Acts 14:17

17 and yet He did not leave Himself without witness, in that He did good and gave you rains from heaven and fruitful seasons, satisfying your hearts with food and gladness."

I then conclude that the part of the curse relating to the ground was abolished after the flood. So far, we have Satan cursed, the woman frustrated in her mandate to multiply descendants, and Adam's curse on the ground abolished after the flood. But let us not trivialize what happened in the Fall. Although I have argued thus far that the earth outside of the garden did not substantially change in its operation, what happened here was of cosmic significance. And to unpack why, we need to look back at Adam.

Romans 5:12-14

12 Therefore, just as through one man sin entered into the world, and death through sin, and so death spread to all men, because all sinned— 13 for until the Law sin was in the world, but sin is not imputed when there is no law. 14 Nevertheless death reigned from Adam until Moses, even over those who had not sinned in the likeness of the offense of Adam, who is a type of Him who was to come.

Paul tells us that sin entered the world through Adam. We have defined sin as being a transgression against, or lack of conformity to, the law of God. God had given Adam a clear law in special revelation to not eat from the tree and Adam broke this law. And when that happened Adam's sin was imputed to all men and death spread to all men. Sin was in the world before the Law of Moses (Sinai covenant), but people did not have this special revelation from God to sin against like Adam had. Nevertheless, death reigned in the time between Adam and Moses. What kind of death is this?

Romans 5:15-17

15 But the free gift is not like the transgression. For if by the transgression of the one the many died, much more did the grace of God and the gift by the grace of the one Man, Jesus Christ, abound to the many. 16 The gift is not like that which came through the one who sinned; for on the one hand the judgment arose from one transgression resulting in condemnation, but on the other hand the free gift arose from many transgressions resulting in justification. 17 For if by the transgression of the one, death reigned through the one, much more those who receive the abundance of grace and of the gift of righteousness will reign in life through the One, Jesus Christ.

Here Paul further defines what this kind of death this is. The many died because of the one transgression of Adam. The judgement on Adam was condemnation, and those who receive the free gift and grace of justification will reign in life. In what kind of life? They will reign in everlasting life with God, although they die physically in this life. They will reign first with the spiritual ego and consciousness without a body, and then with a glorified body with the second coming of Christ and final judgement. Those in Adam were imputed the sin of Adam resulting in condemnation (spiritual death). And this spiritual death reigned from Adam to Moses, even though the Mosaic law was not yet given. Those in Christ are justified by the imputation of our sins to Christ, and by the imputation of His righteousness to us. They reign in spiritual life with Christ. Spiritual death is being contrasted with spiritual life in these verses.

Everyone included in the original covenant of works, with Adam as the head, was imputed Adam's first sin. (Hereafter I will just refer to it as the covenant of works, but the adjective original is to still be assumed. Many theologians would consider what Jesus was doing during His life as a second covenant of works. Jesus was doing what Adam failed to do, obey the law of God and not sin. In other words, Christians are saved by works but they are not our own, they are Christ's works and rewards imputed to us by the Holy Spirit through faith). And we were given a sentence of condemnation for that sin. This is where it is imperative to understand the covenantal nature of Adam with mankind. He is our representative. And that's what Romans 5:14 means that Adam was a type or foreshadowing of Him to come. Adam was a type of Christ in that both are heads of a covenant that represent others. What Adam did, we did. He ate the fruit; we ate the fruit. He was condemned, so are we. He was given a depraved nature, so we have a depraved nature. If we think that this somehow impugns the justice and fairness of God, just remember that by the same basis we are justified by Christ. This is Paul's argument. And it is also the glorious, good news of the gospel of Jesus Christ.

Romans 8:1

8 Therefore there is now no condemnation for those who are in Christ Jesus.

Paul is setting up the analogy that in the same way we are condemned through the act of another, Adam, we are also freely justified (the opposite of condemnation) by the acts of another, Jesus. If we think that we should only be judged based on our own actions and sins, then we also have to say that we should be justified by our own merits. We would have to hope that all our goodness in the end is enough to merit us God's favor and eternal life. But that would negate Paul's whole argument up to this point in Romans. All of us are under sin, fall short of the glory of God, and can never be justified by the works of the law. We need an alien righteous apart from our own works to be imputed to us so that we can be reconciled to God and declared justified.

Let's now move on to the next verses about Adam in 1 Corinthians.

1 Corinthians 15:20-26

20 But now Christ has been raised from the dead, the first fruits of those who are asleep. 21 For since by a man came death, by a man also came the resurrection of the dead. 22 For as in Adam all die, so also in Christ all will be made alive. 23 But each in his own order: Christ the first fruits, after that those who are Christ's at His coming, 24 then comes the end, when He hands over the kingdom to the God and Father, when He has abolished all rule and all authority and power. 25 For He must reign until He has put all His enemies under His feet. 26 The last enemy that will be abolished is death.

After laying out the historical reality of Christ's resurrection in the beginning of this chapter, Paul tells us that Christ is the first of His body, the church, to rise with a resurrected body. Death came by Adam, the resurrection from the dead by Christ. What kind of death and how? Paul again reiterates the point he was making in Romans 5, that we have two covenant heads, Adam and Christ. We all fell in Adam in the covenant of works and the only way to avoid the consequences of being in Adam is to be found in Christ. We are either in union with Adam or we are in union with Christ. But this time the contrast is not between condemnation and justification like Romans 5. This time it is between physical death and physical life in a resurrected body. Now, many theologians at this point would want to stop us and say that condemnation as a penalty for sin involves both spiritual and physical death. And I think they are right. But I think Paul is mainly contrasting spiritual death and spiritual life in Romans 5 and physical death and resurrected life in 1 Corinthians 15. But we need to go a little further to unpack this.

We all die in Adam, and all who are Christ's will be made alive (resurrected) with imperishable spiritual bodies at the second coming of Christ. The kingdom of God has been inaugurated with the ministry of Jesus but has yet to be completed, and Jesus is now reigning until God the Father puts all His enemies under His feet (Psalm 110:1).

1 Corinthians 15:42-49

42 So also is the resurrection of the dead. It is sown a perishable body, it is raised an imperishable body; 43 it is sown in dishonor, it is raised in glory; it is sown in weakness, it is raised in power; 44 it is sown a natural body, it is raised a spiritual body. If there is a natural body, there is also a spiritual body. 45 So also it is written, "The first man, Adam, became a living soul." The last Adam became a life-giving spirit. 46 However, the spiritual is not first, but the natural; then the spiritual. 47 The first man is from the earth, earthy; the second man is from heaven. 48 As is the earthy, so also are those who are earthy; and as is the heavenly, so also are those who are heavenly. 49 Just as we have borne the image of the earthy, we will also bear the image of the heavenly.

Paul continues his argument of either being in Adam or being in Christ by comparing the natural, earthy, corruptible flesh of Adam and the rest of us with the spiritual, incorruptible bodies of Christ and the resurrected saints. The natural body is one prone to decay, corruption, and perishing. Paul explains by way of comparison that the first man, Adam, had a natural body and the last Man, Christ, became a life-giving spirit. The natural man comes first, from the earth, and is earthy and corruptible. The last Man is from heaven. All those in Adam are earthy in nature, those who are in Christ will be spiritual in nature. We must put on spiritual, incorruptible bodies to be citizens of heaven. But a significant point germane to our discussion of the Fall and original sin is verse 45 where Paul quotes Genesis 2:7.

1 Corinthians 15:45

45 So also it is written, "The first man, Adam, became a living soul." The last Adam became a life-giving spirit.

Genesis 2:7

7 Then the Lord God formed man of dust from the ground, and breathed into his nostrils the breath of life; and man became a living being.

Living being in Genesis 2:7 can also be rendered living soul. And the important point is that Paul is talking about Adam ***before*** the Fall. Adam had an earthy, corruptible body before the Fall happened. And even Adam would be required to put on the imperishable before he could enter the kingdom of heaven. We have all born the image of the earthy, like Adam, and those who are in Christ will bear the image of the heavenly, like Jesus. This point of Adam being earthy is very significant and takes us back to Genesis and the curse on Adam.

Genesis 3:19

19 By the sweat of your face

You will eat bread,

Till you return to the ground,

Because from it you were taken;

For you are dust,

And to dust you shall return."

This being the last part of the curse proclaimed on Adam after the Fall, God tells him that he will return to the ground because he was taken from the ground. Most traditional doctrines of the Fall say that Adam was born immortal, both in soul and body, and he would not have died a natural death if he had not sinned. Many theologians do allow that he would have had to be transformed before dwelling in heaven, but still claim that he would not have died naturally without sin. I don't think we can get that doctrine from either the Old or New Testament passages. It is likely, based on inferences from the text, that Adam and Eve would have continued to partake of the sacrament of the tree of life and not tasted physical death, but it seems clear that both he and Eve would have aged and experienced decay in their mortal flesh. Eternal life is the implied blessing for not breaking God's law in the garden, but this blessing would have required a changing of Adam's nature to inherit the kingdom of heaven.

And we need to remember that life in its fullest sense in the Bible is defined as living with God in the light of His countenance. For Him to turn His face away entails the absence of life. We are so accustomed to thinking of death only in terms of physical life. But the word of God sees spiritual death as much more serious than physical death. For the first is eternal separation from God and His favor. And this is how Jesus rendered powerless the devil, who frightened men with separation from God and death all their lives (Hebrews 2:14-18). Jesus tasted death for all those who trust and follow Him (Hebrews 2:9) and He was crowned with glory and honor. And his disciples no longer have to fear death because God promises the same for them.

I have tried to show that Adam was indeed born with a perishable body, made from the earth, and that he would have been prone to natural death regardless of whether he passed some inferred probationary period or not. I believe the difference is that he would have been allowed to continue to eat from the tree of life as a sacrament of the covenant of works (also called the covenant of creation). This covenant is important to discuss so that we can understand in what relation Adam stood to other humans. And for comprehending the transmission of sin, guilt, and depravity. By Adam and Eve living in their original righteousness and partaking of the tree of life, they would not have been under spiritual condemnation and the fear of death their entire lives, knowing that either upon their physical death or being translated into heaven, they would have been given spiritual, incorruptible bodies. Fully clothed to be citizens of heaven. So, in Adam, all die spiritually. And all die physically because we are born with mortal flesh prone to corruption by nature, and because the tree of life is no longer accessible due to Adam's sin.

I believe further confirmation is given to this theology when we look at the human nature of Jesus. Jesus had to be like man in every respect (Hebrews 2:17–18). To atone for the sins of mankind, the eternal Son of God took on a human nature, body and soul, just like ours. Everything about Jesus' humanity was like ours, except He was without sin. Although He had no earthly father and was conceived by the power of the Holy Spirit, He was born of a woman in the ordinary way of maternal delivery (Luke 2:6-7). He developed and grew in both his body and mind like a normal child (Luke 2:40). He got hungry, thirsty, tired, weak, and experienced emotions. He was tempted and had to resist temptation, so he can sympathize with our temptations, and he suffered as a result of them (Hebrews 2:18). Jesus appeared to age normally and hypothetically considering what would have happened if he had not been crucified, he would have presumably died a natural death, although God would not allow His Holy One to undergo decay (Psalms 16:10, Acts 13:35). Now, for the reason of dying to save His people from their sins He came to Earth and was raised again to not undergo decay, but it is simply to make the point of the humanity of Jesus.

Now, most theologians don't deny any of that, except maybe for the hypothetical above. They affirm that Jesus took on the finitude and weakness of fallen man so that He could atone for fallen man in his woeful condition. But I think it raises issues if we believe that corruption of the human body that was now subject to physical death for the first time was part of the penalty for Adam's sin. Jesus was sinless in His life, and He was imputed the sins of those who would believe on the cross. Just like we are imputed the righteousness of Christ by the Holy Spirit when we are justified. We are not made inherently righteous in our own nature when we are justified, although our nature does come to life and is changed in regeneration and becomes more righteous through sanctification. But a penalty can only be just if it is a liability for guilt and that guilt corresponds to an infraction of the law. Therefore, it seems to me that as Jesus was guiltless in His life and not a party to the covenant of works, if corruptible human flesh was indeed a penalty for Adam's sin that was imposed on Jesus prior to bearing the wrath of God on the cross, then it would be unjust. We will return to the implications of Jesus's humanity in a moment, but let's first look at how Adam stands in relation to the human race.

How and why is this sin, guilt, and condemnation transferred to others besides Adam and Eve?

Hosea 6:7

7 But like Adam they have transgressed the covenant;

There they have dealt treacherously against Me.

Some theologians deny that there is a covenant in the first few chapters of Genesis. Some of the reasons they give are because it is not specifically mentioned by the name covenant, there are no explicit agreements between parties, and the blessings and curses are not named. But a covenant is simply an arrangement or contract between two or more parties that has conditions. And this is clearly in the text. The parties, like most covenants in the Bible, involves God as one party and human beings as the other. The exception being the covenant with Noah, his family, his descendants, and the animals.

Although the garden of Eden is five millennia before writing, covenants in the ancient near east during certain time periods involved a preamble describing what the superior party had done and accomplished for the lesser party before laying out the conditions for the covenant. In the covenant of works with Adam, the preamble is assumed by God being the Creator of all things, including the man and the bounty of the temple garden. And then came the blessings and the curses for obedience and transgression. Look at the beginning of the Ten Commandments for the covenant with Israel at Mt. Sinai, which K.A. Kitchen has argued convincingly is to be dated to the 14th-13th century BC. (Kitchen, 2003)

Exodus 20:1-2

Then God spoke all these words, saying,

2 "I am the Lord your God, who brought you out of the land of Egypt, out of the house of slavery.

Biblical covenants sometimes involved God explicitly telling the other party what He has done for them in redemptive history before laying out the terms of the covenant. Grace always precedes law. I think it is no different in Genesis 2. Adam owes everything to God, including his supernatural creation, and God makes an agreement with Adam that he can eat from all the trees of the garden, presumably also from the tree of life. He can live in communion and fellowship with God Himself, which is true life. The conditions of the covenant are to serve and guard the garden temple, keeping it holy and undefiled, to extend its boundaries by exercising dominion and subduing the earth, spreading the glory and kingdom of the Lord to the ends of the earth, and not eating from the tree of the knowledge of good and evil. The only condition given explicit mention in the text is about not eating from the tree of the knowledge of good and evil. The explicit penalty and curse for breaking God's covenant is that **"for in the day that you eat from it you will surely die."** (Genesis 2:17) The implied blessings for obedience are living in a blissful relationship with God in eternal life, presumably not to face physical death and be given a glorified body to dwell in heaven after a time of probation, and to see the glory of the Lord and His kingdom expanded throughout the world by Adam and his godly offspring.

Satan (the adversary) tempts Eve, Adam willingly follows along, and they violate the conditions of the covenant. Adam and Eve are condemned, and they both surely died that day. And then additional curses for breaking the covenant follow before they are expelled from the garden. I then think we can confidently assert that there was a covenant of works from the immediate context itself. This is further supported by our explicit verse above from Hosea 6:7 and Paul's treatment of Adam in the New Testament as the federal head of humanity. And we can see that both guilt and depravity are present in all of Adam's descendants.

Following the breaking of this covenant, God in His justice deals out the penalties in the form of spiritual and physical death, which were the just deserts of Adam and Eve and what was promised for disobedience. Spiritual death immediately occurred for Adam and Eve in the form of guilt and its penalty of condemnation and sin thoroughly corrupted their bodies and souls, marring and defacing the image of God. And physical death was now inevitable because they were expelled from the garden and could no longer partake of the sacrament of life. God could have annihilated them, but in His mercy and longsuffering He instead gives them the first promise of the covenant of grace, that of a Redeemer and Savior. And then God sheds blood and covers them with animal skins in place of their meager fig leaves and expels them from the garden so they are not confirmed in their damnation by eating from the tree of life. From this moment on, people represented by Adam will need atonement for sin and a Savior.

All of us are imputed the sin, guilt, and penalty of condemnation of Adam because of his transgression. So how does this covenantal relationship work? As unfair as it may seem to us initially, we are quick to accept the covenantal relationship with respect to Christ. We rejoice in the fact that He had the sins of those of us who trust in Him imputed to Him on the cross and then bore the wrath of God on our behalf which we deserved. And we sing His praises that we are imputed His righteousness by faith. And this imputed righteousness gives us a standing, reward, and benefits that we did not earn. The same concept applies to those in Adam. His first transgression was imputed to us directly because he represents us. Guilt is defined by its relation to law and subject to penalty or sanction. And the penalty for Adam's transgression was condemnation. In the day that he ate of the fruit, we surely died. We are therefore all guilty and under a penalty of condemnation because of Adam's transgression. But we are also born with a corrupt nature prone to sinning. This corruption of our nature (depravity) is itself worthy of guilt and condemnation, and it does not take us long to commit our own transgressions. Is this corruption or pollution of our nature also imputed or is it transmitted some other way?

Many orthodox Christian theologians and scholars want to argue that if Adam and Eve were not the first human beings on Earth and the parents of the entire human race, then there is no ontological basis for our covenantal relationship with Adam. They claim that God would seem to be arbitrary if the covenantal relationship was not also grounded in something real, like biological ancestry. They say that the imputation of Christ's righteousness to us is by means of the Holy Spirit, providing a real spiritual connection between the Christian and Jesus Christ. And therefore, Christ can really represent us in a covenantal relationship. But I'm not sure it's that simple.

If fact, we are confronted with several questions if ordinary sexual generation is the real basis on which our guilt and corruption in Adam is grounded. Such as, why are we only guilty of Adam's first sin and not the subsequent ones? Why not the sins of all our forefathers? Why would it be just to punish us for simply being descendants? If Jesus is ultimately a son of Adam (Luke 3:38), how is He not guilty or born with a corrupted nature? (Berkhof, 1996) The traditional answer to that last question has been that original sin is transferred by the father to the children, which would initially seem to protect the human nature of Jesus, despite the strange implication that children do not also inherit their mother's nature. But even with this doctrine of the paternal transmission of corruption, that leaves open the first questions. Then it is usually said that Adam is not only in biological relationship with his descendants but also in federal headship. Let's look at this more closely.

Is the corruption of our nature also imputed like guilt and condemnation or it is transferred by some other mechanism? There is no denying that the Bible teaches that we are all born in sin and the reason we commit actual transgressions is that our nature is inclined that way. And reason and experience verify this reality for all of us. We don't have to teach children to lie and disobey, to hit their siblings, or be selfish. Jesus says that out of heart comes the things that defile a person (Matthew 15:18-19). The mess of this sinful world confirms this to us daily. And surely if it was not innate, some remote society in the history of the world would have remained pure with no bad examples to imitate. I think it is clear from the Bible that all of us are born with a sinful nature, but I will argue ***not*** according to the traditional doctrine of ordinary generation.

Is original sin (a corrupted sinful nature, depravity) passed through ordinary generation (human sexual reproduction) like codified in the Westminster Confession and Catechisms? This is one of the few points in which I disagree or have nuance with the Westminster Standards. I am only using the Westminster Standards because I have been persuaded that reformed theology is biblical theology for many years now, but I could have chosen other orthodox confessions and creeds. But I think the doctrine of ordinary generation as the mechanism of transmitting original sin has been and still is the orthodox Christian position.

I think this doctrine first rests on the assumption that Adam and Eve were the sole progenitors of the whole human race. But even if we believed the word of God unequivocally taught this (which I have tried to show otherwise), I believe it is unnecessary to resort to as an explanation. Let's first look at some verses from God's word to determine if we are indeed born with original sin before we examine how it may be transmitted.

Genesis 6:5

5 Then the Lord saw that the wickedness of man was great on the earth, and that every intent of the thoughts of his heart was only evil continually.

Genesis 8:21

21 The Lord smelled the soothing aroma; and the Lord said to Himself, "I will never again curse the ground on account of man, for the intent of man's heart is evil from his youth; and I will never again destroy every living thing, as I have done.

Jeremiah 17:9

9 "The heart is more deceitful than all else

And is desperately sick;

Who can understand it?

Mark 7:21-23

21 For from within, out of the heart of men, proceed the evil thoughts, fornications, thefts, murders, adulteries, 22 deeds of coveting and wickedness, as well as deceit, sensuality, envy, slander, pride and foolishness. 23 All these evil things proceed from within and defile the man."

Romans 3:23

23 for all have sinned and fall short of the glory of God,

Romans 3:9-10

9 What then? Are we better than they? Not at all; for we have already charged that both Jews and Greeks are all under sin; 10 as it is written,

"There is none righteous, not even one;

Ecclesiastes 9:3

3 This is an evil in all that is done under the sun, that there is one fate for all men. Furthermore, the hearts of the sons of men are full of evil and insanity is in their hearts throughout their lives. Afterwards they go to the dead.

Ephesians 2:1-3

And you were dead in your trespasses and sins, 2 in which you formerly walked according to the course of this world, according to the prince of the power of the air, of the spirit that is now working in the sons of disobedience. 3 Among them we too all formerly lived in the lusts of our flesh, indulging the desires of the flesh and of the mind, and were by nature children of wrath, even as the rest.

We could multiply verse after verse, but we get the point. We are by nature children of wrath, born in sin and corruption, spiritually dead. Let's examine some of the proof texts from the Westminster Confession of Faith in the section on the corrupted nature being conveyed through ordinary generation.

Acts 17:26

26 and He made from one man every nation of mankind to live on all the face of the earth, having determined their appointed times and the boundaries of their habitation,

They start with the assumption that all people on the planet descended from Adam (or Noah, and by extension Adam, after the flood). I have argued in a previous chapter that all the face of the earth here is not the entire planet. Even with that being the case, I still believe it is true that all the ancient near east was eventually directly or indirectly related to Noah's three sons.

They then use Romans 5, which we have looked at extensively. In this chapter both guilt and depravity are in view, as we will see. But this chapter in Romans does not actually support inherited depravity through ordinary generation. In fact, it doesn't imply any other mechanism apart from the direct imputation of sin, its penalty, and its effects on our nature. Then they go to 1 Corinthians 15, which we are by now familiar with. In the context of resurrection, all die in Adam, all will be made alive in Christ. Just as we have born the image of the earthy, we will also bear the image of the heavenly. This is then applied to the following two verses which they use as proof texts to make their argument.

John 3:6

6 That which is born of the flesh is flesh, and that which is born of the Spirit is spirit.

Genesis 5:3

3 When Adam had lived one hundred and thirty years, he became the father of a son in his own likeness, according to his image, and named him Seth.

Like things beget like things. Flesh begets flesh. Adam begets a son in his own likeness and image. We who are earthy bear the same image (1 Corinthians 15:48-49). But all that is being reinforced here does not prove the case for ordinary generation as the means of transmitting moral corruption. While it is true that we must be born again because flesh cannot beget spirit, it is not necessarily addressing depravity through ordinary generation. Now, some could argue that our Lord is using the Greek word Sarx for flesh in John 3:6 to say that those with a sinful nature beget others with a sinful nature. But the same Greek word for flesh is used in John 1:14 **"the Word became flesh, and dwelt among us"**. And surely, we would not want to apply the sinful nature to Jesus. I think Jesus is just making the point to Nicodemus that we must be born of the Spirit to see and enter the kingdom of God. Although I think there is some truth to this idea that like begets like and can include moral qualities.

The Bible is full of instances where the children suffer the consequences of the choices of the parents (Exodus 20:5), and nations and people groups recapitulate the sins of the fathers. Think of Cain and his descendants. Or the 400 years God waited to judge the Amorites because their iniquity was not yet complete (Genesis 15:16). And Jesus tells the Pharisees that they are acting like their father, the devil (John 8:44). And many curses are given to peoples and nations descended from a sinful actor. Think of the curses on Canaan because of what Ham did to his father Noah (Genesis 9:24-27). And how the blessings of covenants are reserved for biological descendants in the immediate context. For Abraham's line, for David's line, etc. So then, the concept of biological descendants as the grounding for inclusion in a covenant, with its blessings and curses is well attested in the Bible.

But all these covenants find their fulfillment in Jesus Christ. All the promises are yes and amen in Him. All Christians inherit the promises and blessings of Abraham, regardless of if they are Jew or Gentile. There are far more people living on the planet that do not descend biologically from Abraham compared to the ones that do. Therefore, the parties involved in the covenant with Abraham are not those who have Abraham as a biological ancestor, but those who are of the faith of Abraham (Galatians 3:6-9).

The concept of sons being like their fathers is also prevalent in the scripture. And I think that is what is going on for Adam and Seth. Seth being born in the likeness and image of Adam is confirmation in the immediate context that the image of God still exists after the Fall. But I believe it is also there to show us that Seth is like his father Adam in character and nature. Adam acted in faith after the curses by naming his wife Eve (Genesis 3:20). Eve acts in faith by naming Seth (Genesis 4:25). And this lineage began to call on the name of the Lord (Genesis 4:26). Moses could have told us that Cain or Abel was made in the likeness and image of Adam, but he doesn't. After Adam disobeyed and was expelled from the garden, Moses gives us every reason to believe that he repented, acted in faith, and followed God. Seth is like his father and produces godly offspring that call upon the name of the Lord. And this concept of sons acting like their fathers we know from experience, reason, and science. People often have the behavioral traits of their parents. We can often see ourselves in our parents and they can see themselves in us. We not only share physical traits which show a likeness and image, but in the way we act as well. And science has borne this out, especially in the study of identical twins. Biology and genetics play a large role in behavioral traits. Two biological twins separated at birth and raised in different households will be far more alike than the siblings they grew up with. Like does indeed beget like.

But I think that Adam is unique from the standpoint of how the Bible portrays and names him, and this further supports our thesis and alleviates any protest some may have with not having a "real" connection with Adam. In Genesis 1:26 'adam stands for mankind and then the Hebrew grammar through these first few chapters usually designates 'adam not as a proper name, but with reference to a masculine pronoun to let us know we are talking about "the" man. In verse 1:27 'adam is referenced first by a masculine pronoun and then as them, referring to mankind in general. And then this man, who also stands for mankind, male and female, is given the same proper name to reflect his identity. This man, who represents mankind and is named after them, is from the earth and is earthy in his human nature. Adam is the head of all earthy mankind.

Adam, who had lost his positive righteousness by the time of the birth of Seth, begot Seth into his image and likeness. We know from our study so far that Seth was directly imputed Adam's sin and condemnation. For only the sin and guilt or righteousness of someone else can be imputed. It doesn't make sense to say that we inherited guilt, for guilt is liability to the law for a specific infraction. It does at least make sense in principle that we could inherit depravity. For depravity has to do with our nature and that is determined by several factors, including our genetic makeup, behavioral traits, and the presence of sin.

We can then see that Adam represents humanity in general. This mechanism of inherited corruption based on ordinary generation assumes Adam is the first human being and that it is the sole means of transmitting depravity. But this depravity could have its grounds in direct imputation or some other mechanism instead of ordinary generation, even for those outside of Adam's biological descendants. Adam can stand in solidarity and as covenant head with other human beings outside of his family and lineage because we are like him in our nature and God created him to represent us. Just like Christians are parties to the Abrahamic covenant because we are of the same faith as Abraham, we can be parties to the covenant of works represented by Adam because we are of the same nature as Adam, both with earthy flesh and the image of God.

Back to the confession, where they are reasoning about ordinary generation with the following proof texts.

Job 15:14

14 "What is man, that he should be pure,

Or he who is born of a woman, that he should be righteous?

Psalm 51:5

5 Behold, I was brought forth in iniquity,

And in sin my mother conceived me.

If these verses are used for proof texts about depravity conferred through the birth of a woman, it has heretical implications for Christ's sinless human nature. Apart from some doctrines affirmed in the wider Chistian faith, Mary the mother of Jesus was not sinless. She confessed her personal need for a Savior in The Magnificat (Luke 1:46-55), and she was not a perpetual virgin (Matthew 1:25). She and Joseph had other children after the birth of Jesus, the brothers and sisters of Jesus (Matthew 12:46, Matthew 13:55-56, Mark 3:31, Luke 8:19, John 7:1-10, Acts 1:14). Not that sex within marriage is sinful, but often these doctrines of the sinlessness and perpetual virginity of Mary go hand in hand. Mary was also imputed Adam's sin and had moral corruption in her nature. Jesus came to save those in Adam, including Mary.

Setting aside the fact that a lot of what Job's friends say is wrong, the above verse from Job is just describing the condition of natural man. Not establishing a causal relationship between sex, embryo development, and birth to original sin.

The rest of the proof texts simply point to the sinful nature of natural man in his flesh. (Westminster, 1646)

This brings us to one of the reasons I think that the Westminster Standards makes specific mention of ***ordinary*** generation. I believe it is to make a distinction of the virgin birth of Jesus. It would be heretical to think that because He condescended to our nature and emptied Himself of His Godly form and prerogatives to become man, that He was born with original sin, or had a marred image of God. I believe the confessions and creeds are trying to protect our Lord from inherited depravity but doing it in a way that is unnecessary. We can clearly see in the gospel accounts of the birth of Jesus that He was conceived of the Holy Spirit (Matthew 1:18-20, Luke 1:35). He was born of a woman and had two distinct natures of God and man in one person, so that He could reconcile man to God. I believe it only makes sense that He shared Mary's DNA (or substance like the older theologians used to say), so we could say that Jesus shared in Mary's human nature. Jesus had a normal gestation period, with normal biology and development like any human baby. Otherwise, people could try to claim that Jesus was not truly human and call into question the nature of the God-man. To say that God was implanted in the womb brings us into all kinds of problems about the true humanity of Christ. I think this is more evidence that moral corruption does not have to be transmitted through ordinary generation, even if we take the somewhat awkward position that only the sinful nature of the male gets propagated.

Beyond these things, I see it unfruitful to speculate. The main point regarding our thesis is that ordinary sexual generation does not have to be the mechanism to transmit moral depravity. Jesus was born without a sinful human nature and with positive righteousness because He is the final Adam and the eternal Logos who took on human nature, not anything to do with ordinary generation. Jesus was born a real flesh and blood human being because He is the Son of Man. But the image of God that Jesus bore in His human nature I believe to be the same as Adam's. A rational soul, moral conscience, and positive righteousness. Born in true knowledge, righteousness, and holiness.

Jesus fulfilled the righteous requirements of the law that Adam had failed to uphold. Adam was a type and foreshadowing of Jesus. Both were born without a fallen image, guilt, or original sin. Both had wills that were subject to temptation. But Jesus, unlike Adam, was the eternal Son of God who took on human flesh, our nature, and dwelt among us. The transcendent God became immanent and tabernacled among us once again. Adam gave in to temptation and sinned, while Jesus resisted temptation and fulfilled the righteous requirements of the Law, always delighting to do the Father's will. I don't think then that denying the idea that original sin is passed through ordinary generation puts Jesus's sinless humanity in question. Jesus was born without imputed sin, guilt, or moral corruption because He was not a member of the party in the covenant of works. He was not represented by Adam in this first covenant. Jesus is the head of a new covenant.

Jesus is clearly a descendant of David, Abraham, and Adam on both Joseph and Mary's sides. That is clear from the genealogies of Matthew and Luke, not to mention all the Messianic prophecies of the Old Testament and their clear fulfillment in Christ. He is the federal head of the new covenant, and He represents those whom the Father has given Him. Those who repent, trust in Him, and follow Him. As well as those Old Testament saints, including Adam himself, who were justified by faith and lived in light of the promises of the seed of the woman who would crush the head of the serpent. Those who could see the types and shadows pointing to Christ and believed God. Jesus is not influenced by Adam; Adam needs the eternal divine Son of God who took on human nature to save him.

So, the biblical texts make clear that we are born in condemnation due to the direct imputation of Adam's sin and guilt of his first transgression. All die in Adam (Romans 5:15, 18, 1 Corinthians 15:22). How then are we born with a corrupt nature, if not by ordinary generation? In the same way, by direct imputation and its effects.

Romans 5:12

12 Therefore, just as through one man sin entered into the world, and death through sin, and so death spread to all men, because all sinned

We are born with moral corruption in every constituent part of body and soul according to Adam's fallen nature because we all sinned in Adam.

Romans 5:16

16 The gift is not like that which came through the one who sinned; for on the one hand the judgment arose from one transgression resulting in condemnation, ...

The imputation of Adam's sin results in condemnation.

Romans 5:19

19 For as through the one man's disobedience the many were made sinners, ...

And because this imputation is real, it corrupts our entire nature, body and soul, resulting in us being born in original sin.

I think the reason many of us have a hard time understanding this doctrine of imputation or why we attempt to come up with other mechanisms is that it doesn't seem real to us. But the Old Testament sacrifices that atoned for sin were very real. The animal really lost all its life blood and was killed as a substitutionary atonement. Our sins were really imputed to Christ and He really suffered the wrath of God because of them. They were real enough that Jesus's sweat became like drops of blood in His agony and asked the Father to remove the cup of His wrath in the Garden of Gethsemane on the Mount of Olives (Luke 22:39-46). This was not some legal fiction; He bore God's wrath in our stead. Likewise, the imputation of Adam's sin is not some legal fiction; it really led to our condemnation and corrupted our nature.

And then I also believe that God chooses to withhold positive righteousness from the image of God for those in the covenant of works as a punishment and penalty for Adam's sin. This is what happened to Adam's children and their descendants. And this is what happens to us when we are born. And the reason that Jesus was not born with imputed guilt and a corrupted nature is because He was not a party to the covenant of works and His human nature was born in the original image of God, with included positive righteousness. Jesus did not suffer the penalties of Adam's transgression because He was not in Adam. Only a sinless substitute with our human nature can vicariously atone for the sins of mankind. And only a divine substitute can offer a substitutionary atonement of infinite worth.

Therefore, Adam's first transgression had at least four consequences. The first is that we are imputed the guilt of Adam's first sin and are therefore born under a sentence of condemnation. The second is that we are born with original sin (depravity) because we are imputed Adam's sin, and this sin corrupts our nature from conception. The third is that God withholds positive righteousness from the image of those in Adam. And the fourth is that we face physical death because we are born with an earthy body and the way to the tree of life is now only reserved for the saints and citizens of the new Jerusalem who can freely eat of it once again.

These moral and spiritual realities occur because Adam is our federal head in the covenant of works, not because he is biologically the father of the whole human race. Adam stood in solidarity with the rest of mankind by way of God's covenantal choosing of Adam to be its head. And therefore, I believe it is neither necessary nor warranted from the Bible that the soul, the image of God, or Adam's sin, guilt, and heart of stone are conferred to the rest of us through sexual reproduction. We are "in" Adam, and he represented us in the covenant of works in a similar manner in which the last Adam, Jesus, represents the spiritual children who are "in" Him. The first relationship is one of the flesh, the latter is one of the Spirit. All those human beings who are spiritually dead are in union with Adam, and all those who are spiritually alive are in union with Christ. We are either in Adam or we are in Christ (Romans 8:1).

I hope that we can see that this covenant of works with Adam as the representative head was of God's own sovereign choosing. Like with Abram, God always chooses and elects based on His good pleasure. Not in anything we bring to the table or that is foreseen in us. If we are to be declared righteous because of the work of Christ on our behalf, we should not think that Adam's imputed sin and our resultant depravity are somehow unjust. The idea of corporate responsibility and covenant relationships are throughout the Bible. God saves Noah's family because he is found to be righteous. God says He will spare Sodom and Gomorrah if ten righteous people are found within its walls. Lot's family is saved on account of Abraham. Israel and Judah are judged and carried into exile because of the sin of the people, although surely there was still a small, faithful remnant walking by faith. But sometimes even the most righteous men can't save the wicked from their consequences (Ezekiel 14:14, 14:20). Although Israel in the

Old Covenant offered experienced the blessings and curses of the covenant corporately, individuals were still held responsible for their own faith and actions (Ezekiel 18). But the eternal destiny of individual men and women and their reconciliation to God has always depended on what they do with Jesus Christ (Romans 5:17, John 3:16), or prior to Jesus if they were justified by faith in the promises of God and had eyes to see (Habakkuk 2:4, Hebrews 11).

So, I think it is unwarranted to say that God had to choose the first human being to be in this role. If Adam was not the first human being, God is free to create a representative in the fullness of time to represent humanity. I contend that not accepting the idea that Adam is the representative head of humanity because not everyone was somehow in the "loins" of Adam, and therefore not somehow sharing in his "divisible humanity" does nothing to lessen his role. He stands as the covenant head because God chose to create him for such a role. And I don't think any allegations that God is being arbitrary for our union with Adam based on the lack of perceived realism or ontological grounds is warranted. For if this realism were the grounds for our union with Adam, it brings its own set of problems with it.

I believe God's word teaches us that Adam would have died naturally regardless of the outcome of his obedience, because he was made with an earthy body that was subject to corruption and decay. He would have continued to eat from the tree of life had he stayed in his state of original righteousness and not sinned. He would have been free from condemnation and the fear of death and would have been given an incorruptible body either before or upon death. Now after the Fall, we face the condemnation of sin, we need atonement for our sin, sin fractures our relationship with God, with other humans, with the animals, and with the earth itself. We all will face physical death unless we are alive at Christ's second coming. But through faith in Christ and our union with Him we are guaranteed the promises of eternal life as children of God and will be allowed to freely eat from the tree of life in the new Jerusalem. (Revelation 2:7, 22:2, 22:14)

Let's now look at the soul and image of God again before trying to tie it all together. We have argued that part of the biblical definition of man is that he is a unity of body and soul. And that being made in the image of God separates him from the other creatures. I argued that the Bible teaches that the soul is inseparable from the body in its normal functioning, and even though the soul can exist apart from the body, it is unnatural and awaits its resurrected body in the final judgement. And then we saw that even though the moral properties of positive righteousness were lost in the Fall and now needed to be renewed in Christ, man is still made in the image of God. We argued that this marred image of God consists of the attributes in which man is like God to exist eternally, to be spiritual, to have morality and make moral decisions, to think, plan, choose, exercise dominion, and be in social relationships.

When thinking about the transmission of the soul and image, we should be constrained by several considerations. The most important one is the Bible's teaching on each. The Old Testament use of the word (nephesh), sometimes translated soul, is mainly that of a living being. Other creatures are said to have it, and God gives it and takes it away. But people don't have a soul (nephesh); people are a soul (nephesh). It is synonymous with the person themself. It is not only a principle that animates him and gives him life, but the essence of what the person is. But we argued that this is the biblical anthropology throughout the Bible. People do not have a body and soul; they are body and soul. Even though I argued (contrary to many) that the Old Testament does not deny the immaterial nature and origin of the soul, it is more clearly taught in the New Testament. But even the New Testament equates it with the very essence of the person.

As far as the image of God, we argued that the rational soul and moral conscience are what separates man as a special creature. The positive righteousness given to Adam and Eve is what was lost and needs to be renewed in Christ. We therefore concluded that God gives and takes away the soul in all creatures, and that the image of God imprinted on the soul is what makes man man. The image of God gives the soul its properties of rationality, morality, and eternal existence after its creation. These properties allow man to exercise dominion and live in social relationships with God and other image bearers. To experience eternal life with God or eternal separation from Him.

When talking about the transmission of the soul and image, we bear these things in mind. Which then leads me to conclude that the soul and image have their ultimate origin from God, and every human baby conceived is because of God giving life. This life (body and soul) is imprinted with the image of God, but

God withholds positive knowledge, righteousness, and holiness as an effect and penalty for Adam's sin. This withholding of the positive moral qualities of the image plus the imputed sin of Adam is what confers corruption upon the individual. God in His justice could withhold soul-giving life to a sinner, but He is gracious to the just and unjust. God could withhold imprinting His image on the soul, but then the human creature would be like any other creature, lacking the capacity and potentiality to relate to God in mind, heart, soul, and strength.

This then leads us to the other human beings alive when Adam and Eve were created. It is estimated there might have been somewhere around 5 million people alive on the Earth around 8000 BC. These people were already human in their intellectual capacity and some moral reasoning. They had in the last couple millennia begun to settle down in communities practicing agriculture and animal husbandry in the Fertile Crescent of the ancient near east. Most of the rest of the world was still in the Paleolithic culture of hunting and gathering. These people owed their existence to the Lord God for creating the universe and for giving them life. And although they already possessed a life-giving soul given by their Creator, it was not imprinted with the image of God until the birth of the generation after Adam's creation. This new generation was then separate and distinct from the other creatures due to this fact. And this is why Genesis 1 can talk about humanity in general being made in God's image. But only three human beings have ever born the full image of God: Adam, Eve, and Jesus.

But doesn't this impugn God's character? If God lets humans evolve, gives them life, and then imprints that life-giving soul with His image (minus positive holiness) when they are intellectually, morally, and socially "mature", isn't God creating something already with sin? We have argued that the Bible teaches that the soul (the animating life) and the image all come from God. But neither of these are intrinsically evil. God is said to give all people life and breath and everything long after the Fall (Acts 17:28). All these people living in Athens in Paul's day in Acts 17 had physical bodies with immortal, rational, and moral souls. But there is nothing innately evil in the body or soul. Jesus had both. Nothing in the body and soul is inherently corrupt, whether that soul is imprinted with the full image or one lacking positive holiness. It is sin that corrupts, blemishes, and spoils. And God is not the author of sin. Therefore, God giving life and breath to all humanity who were parties in the covenant of works does not make God the author of sin.

Then were these people living in 8000 BC already sinners in the eyes of God? Didn't they already have the moral law written on their hearts? Couldn't they be held responsible for not knowing God and His invisible attributes from the creation of the world (Romans 1)? For the people prior to this time in history, the answer must be no. They were not members of the covenant of works, they did not have the image of God imprinted on their souls, and were therefore like the rest of the creatures, without moral culpability. They did not have an immortal soul and therefore perished in death and did not have an existence after death. We know that sin entered the world through Adam. And this is true because sin is any lack of conformity to or transgression of the law of God. And the prior generations did not have the image of God imprinted on their souls, therefore they did not have the moral law of God written on their hearts to sin against. Nor did they have a true rational soul or moral conscience. In other words, they could not perceive God through nature, they did not have a conscience with the moral law written on it, nor did they have special revelation. In other words, there was no sin in the world. For sin is not imputed where there is no law (Romans 5:13). And sin is lawlessness (1 John 3:4).

But those who were born after the time of Adam's creation, when God determined that the fullness of time had come, they had their souls imprinted with the image of God at conception. Because they had natural bodies and the image of God, they now met the conditions for being incorporated as parties in the covenant of works with Adam as their head. Adam and Eve were placed in the temple garden as priests (Beale, 2018) with the pristine image of God to expand the boundaries of the garden and bring the rest of this covenant humanity into the knowledge and worship of the one true living God. Israel had the same commission. Christians in the new covenant have the same commission. Adam was created for this role so that he may lead people to seek and find God and reach the goal of creation, that is, life with God.

Acts 17:26-27

26 and He made from one man every nation of mankind to live on all the face of the earth, having determined their appointed times and the boundaries

of their habitation, 27 that they would seek God, if perhaps they might grope for Him and find Him, though He is not far from each one of us;

But when Adam sinned, his transgression resulted in the imputation of sin and guilt to all his contemporaries with the image of God and their descendants, resulting in condemnation. And because of this imputation of Adam's sin and having the image of God without positive righteousness, this resulted in a corrupted nature that was liable to guilt and led to their own transgressions. These people were now liable for the moral law written on their hearts, for knowing God through His creation, honoring Him, and giving Him thanks. They and subsequent generations were now like Adam. They were guilty and condemned, they had an incomplete and marred image of God, they had a moral conscience that accused or excused them, and they had a dim knowledge of God through what had been created. But most of them exchanged this knowledge for idolatry and suppressed their moral consciences in sin.

Adam and Eve were expelled from the sacred space of the garden and physical death for Adam, Eve, their descendants, and all the members of the covenant of works was now guaranteed because the tree of life was guarded and out of reach.

If someone should protest at this point that it is not fair that they were held responsible because they were unable to keep the law of God due to the imputation of Adam's sin and their own corruption, I would say we need to be careful to avoid the errors of Pelagianism, which taught that people were not guilty or imputed the sin of Adam for eating the fruit, nor are people born with a corrupted nature. It taught that people are born good and have the power in their own wills to act and live righteously. Apart from the overwhelming biblical data that this is not true and our experience of every person and culture in the world showing the effects of sin, our responsibility and culpability before God does not ultimately depend on our ability. We know what God requires and that those who transgress are worthy of death, and yet we find ourselves committing those sins and even giving hearty approval to others for the same choices (Romans 1:32). The unregenerate man is presented in the Bible as a slave of sin (John 8:34-36, Romans 6:6, Romans 6:16-20), yet our liability to punishment remains. But is this true in our experience?

People often say that non-Christians are decent, moral people and that is often true when it comes to relating to other human beings using our own standards of righteousness. But Jesus said that the greatest commandment is: **"and you shall love the Lord your God with all your heart, and with all your soul, and with all your mind, and with all your strength."** (Mark 12:30). We can therefore see that non-Christians break the greatest commandment every day. And we read from Romans 1 that this is one of the main reasons God gives people over to sin, because they fail to honor Him, give Him thanks, and instead exchange this knowledge to practice idolatry. Unbelief is more than just lacking something a Christian has, it is essentially calling God a liar for what He has said. Original sin (depravity) does not mean that we are as morally bad as we could be, as God has common grace for all His creatures. It means that sin has infiltrated all aspects of our nature, body and soul, so that we cannot know and please God like we were created to do. But that's why we need grace. And that grace is abundant in Jesus Christ.

But why wait until 8000 BC to create Adam and Eve? Why now include the humans in the covenant of works? Because by now man had developed the necessary capacities to have souls that could be imprinted with the image of God to display spirituality, rationality, morality, language, society, and culture to reflect Him and have dominion. And we can see the results of this dominion clearly between 8000 BC and today.

Up until the creation of the covenant head of humanity, Homo Sapiens were to be regarded as brutes, and sin did not exist because there was no law. The fullness of time had come for a man to be made in the pristine image of God with positive knowledge, righteousness, and holiness to represent humanity. Because now that mankind had the capacities, he had achieved the teleology of God as image bearers. But now they were without hope in the world. Listen to Paul in the book of Ephesians talking about the hopelessness of the Gentile world prior to Christ.

Ephesians 2:12-13

12 remember that you were at that time separate from Christ, excluded from the commonwealth of Israel, and strangers to the covenants of promise,

having no hope and without God in the world. 13 But now in Christ Jesus you who formerly were far off have been brought near by the blood of Christ.

These so-to-speak "mature" human beings, who were now without hope and without God in the world, needed general and special revelation and to be parties to the covenant of works so that they could enjoy potential communion with their Maker. The question one step farther back is why does the sovereign Creator have dealings with this brute who has progressed to a human being in the image of God?

Hebrews 2:5-7

5 For He did not subject to angels the world to come, concerning which we are speaking. 6 But one has testified somewhere, saying,

"What is man, that You remember him?

Or the son of man, that You are concerned about him?

7 "You have made him for a little while lower than the angels;

You have crowned him with glory and honor,

And have appointed him over the works of Your hands;

God miraculously created Adam and Eve to be like these image bearers on the Earth so that Adam could be their covenant head. Just like Jesus had to be like those whom He represented. If Adam and Eve had not sinned, those in the covenant of works with them would not have been imputed Adam's sin and they could have worshipped the one true living God and been given access to the tree of life. They may have been required to approach God and this sacrament through the priesthood of Adam and Eve and their posterity born with positive holiness, but their covenant head could have been their mediator had he not sinned. And his righteousness could have been imputed to them as their covenant head who did the Father's will. But Adam did sin and took us with him. But praise be to God the Father who loved the world so much that He sent His Son into the world, the One who did keep God's law, Whose righteous works are imputed to those of us who believe, in order that we may be reconciled to God.

What about Romans 8 then? God's purpose in creation is ultimately His glory. And He subjected the original creation to futility in order that it would not be redeemed apart from the sons of God.

Romans 8:18-23

18 For I consider that the sufferings of this present time are not worthy to be compared with the glory that is to be revealed to us. 19 For the anxious longing of the creation waits eagerly for the revealing of the sons of God. 20 For the creation was subjected to futility, not willingly, but because of Him who subjected it, in hope 21 that the creation itself also will be set free from its slavery to corruption into the freedom of the glory of the children of God. 22 For we know that the whole creation groans and suffers the pains of childbirth together until now. 23 And not only this, but also we ourselves, having the first fruits of the Spirit, even we ourselves groan within ourselves, waiting eagerly for our adoption as sons, the redemption of our body.

The creation itself then is waiting for the glorification of the saints so that it too may be redeemed. The world was never perfect in the sense of being created with no forces opposed to life. It was created with darkness and the seas as a part of it. It had predation, disease, natural disasters, extinctions, and death because God did not want to redeem it from corruption apart from the sons of God. None of these things are sin, and therefore God is not the author of sin.

Like Jonathan Edwards reminded us long ago, the chief purpose for creating the universe is God's glory. And this is plain throughout the Bible. I then contend that the ultimate purpose of God was creating a type of world where his creatures would give Him thanks and praise, glorify and delight in their Maker, and enjoy His presence forever. And that God gets the most glory by creating this kind of world that is only redeemed after the children of God. And He will redeem this world in His timing when the full number of the saints come in, God is glorified for His grace, mercy, and justice, and then He becomes all in all.

1 Corinthians 15:23-28

23 But each in his own order: Christ the first fruits, after that those who are Christ's at His coming, 24 then comes the end, when He hands over the kingdom to the God and Father, when He has abolished all rule and all authority

and power. 25 For He must reign until He has put all His enemies under His feet. 26 The last enemy that will be abolished is death. 27 For He has put all things in subjection under His feet. But when He says, "All things are put in subjection," it is evident that He is excepted who put all things in subjection to Him. 28 When all things are subjected to Him, then the Son Himself also will be subjected to the One who subjected all things to Him, so that God may be all in all.

In the final chapter we will attempt to tell the full story.

Chapter 14

Bringing it all together

We have been through a lot together to reach this point. With all the information and data, we need to consolidate a narrative of this creation and God's plan of redemption. Humans are moral and religious beings. We can't help but think there is purpose to our lives. Why should the human species be so interested in knowing about the world in which we live? Why should we be so concerned about justice and morality? Why should this universe make sense and why are we so preoccupied with making sense of it? No animal loses sleep at night over metaphysical and epistemological questions. No animal feels like they have dignity and worth above nature itself. No animal feels like they have fallen from glory and that they have lost something that needs to be regained. If purpose, meaning, and objective morality did not exist, why should we always be searching for them?

The answer is, in the beginning God created the heavens and the earth. And that we are created in the image of God with infinite worth and value. We long for heaven because we are spiritual beings. We long for the garden because we have a sense that we were made for more and have fallen from our created purpose. We and the universe are no accidents of chance or fate. An eternal, transcendent Creator who is good decided to create for His glory and our eternal joy.

One thing I have tried to make clear throughout this book is that God's word is the ultimate authority. But special and general revelation cannot contradict each other if they are meant to be interpreted in the same sense. And it is important that we qualify it in those terms because literature should be interpreted in the sense that it was intended for. We have argued repeatedly that Genesis chapter 1 was not meant to be interpreted as historical narrative because the literature itself does not lend itself to that genre. Just like we would be amiss if we interpreted Revelation or many of the Psalms as literal, historical narrative.

Highly stylized and magisterial prose should be interpreted as such, apocalyptic literature should be interpreted in light of metaphor and symbolism, and poetry should be interpreted as poetry. That does not mean that these different types of literature cannot teach both truth and history, but we need to be careful that we are getting the truth and history out of them correctly by proper methods of interpretation.

But the Bible and nature must correspond to the same truth when discussing the material universe and physical events that happened in space and time. Some neo-orthodox theologians claimed that contradiction was a hallmark of mature Christian faith because they were trying to protect the Christian faith from falsifiability by rooting it entirely in subjective experience. I do not think that God gave us revelation in historical narrative literature that was meant to be solely rooted in subjective experience, but that it also corresponds to actual events in space and time.

Although the Bible is much more than a history book, it is not less than one. It is a living book, and we need spiritual perception to believe and apply it, but God created this universe and works out redemptive history on this planet. God was governing the events and miracles of the Exodus, but He continually reminds Israel of how He redeemed them out of slavery from Egypt at a certain moment in history, under a certain pharaoh, using actual locations, place names, and timing. And then He commands them to remember these events and to teach them to their children. And most Christians intuitively believe this. They know that Christianity claims to be rooted in historical events. Like Paul says, if Jesus did not actually rise from the dead, our faith and hope are in vain (1 Corinthians 15:16-19).

I think that if we properly interpret the scripture, it will not lead to contradictions when studying nature. But (and this is a big but) our study of nature requires interpretation as well. To start with, all our observations and data must be systematized by our rational minds to reach any conclusions. Some scientists seem to imply that empirical observations and the conclusions that follow from them are self-evident. But all sensory input and the thoughts corresponding to them need to be systematized by our rational minds to produce a coherent picture of the world. Otherwise, we just have a bunch of discrete input data that is unrelated. So, some very intelligent people can study the heavens, the earth, or the life on this planet, but all the observations and data need to be placed into a coherent story to give it any explanatory power. Secondly, there are multiple ways to interpret data and observations that lead to various conclusions. And last, if we begin with the assumption that only physical observations and reason can give us truth and knowledge, we have tacitly supported the materialism and atheism of the modern world before we even begin. No one, no matter how brilliant, has verified everything they believe through the methodology of science. Most of what people believe they have not discovered or verified for themselves. In fact, the methodology of science is ill suited to all kinds of knowledge that is critical to us. Most of written human history and thought could not be reconstructed through science. What you ate yesterday for breakfast is not revealed by the scientific method. Sense perception, memory, reason, logic, mathematics, and beliefs are more fundamental than the scientific method.

This leads us to the idea of models in science. A model is a framework that tries to best explain the data. It can also be called a paradigm and is also sometimes synonymous with theory. Anomalies in observations and data from what the model or paradigm proposes to explain lessens our confidence that the model is a good one. A good model explains the observations and data without resorting to elaborate or unnecessary mechanisms. This last criterion is admittingly kind of subjective, especially because the more we learn about nature, the less simplistic our explanations tend to be to account for the data.

Before the rise of modern science, people could speculate and form models for the laws of physics, what the heavens were composed of and how they operated, the processes of earth sciences, and how biology works. But now we have complicated mathematical models for physics like the standard model of particle physics, the general theory of relativity, and quantum physics. Complicated mathematical models of cosmology like the lambda cold dark matter model, complicated models of plate tectonics, chemistry, organic chemistry, rock and water cycles, nitrogen cycles, etc. And complicated models of biology, genetics, development, and biochemistry.

It turns out that nature is complicated. Even in a modern field like biology, the relatively straightforward idea that DNA contains a sequence of nitrogenous bases that are translated into proteins and that these genes change based on single point mutations in those bases is a vast oversimplification the more we learn. Much of what used to be considered "junk" DNA turns out to have function, genes modify other genes, they control what is turned on or off, biomolecules influence genes and genes influence biomolecules, the DNA, chromosomes, and proteins all take three dimensional shapes that influence how they operate and when and how they are expressed. In other words, biology is extremely complicated. And the more we learn the less simplistic explanations and stories are persuasive. And there is still a lot to learn about the natural world. But a good model should also be able to make predictions. There are many ways to explain various phenomena, but only accurate models will be able to generalize and make predictions that can be tested.

I have tried to be charitable throughout this book in not presenting scientific ideas in a way that experts themselves would not own. No single person can be an expert in everything, but I have tried my best to represent the best models. And I have aimed to be as accurate as possible with respect to the consensus theories of our day. I never had any desire to present strawman arguments or make people's ideas look too simplistic or easy to demolish. The same thing goes for Christian theories and models of creation science. I have argued that many scientists are too bold and confident in their assertions of what science can and has explained. We have tried to show where the theories from cosmology, astronomy, physics, geology, and biology are sometimes making bold assertions where there should be humility, mystery, and open questions. And I have tried to show where I believe some Christians have made poor interpretations of God's word and extrapolated those to dogmas. And I am certain that others who read my book will have criticisms of the ideas that I put forth.

But that still leaves us needing models that best explain the world and what we observe. I have heard extremely intelligent Christians, many of whom have doctorates in scientific fields, who believe in a young Earth and interpret Genesis 1 as historical narrative, that present their own models of creation, geology, and biology. And these models purport to explain the data, using the early chapters of Genesis as their framework. And many of them seem quite well thought out and convincing. They also make some great points (many of which I have argued in this book) about the improbability of materialist explanations to account for the universe, history of life, and psychology and spirituality of man.

Many creation scientists place the biblical concept of animal kinds somewhere around the taxonomy of animal families, instead of species, in the taxonomy of Linnaeus. So, using that categorization of plants and animals, they would argue that random mutations with undesigned and unguided genetic mechanisms plus selection pressures cannot account for one kind (family) of organisms evolving into a different kind of organism. Many would argue that organisms can evolve within their families but not split off to make a different family. And I have sympathy for this critique. I think too many biologists make confident assertions where there is mystery, and that a lot of things must go just right at the genetic level in a population for a family of organisms to be a common ancestor of a different family. But in their defense, most biologists can't see a better alternative to the model of biological evolution with common descent. And the relatively recent science of genetics makes a lot of things harder to explain without the theory of evolution.

And that brings me to an important point. As Christians we can say that the world was created in six days 10,000 years ago, but how much data does a model like that account for? Does it have a lot of untestable and unreasonable assumptions? Does it make predictions that can be verified? Can it account for observations from various fields of study like cosmology, geology, archaeology, anthropology, paleontology, and biology? And if we propose a different creation model, are we correctly interpreting the text of scripture based on sound principles of exegesis? Does that new model account for the data?

I have found that many Christian scientists present their models in such a way that the Earth's systems and life are in harmony and each species is looking out for the good of their ecosystem and neighboring species. They give examples of all the mutualism on Earth and its life systems and how each component of these systems depends on each other. Many secular biologists scoff at this and present examples of what they see as examples of poor design, predation and death, viruses, bacteria, and parasites trying to kill us, competition for resources, disease, and developmental and birth defects. Most creation scientists counter with the Fall and the entrance of sin into the world to explain the darker sides of life. But as we have argued before, the assertion that evolution is unguided and without purpose is a philosophical statement, not a scientific one. Science cannot tell us that there is no design or purpose, even in principle. And in my view, the study of science in general definitively supports the argument from design.

Based on all these considerations, when I look at the genre of Genesis 1 and the data, observations, and models, I think most modern scientific models and theories generally do the best job of explaining the most data about nature, while also having predictive power. Despite many times overplaying their hands, being confident where it is unwarranted, and inserting philosophical claims into their explanations. Although what we actually find are that many theories that claim to be predictive have actually been tweaked and changed based on observations to better fit the data, so I'm not sure that we can call many of them predictive. But I am not too wed to any of these theories and explanations to think they are infallible. I just think they do a better job of explaining nature and its processes from a human perspective compared to the other alternatives that I have seen.

But modern science doesn't stop with theories, models, and paradigms. They have stories. They produce narratives about why we are here, how we got here, how things came to be, and then make inferences from the stories about the value and meaning of our lives and the things around us. Mesopotamia and Egypt had their stories about how the gods were created, how the cosmos was ordered, how man was created, and what man's place in that cosmos was. Modern science also has a story. It goes something like this.

There is no transcendent being or creator. There is no spiritual realm, only material existence. In the beginning we don't know what to say, but it is material. There was a hot Big Bang that was probably infinite in extent where the energy in the universe was at Planck density. We don't know much about these initial conditions, or the physics involved. But there was at least one region of space that had most of its energy in a metastable scalar field. And because of this, after a fraction of a fraction of a second this scalar field inflated the universe by 10^1,000,000 in a trillionth of a trillionth of a trillionth of a second. This process effectively caused space to have zero particle density and temperature. But when this unstable scalar field decayed, the energy of the field reheated all space, and tiny quantum fluctuations in the scalar field right before it decayed produced small differences in density. The result was extremely hot and dense, but not enough energy was available to once again produce magnetic monopoles. This energetic, hot, and dense universe produced all kinds of exotic fields and particles, but almost all of it was transformed into electromagnetic radiation because of the unstable nature of the energy and particles and due to matter-antimatter annihilation. For some reason there was a little bit more matter than antimatter and because of this the universe (and we) exist. But it did produce other effects that make the exquisitely fine-tuned initial conditions of the universe without inflation more explainable. Like why the universe is so uniform, why the geometry is flat, and why we don't see magnetic monopoles. It doesn't explain why a region of space had most of its energy in a scalar field, if this is even possible, or why this metastable region began expanding or didn't start expanding sooner or later. It doesn't explain why the differences in density are where they are, why matter and antimatter were asymmetric, why the laws of physics are what they are, how the fundamental forces of nature separated, and why the free parameters have the values they have. Whatever the reasons, we exist and can observe the universe because of a field and quantum fluctuations.

The energy and matter left over after annihilation were a hot plasma of subatomic particles that quickly formed protons and neutrons that could make atomic nuclei as the universe expanded and cooled. This energy and countless atomic nuclei existed for hundreds of thousands of years and then cooled enough to allow the formation of neutral atoms. Through billions of years of the expansion and cooling of this universe, this energy and matter combined by the strong and weak nuclear forces, electromagnetism, and gravity, producing stars, planets, and galaxies.

After billions of years a certain galaxy was formed that we call the Milky Way. Within that galaxy, a certain solar system formed, and approximately 14.3 billion years after the Big Bang the planet we call Earth was formed. This planet was molten and unfit for life. But as the planet cooled and water from within or without formed oceans, life arose from non-life. And most of the elements for that life to exist ultimately came from exploding stars in the past. But this pale blue dot of a planet is just a speck in a universe that is so vast as to make anything on this planet insignificant in the context of the whole universe.

But despite this triviality, that simple life was capable of reproducing, and increased in variation and complexity through random processes like mutations and genetic drift and by non-random natural selection and sexual selection which picked and chose organisms that were best adapted to survive and reproduce. These processes over billions of years led to biodiversity, which was hindered at times by numerous extinction events. Until eventually ending up with the evolution of one species that was much more intelligent than the others, Homo Sapiens. But this species is made of stardust and biochemistry like the rest of them.

The Homo Sapiens proliferated throughout the various ecosystems of the world because of their big brains and hands that are dexterous and work well at producing things. They were also successful because evolution had shaped their brains to develop and use language and to have motivating emotions that make them work well with others of their kind to modify their environment. These emotions and intuitions that they called morality and human dignity are not based on anything objective but did make them a very successful species. To the point where they modified the entire biosphere. This species is a blip on the radar screen of cosmic timescales and if they should avoid extinction like 99% of the species who have ever existed, they too will meet their demise when their unexceptional star runs out of nuclear fuel and doesn't have the light and heat to sustain the planet. But it is a moot point because as the expansion of the universe continues over time, with dark energy pushing everything apart, and the nuclear fuel runs out for each star, everything in the universe will be so unfathomably far apart that all that is left is a cold, dark, empty universe.

Kind of fills you with hope and joy, right? Sorry to be a downer but that is the story that science tells. It is the modern myth of origins and purpose. It is the story of an indifferent universe made up of impersonal matter and energy, inviolable laws of physics and chemistry, creating life and people who were created without purpose, meaning, or worth.

I argued previously in chapter 8 that this story of the universe has many difficulties. First and most importantly, the atheistic or agnostic philosophical worldview that is not deduced from the scientific method. And then there are the conjectures proposed for the beginning of the universe, using explanations like multiverses, cyclical universes, no boundary universes, a wavefunction for the universe, or a universe from nothing employing virtual particles or quantum tunneling. And then we looked at critiques for the theory of inflation. Why did this region of space exist at all, why did it start expanding, why did it have this energy density, why did it exit at just the right scale factor of expansion to allow for the evolution of our universe? Then there are the questions that don't get answered, like why the laws of physics are what they are, why do the constants and free parameters of nature have the values they have, why is the expansion rate just right, and on and on.

Then we probed the improbability of truly random mutations producing and expressing beneficial genes to make exquisitely adapted creatures. And even more importantly, why man has morality, a feeling that he is a superior creature created for a reason, one who screams at injustice and suffering, who has an insatiable need to understand his world, a need for meaning, a feeling he has lost something, a longing for the transcendent, and one who clings tightly to ideas like human rights and dignity that do not naturally follow from his materialistic worldview? I think in the final analysis, the scientist who tells the above dreary origins myth must resort to either the anthropic principle, which states that we are observing these things and talking about them because we just happen to be where all these things came together correctly. And if they didn't come together correctly, then we wouldn't be talking about them. But this anthropic principle only works by resorting to the multiverse to employ the law of large numbers, otherwise this story looks like it has been fixed if there is only one universe. But this is a vanishingly improbable explanation compared to the Christian story and still leaves them with no rational reasons to believe in objective morality, meaning, and purpose.

So, we have the caveats in place about what the models of modern science claim to explain, and what they are incapable of explaining. But before I begin to tell what I believe is the true story, I must remind us again that science is provisional. If new data, observations, theories, and arguments come along in the future that change scientific theories that are current in the year 2024, we will still have the word of God. We will always know that God created this universe, that He created mankind for a purpose, that morality is objective and real, that the Bible is thoroughly confirmed by history, that tons of prophecies have been fulfilled, that Jesus lived, died, and rose from the dead, and that His apostles and inspired authors gave us the New Testament writings for our faith and practice. It will always be a living book that tells us who God is, what He has done, who we are, and points us to the Savior. It will always be a book that changes lives, takes our eyes off material explanations, and gets them back on the One who controls all things and has been pleased to reveal Himself to us.

With that being said, let us begin the story I think is true.

The triune Lord God, Yahweh Elohim, created the universe out of nothing by His will and power, for His glory, and for our everlasting joy.

Although science is always provisional, it makes a convincing case that the universe is around 13.8 billion years old, and that the age of the planet Earth is around 4.5 billion years old. This universe has laws of physics, values of free parameters, and initial conditions that make biological life possible. But God is not a hands-off God who winds things up and then lets them play out according to His fixed laws. He governs the material universe that He created, down to the laws of physics and chemistry, every molecule, atom, subatomic particle, and biology itself. God has purposes and goals for His creation and man within it, and He always makes those things come to pass. Nothing in the evolution of the universe was outside of His control and providence. He made sure that the Milky Way was formed, that our solar system was made, and that planet Earth was here and ideal for life. And this is not some idea that we are imposing on the Bible, it is the God revealed to us in scripture.

According to our best current theories of science, about 100 million years after the planet had formed, the Earth already had substantial water. Science is not sure where it comes from. The leading theories hypothesize that it came from the rock that formed the planet, from asteroids or comets, or a combination of both. About 200 million years later, the Earth may have been largely covered by oceans. We can begin to see why some Christians like to interpret Genesis 1 as God giving us a historical account of material creation. We have the Earth covered by water and it is formless and void.

Possibly about 600 million years after the Earth was formed, biological life somehow emerged on this planet. As of the time of this writing, no one knows how. They don't know if it happened just once or multiple times. In fact, scientists aren't sure if several of the first steps of simple life evolved only once or multiple times. But God was still active in providence in all aspects of His creation. The life on Earth at this time was composed of simple prokaryotic single-celled organisms, and the earliest solid fossil evidence dates to about 3.5 billion years ago. At which point, prokaryotes seem to have already come together to form simple multicellular structures. About 2.7 billion years ago, some bacteria evolved a form of energy metabolism that released oxygen as a byproduct. This oxygen built up in the atmosphere over the next couple billion years. About 2 billion years ago, eukaryotic cells emerged. These types of cells eventually allowed for more complex multicellular life, including plants and animals.

About 1.5 billion years ago, multicellular eukaryotic life diverged to begin evolving separately as plants, animals, and fungi. But to say they are plants, animals, and fungi in the sense that we think of them is a gross over exaggeration. Think very simple multicellular organisms who can live in a relatively oxygen poor environment with a fair amount of methane in the atmosphere. In fact, the evidence for truly complex multicellularity in eukaryotes did not appear until around 1 billion years ago.

About 800 million years ago, life on Earth appeared to consist of mainly bacteria, archaea, fungi, sponges, and seaweed. This was approximately 800 million years ago. So, we are talking about 3.7 billion years since the Earth was formed. If we went to the zoo at this time in Earth's history to see the critters, we would see small simple squishy stuff and microscopic single-celled organisms.

The climate of the planet varied over the eons from very warm with little snow and ice and warm, shallow seas covering the land, to snowball Earths where most of the planet was in a deep freeze. The oxygen in the atmosphere was continually increasing, making it more likely for organisms who can live on the land and in the sea, although it was still much lower than today's levels. This environment led to the animals at the time consisting many of sponges, who could live with low levels of oxygen. These sponges seemed to dominate for another 200 million years; at which time several other animals appear in the fossil record resembling worms. And others with strange body plans living on the seafloor.

About 540 million years ago, what scientists call the Cambrian explosion occurred. This was a vast explosion in new animals with new body plans and hard parts, including shells and spines. This would be the first time most of us would recognize something as an "animal". Although many biologists like to downplay this relatively short era and point to evidence of prior animals, this is hard to explain as the gradual intermediate forms progressing by imperceptible changes in traits on the tree of life envisioned by Darwin. In fact, it is still true that the fossil record is largely devoid, but not absent, of intermediate species. And rather than showing a nice progression of branching and intermediate forms, it generally shows periods of long stasis followed but relatively short periods of dramatic change.

This is not to say that there is no fossil evidence of creatures with traits belonging to different species or that we are talking about the changes in terms of days or weeks. But we should be wary of any scientist who is too bold about the data available from the fossil record showing slow, gradual changes and plenty of intermediate species. But to be fair to evolutionary biologists, the theory of evolution has come a long way since Darwin. And many acknowledge the data from fossils and genomes and acknowledge genetic mechanisms that allow for relatively large and quick changes. But most of them now construct the tree of life based on genetic data, and then look to other lines of evidence to support their hypotheses. With that caveat, much of the tree of life, speciation events, and how it happened is still a black box. And it is assumed rather than explained.

Back to the Cambrian explosion. It is during this time that true vertebrates appeared. And many animals had a clear distinction between the head and tail that showed locomotion in a certain direction for chasing prey. Trilobites, a kind of arthropod, were one of the dominate animals and lived in the oceans for another 200 million years before going extinct. There was an unprecedented diversity of life in the seas at this point, with various creatures occupying different niches in the ocean environment. Worms, somethings like eels, and new reef-building organisms called brachiopods also developed.

By the end of the Cambrian (about 490 million years ago) most of the phyla (groups) of animals we know today were present. Starting around this period, marine biodiversity exploded, and most animal groups were refined and changing, versus developing new body plans. Many new varieties within these groups appeared. Biologists propose that green algae that existed in the oceans appear to be the root for the evolution of the first land plants, which began to colonize the land at this point in history. Red and green algae still exist today. Critters like starfish and brittle stars are attested to. Cephalopods (relatives of squid) and the first creature who looked like a fish and breathed through gills emerged in this period. Animals related to millipedes and centipedes may have begun to explore the dry land. Around 445 million years ago, a couple of mass extinction events caused massive loss of fauna.

Around 430 million years ago, fish split off into bony and non-bony fishes. Some of these fish had the first moveable jaws. The non-bony fish have skeletons of cartilage instead of bone and are related to modern sharks and rays. Vertebrate animals were still rare, and invertebrates made up most of the fauna. We find forests of moss on the shorelines of bodies of water. We still find trilobites roaming the seas, along with animals very similar to modern snails and sea scorpions. Bony fish diverged into lobe-finned fish and ray-finned fish. Evolutionary biologists claim that the lobe-finned fish were the common ancestors of mammals, birds, reptiles, and amphibians. Coral reefs became increasingly important for the base of the ocean ecosystems.

By about 360 million years ago, sharks and bony fish patrolled the seas, being the dominate predators. While many other fish found niches in freshwater. Various vascular plants and many trees were found on the land creating new ecosystems. A type of fish known as a coelacanth was present and is still alive today. That is close to 400 million years of being almost unchanged! Evolutionary biologists claim that four-legged (tetrapod) land animals branched off from a common ancestor who was a fish that could breathe air through both gills and lungs. Something similar to a lung fish, that had fins that it also used as feet to temporarily explore the land. We would have also seen insects, spiders, and scorpions. Another mass extinction event occurs over millions of years leading to the extinction of about ¾ of the species alive at the time.

10 million years later, the amphibians branch off from the tetrapods. The land fauna included winged insects, beetles, reptiles, millipedes, and various amphibians. Flora on the land began to diversify and included ferns, and various

plants and trees. The seas were dominated by sea lilies, corals, and various animals like sharks and fish. Shallow, warm seas covered most of the land masses and the skeletons of animals rich in calcium carbonate led to large limestone deposits.

The tectonic plates continued their movement over the eons and by 300 million years ago the land masses of Earth were largely connected. The climate of the time was quite like today, with ice covering the poles, warm areas around the equator, and temperature regions in between. This left the rest of the planet covered by essentially one giant ocean. Because of a cooling climate, the warm shallow seas were replaced by more temperature regions and more distinct ecosystems. One of which was vast swampy tropical forests that scientists claim were the source of coal and fossil fuel deposits. The tetrapods continued to diverge with one with another, leading evolutionary biologists to claim that one lineage was the common ancestor of reptiles, dinosaurs, and birds. And the other leading to another lineage, with some of them evolving into mammals. The oxygen in the atmosphere became very high during this era, a main contributing factor leading to giant animals, including insects. Arthropods (some quite huge) like insects, beetles, millipedes, spiders, and scorpions filled the land, while others like the successful trilobites are becoming less numerous. They also shared territory with numerous amphibians.

From about 300-250 million years ago, there was a major diversification of sea animals and an order of mammal-like reptiles called therapsids, attested in the fossil record. Evolutionary biologists claim this animal was a common ancestor of mammals. Numerous and varied therapsids, reptiles, amphibians, and insects filled the ecosystems. Due to selection pressures from a changing environment, herbivores, and intra-species competition, land plant biodiversity ramped up. There were large vertebrate herbivores for the first time. At the end of this period, scientists tell us the biggest mass extinction in the history of Earth is estimated to wipe out up to 90 percent of the species worldwide. Including up to 95 percent of the species in the sea. Within about 5 million years after this catastrophic event that occurred over thousands of years (possibly due to massive volcanism), animals closer to the top of the food chain seemed to have emerged first in the sea. But like many open questions in science, this remains unsure.

Whatever the chain of events that followed, the recovery of ecosystems and the diversity of flora and fauna took at least 10 million years, with microbes dominating and widespread, but a much lower biodiversity overall. The land masses were all together in a giant single landmass that scientists call Pangea at this time. There were vast conifer forests covering the land with mammal-like reptiles roaming about, along with small, nocturnal therapsids. The freshwater ecosystems were filled with amphibians of various sorts, including animals resembling crocodiles. The seas were once again filled with stromatolite reefs and cousins of the octopus and nautilus. A dolphin-like creature, that was actually a reptile, called ichthyosaurs dominated the oceans. About 240 million years ago, the famous dinosaurs first appeared, as attested in the fossil record. These began as small creatures. The first vertebrate to evolve flight is said to have been a cousin of the dinosaurs, starting out with a modest size to evolve into giant winged creatures.

Other non-mammal-like reptiles, like the dinosaurs, began to gain prominence, and by the end of this period (about 200 million years ago) they were more numerous and dominant than mammal-like reptiles. It is also by this time that the dinosaurs dominated the terrestrial ecosystems, along with pterosaurs and crocodiles. Biologists claim that by this time warm-bloodedness had evolved. Whether some or all the dinosaurs were warm-blooded remains an open question. Another mass extinction event occurred around this time leading to around a quarter of all animal families going extinct.

After this extinction event, dinosaurs rose to prominence as the king of the terrestrial ecosystems, while the common ancestors of the mammals try to avoid being eaten by them. These terrestrial ecosystems were largely covered by conifer forests with underbrush of ferns and other various plants. Dinosaur fossils are robustly attested in the fossil record from this time. They are found on every continent and represented by diverse genera and species. This is attributed to their long dominance spanning about 180 million years, and their diversity and adaptability to different ecosystems and niches. They ranged from tiny to enormous. Some of them, which evolutionary biologists claim evolved into birds, had feathers. Icthyosaurs, sharks, crocodiles, bony fish, oysters, lobsters, and crabs made up some of the fauna in the seas, which had large reefs made from the bodies of corals and sponges. Medium to very large pterosaurs flew in the skies above.

By around 180 million years ago, common ancestors of mammals diverged from ones who laid eggs to those who give birth to live young.

10-20 million years later, there is evidence of different feathered dinosaurs and a dinosaur with winged membranes like a bat. Mammals may have diverged at this time between the marsupials and the placental mammals.

By about 145 million years ago, there is evidence of the famous Archaeopteryx, which scientists claim was an intermediate form between the dinosaurs and birds. By this time the first true birds also show up in the fossil record. Lizards were also on the land, eating insects such as beetles, wasps, and the first pollinating insects.

By 125 million years ago, flowering plants are found in the fossil record, but grasses have still not evolved. So, there would be no ecosystems like the grass plains of today. Sea levels were quite high in this period and would have produced vast shallow sea environments resulting in large marine biodiversity of flora and fauna.

By 90 million years ago, placental mammals appeared to split into their major distinct groups. But this is still the age of the reptiles. Dinosaurs seemed to have reached their peak sizes by this period, with animals like T. Rex roaming the earth, and other giant dinosaurs dominating the seas and air.

Around 70 million years ago, grasses have evolved, and mammals continued to diversify, including a lineage that will become the primates. The land masses have broken apart and most of us could have recognized the modern continents that were gradually moving away from each other.

66 million years ago, another extinction event occurred, which most scientists believe was caused by a giant asteroid impact. It is hypothesized that the debris from the blast clouded the skies with dust and material for a prolonged period, blocking sunlight from reaching the surface of the earth and killing the base of many food chains. It may have also acidified the oceans, significantly contributing to the demise of ecosystems, plants, and animals. It is believed that up to 75% of the species went extinct, including many of the large reptiles who

had ruled Earth for millions of years by then. All the large dinosaurs, including the giants on the land, beasts of the seas, and the related pterosaurs who had dominated the skies all went extinct.

Fossils from non-avian dinosaurs have been found in abundance on every continent, but none in sedimentary rocks after this period. Also lost in the geologically very quick extinction event were various mammals, birds, reptiles, lizards, insects, plants, and many species of phytoplankton. What seems to have survived on land were some animals that were omnivores and less than about 60 pounds. This left fauna of mainly smaller animals including mammals, lizards, amphibians, turtles, crocodiles, avian dinosaurs, birds, many bony fish, octopus, nautilus, and some sharks. Some marine and terrestrial plant species survived but in much less diversity and numbers. The farther away from the blast site, the greater number of species of these plants and small animals that seemed to survive.

Scientists claim that it took a few million years for ecosystems to recover, but when they did there was an expansion and diversification of animal species due to empty ecosystems and niches in the environment. Mammals, along with the others that survived the extinction event, were mainly quite small critters. Most about the size of small rodents. These small primitive mammals and birds diversified and filled the land habitats. The age of the reptiles was over; now was the age of the mammals.

Around 55 million years ago, conifer forests, flowering plants, and grasses carpeted the terrestrial ecosystems, although grasses still had some way to go to become habitats and the sole source of food for herbivores. Deciduous trees also became an important component of various ecologies. Primates split into various lineages and insects became more abundant again. Modern-type fishes and sharks swam in the seas, along with crocodiles and turtles in similar aquatic habitats. There was also a dramatic rise in global temperatures around this time, hitting deep water species the hardest. Due to this climate, the world looked much different than today, with trees and crocodiles in the Artic circle and tropical forests in mid latitudes.

By 45 million years ago, a group of animals that were ungulates, and from whom deer and bison would later evolve, are attested in the fossil record.

Biologists claim that this group of animals branched out and one of the species led to whales. This early intermediate species was a land mammal and carnivore. Over time, this species (or a related one) began to have both a terrestrial and aquatic lifestyle, eventually evolving into a fully aquatic lifestyle. They believe that these more aquatic creatures would return to land to give birth to young.

By 33 million years ago, there were terrestrial animals similar to dogs, cats, rhinos, horses, and camels. Although they varied in size much more than today. From small dog-like body sizes to true giants. Higher primates, like the new world monkeys, had diverged from other primates. We find evidence of them in what is today South America, Asia, and Africa. And we find bird fossils that are very much like those of today.

By 25 million years ago, apes split from the monkeys and continued to evolve. The climate cooled and grasslands expanded. Primitive ancestors of modern dogs, cats, horses, and rhinos roamed the landscape. Lots of prey and lots of predators filled the forest and grass ecosystems. The oceans still had fish and sharks, and dolphins and whales were abundant as well.

By 14 million years ago, the common ancestors of gibbons and orangutans had separated. And so has the group of ancestors between orangutans and hominins, chimpanzees, and bonobos. Many marine and terrestrial organisms were unchanged from today. There was another long period of global warming during this time. This global warming, together with Africa and Eurasia becoming connected again, led to the dispersal and diversification of primate species into new habitats, expanding from what was previously just living in tropical forests and woodlands. Apes also diversified and expanded.

By 9 million years ago, we find the last common ancestor of gorillas, hominins, chimpanzees, and bonobos. About 2 million years later, the last common ancestor of hominins, chimpanzees, and bonobos.

Shifts in climate during this time led to cooler and drier conditions, and by 5 million years ago, most of the diverse ape species went extinct. Oceans currents were also greatly changed by this period, which led to changes in climate and ecosystems around the globe.

So, where are we at in our story? It is all too easy to revert to what we in the modern west have been conditioned to believe in our retelling of the consensus story of Earth's history. That everything is easy to explain in materialistic terms. The universe just expands and cools, the evolution of the universe continues through blind deterministic laws of physics, planets form, life begins, life evolves through mutations and selection pressures, and poof, here we are. Easy peasy. Apart from the fact that this narrative is firmly rooted in atheism, so many things are brushed under the rug. The beginning of the universe itself, the fine tuning, and the assumptions of all the forces, matter, and energy evolving according to fixed laws without the governing of a rational Mind. The emergence of life from non-life, and the design and precision of biochemistry and mutations needed to produce the wonderful diversity of organisms. The evolution of master control genes for gene expression and development, the information and control for different types of cells expressing the right proteins in the right cells at the right times and influenced by epigenetics and the three-dimensional structures of chromosomes and proteins. The specific environmental conditions acting as selection pressures, the local ecosystems and habitats, the mutualism, symbiosis, competition, extinction events making some species dominate over others, and on and on we could go. We need to remind ourselves of these things before we expand and move on to the human story. And the redemption story with the creation of Adam and Eve.

God created the universe and all things in it by His will and power. He had a plan for all of creation and governed the universe to see that plan through. He guided the evolution of life from the earliest microbes to our present stopping point. We know that God's cares for the creation and the animals within it. He created all things and filled the spheres of the heavens and earth for His glory, His pleasure, and our redemption and eternal joy.

1 Chronicles 29:11

11 Yours, O Lord, is the greatness and the power and the glory and the victory and the majesty, indeed everything that is in the heavens and the earth; Yours is the dominion, O Lord, and You exalt Yourself as head over all.

Nehemiah 9:6

6 "You alone are the Lord.

You have made the heavens,

The heaven of heavens with all their host,

The earth and all that is on it,

The seas and all that is in them.

You give life to all of them

And the heavenly host bows down before You.

Job 28:24

24 "For He looks to the ends of the earth

And sees everything under the heavens.

Job 35:11

11 Who teaches us more than the beasts of the earth

And makes us wiser than the birds of the heavens?'

Psalm 89:11

11 The heavens are Yours, the earth also is Yours;

The world and all it contains, You have founded them.

Isaiah 40:12

12 Who has measured the waters in the hollow of His hand,

And marked off the heavens by the span,

And calculated the dust of the earth by the measure,

And weighed the mountains in a balance

And the hills in a pair of scales?

Isaiah 42:5

5 Thus says God the Lord,

Who created the heavens and stretched them out,

Who spread out the earth and its offspring,

Who gives breath to the people on it

And spirit to those who walk in it,

Isaiah 45:12

12 "It is I who made the earth, and created man upon it.

I stretched out the heavens with My hands

And I ordained all their host.

Isaiah 45:18

18 For thus says the Lord, who created the heavens (He is the God who formed the earth and made it, He established it and did not create it a waste place, but formed it to be inhabited),

"I am the Lord, and there is none else.

Now that we have readjusted our perspective and given God the glory that He is due, let us return to our story, picking up around 6 million years ago. It is during this time that the lineages between the hominins and chimpanzees is believed to have diverged.

Ardipithecus, the earliest known genus of the animals that make up the family lineage separate from chimps, lived from around 5.8-4.5 million years ago. Scientists often refer to this as Ardi. Around 100 specimens of Ardi have been discovered in Africa. This was a mostly-ape looking creature with teeth that are a little less ape-like. There is debate on whether it walked on two feet or not but based on the context of the fossils and the anatomy, it probably spent significant time in the trees.

Australopithecus anamensis, a different genus, also lived in Eastern Africa from around 4.2-3.8 million years ago. It was apelike in the features of its head but had a lower body anatomy strongly suggestive of walking on two legs for at least part of the time. It is found in ancient habitats of forests and woodlands, so probably spent a lot of time in the trees as well.

Australopithecus afarensis lived in Eastern Africa from about 3.8-2.9 million years ago. This long-lived species is attested to by about 300 specimens. It is nicknamed Lucy, a name given to it by its discovers Donald Johanson and colleagues in 1974. This creature had an apelike head with a small brain, but its anatomy shows that it was capable of walking upright for substantial periods of time. It is still believed that it spent a good deal of time in the trees though. It had a mainly plant based diet but was still an omnivore.

Australopithecus africanus lived between 3.2-2 million years ago in South Africa. It had a slightly less apelike cranium with a brain that was still small but larger than a chimpanzee. The place where the brainstem enters the skull and the lower body anatomy point to bipedal walking, although again this species probably lived in the trees. One thing that became apparent with these fossil discoveries was that walking upright preceded large brains.

Paranthropus aethiopicus was a largely enigmatic species which lived between 2.7 and 2.3 million years ago in Eastern Africa. It was the first of the robust genus and had huge face and jaw muscles. It also walked upright but the lack of fossils makes it difficult to directly determine much about its anatomy, although findings from other members of the Paranthropus genus can give us clues.

Paranthropus boisei lived in Eastern Africa from around 2.5-1.1 million years ago. It was the one of the most heavy and thick of the genus and appeared to have differed substantially in size between males and females. This species lived for over a million years.

Paranthropus robustus lived around 2-1 million years ago in South Africa. It was found in what was mixed and open woodlands at the time. It is the most robust of the genus discovered so far, averaging a little under 4 feet and weighing around 120 pounds.

Australopithecus garhi fossils have all been found in Ethiopia and have been dated to 2.5 million years old. It had a robust, apelike cranium and a brain size similar to other australopithecines.

The oldest discovered creature from the genus Homo is Homo habilis. They lived from about 2.3-1.6 million years ago in East and South Africa. Its name means “handy man” because many paleoanthropologists think it was the first to use stone tools. The earliest stone tools discovered predate the earliest habilis fossils slightly. It had a larger brain and less sloping forehead than the earlier hominins and different than the apes. But some fossils showed that it had a more apelike anatomy with long arms and shorter legs than was originally thought. But it still spent most of its time walking upright and had dexterous hands for possible tool usage.

Homo rudolfensis lived in Eastern Africa from about 2.4-1.8 million years ago. There is debate among experts, as is the case for many details of hominin evolution, on the classification of this species. It had a larger brain on average than Homo habilis, but despite very few fossil specimens, had other features like australopithecines. So, most are not sure where it fits.

Homo erectus appeared in Africa about 2 million years ago and spread into Asia within 100,000 years of its African beginnings. It had more humanlike proportions with longer legs and shorter arms, which indicates that it spent its time walking upright on two legs like us. It also appears to have had a similar gait to us compared to other early hominin species. This species spanned more than 1.5 million years, possibly as long as 1.8 million-110,000 years ago. And because of this longevity, it varied a lot within the species. They still had craniums with strong brow ridges and large faces. Their skulls still had low, sloping foreheads but there was considerable variation in skull shape and brain size. Most scientists believe they were the first to use new stone tool technology. Among this technology were tools they call hand axes.

So, we have four species of hominin living in Eastern Africa together beginning around 2 million years ago and coexisting for about half a million years: Homo rudolfensis, Homo habilis, Homo erectus, and Paranthropus boisei.

Homo heidelbergensis lived somewhere between 800,000-200,000 years ago, depending on which fossil specimens are placed within this species. They lived in Africa, Europe, and possibly Asia. The general consensus hypothesis seems to be that this species was the last common ancestor of both Homo neanderthalensis and Homo Sapiens. That the European population of Homo heidelbergensis were the direct ancestors of Neanderthals, while the African population of Homo Sapiens. And that this speciation event into Homo Sapiens occurred in Africa prior to 300,000 years ago. Some experts disagree. But whatever the case is, Homo heidelbergensis appears to have lived in social groups and hunted cooperatively, using the older stone tool technology with advances happening in later populations for things like wooden spears with stone tips. Their skulls still had a sloping forehead and brow ridge, but they had brains similar in size to modern humans. But they had thicker bones and were more robustly built than modern humans. They may have constructed shelters (instead of just living in natural shelters like caves), and they may have used controlled fire.

Homo neanderthalensis (the Neanderthals) lived in Europe and the ancient near east from somewhere between 400,000-28,000 years ago. Thousands of fossils from hundreds of individuals have been found to date, making it one the best understood species of Homo besides Homo Sapiens. Members of this species were shorter and stockier than the average human. But they were nothing like the early view of them as dimmed witted cave men. Their average brain size was actually larger than modern humans, but brain size alone can't tell us everything. It also depends on how the brain is structured and proportioned. Genetic evidence published in the last 15 years suggests that Homo sapiens interbred with Neanderthals. Contemporary human beings, with a larger percentage in non-Africans, share some of their genome with this species, indicating genetic admixture after some Homo Sapiens had left Africa. The Neanderthals had complex social cultures and may have spoken languages, although this is highly contested. Although experts debate the details, their behavior and cultures were surely different than Homo sapiens. The evidence for abstract thinking and symbolic behavior is much stronger for Homo Sapiens living at the same time as Neanderthals. But they were no gorillas.

They lived in groups, built shelters, used fire, hunted cooperatively, made and wore clothing, and buried their dead. Although many hypotheses about ritualistic behavior associated with their burials are speculative and controversial. They may have developed art and jewelry or simply acquired them from Homo Sapiens. Or they may have just copied what they saw Homo Sapiens doing. But they probably had the brain capacity to do it. They lived in Europe and parts of the ancient near east with the most recent dispersal of Homo sapiens for about 10,000-15,000 years before going extinct. There are several theories of why the Neanderthals went extinct. Most are a combination of hypotheses, including fewer of their children surviving to reproduce, their inability to plan and think like humans, climate change and catastrophic population reductions they could not recover from, to arguments about metabolism and developmental differences between Neanderthals and Homo Sapiens. Whatever the story that unfolded, they went extinct by around 30,000 years ago.

Denisovans, named after the cave in Russia in which they were found, are not yet given a genus and species due to the paucity of fossils found. Most of what we think we know about them comes from DNA analysis from small bone fragments. One hypothesis is that Denisovans and Neanderthals were closely related and descended from a European population of Homo heidelbergensis. With the Neanderthals staying in Europe and the ancient near east and the Denisovans traveling eastward to Asia. Another hypothesis is that the Denisovans were direct descendants of a population of Homo erectus that left Africa much earlier. Whatever the true story, native populations living in Tibet, Melanesia, and Australian share some of their genome with Denisovans. This implies that populations of Homo Sapiens in Asia interbred with Denisovans. They probably went extinct 30,000 years ago. Not much can yet be said about their anatomy, lifestyle, and culture.

Our species, Homo Sapiens, has existed from at least 300,000 years ago to the present day and is the only surviving species of Homo. While evidence points to East and South Africa as the beginning location of our species, it is likely that Homo Sapiens were spread across Africa quite early. While there is a fair amount of diversity in the anatomy of the fossils found, they all seemed to more or less share similar material culture and technology. Although this too varied by region.

Although there are different theories proposed by paleoanthropologists to account for the data, most would agree that our species evolved in Africa and then dispersed through multiple events from there. Some of these events being more successful than others. With some of these populations mixing and interbreeding with both Neanderthals and Denisovans, including the most recent successful waves that led to modern populations. Although the details can be hotly debated, it seems most warranted by the evidence that separate Homo Sapiens populations left Africa, interbred amongst themselves and with Neanderthals and Denisovans, mixing genes and technology. And that the earliest migrations of Homo Sapiens out of Africa did not generally result in populations that survived. Although some may have survived and contributed small amounts to modern genomes.

Homo Sapiens fossils have been found in modern Israel dated to around 180,000 years ago, and they may have been living there for 50,000-80,000 years. Fossil finds in China indicate that some Homo Sapiens were living there by about 100,000 years ago. Stone tool technology associated with them have been found in India and the ancient near east dated to around 80,000 years old. But the big migration event that most experts agree is the main contributor to the ancestry of modern populations was one that happened between 50,000-60,000 years ago. It is unknown how many of them were involved in this(these) migration(s), but it seems an effective population size of around 2000-5000 individuals is usually claimed. But effective population size in not the same as actual population size. It was in actuality probably more than 2000-5000 individuals. But this refers to the effective population size of the population that left Africa around 50,000-60,000 years ago. It is probably three times that accounting for both those in Africa and those that left. In other words, the effective population size for the entire human population is probably closer to 6000-15000, and the actual number of Homo Sapiens who left during the main event was almost certainly more than 2000 individuals. But all these estimates rely on certain assumptions and how far back in time we are looking for an effective population size.

Although theories differ, it is generally believed that this group first crossed the then-dry bottom of the Red Sea near Djibouti or southern Eritrea. They then expanded and separated into various groups, going through the ancient near east and India, eventually into Australia and Melanesia to the east by about 45,000 years ago, and through modern Turkey into Europe by around 40,000 years ago. Other populations spread into China by about 35,000 years ago, continuing up through northeastern Asia, and crossing the then-passable Bering Strait and into North America by 15,000 years ago. Although many have challenged this story and claim there is evidence for people in the Americas before 20,000 years ago. From there, these populations continued down through North America and eventually to the tip of South America approximately 12,000 years ago, but again this is challenged. These theories are bolstered by archaeologic findings and genetic studies that show various founder effects and loss of genetic diversity the farther the populations are found from Africa. All these conclusions are tentative, especially when populations arrived at certain localities, and could be changed with new discoveries.

These groups were all hunters and gatherers, and it is hard to sustain large populations in hunting and gathering lifestyles. So smaller groups broke off from larger ones to maximize the local environment for these people. And although this is long before written records, we can surmise that population size would have been constrained by local resources and the birth and death rates. Probably also by some horrific means, like infant exposure, which has been documented in many cultures. Although until recent times even wanted babies very often did not survive into adulthood.

By around 15,000 BC, the world was populated with various Homo Sapiens populations in hunter gatherer groups and lifestyles. But around 10,000 BC, this began to change. And this brings our perspective back to the world of the Bible. The last ice age was beginning to end and some groups of Homo Sapiens around the ancient near east were beginning to put together the ideas of taking the wild grains that they had been gathering and sowing the seeds themselves. Along with the related idea of trying to domesticate the wild animals they hunted for food. It may have developed in this area independently several times, but I think that is a hard case to make with people socializing, trading, and passing on ideas. But it did arise independently in different parts of the world. There is also some debate on whether people started farming because they had built permanent settlements or whether the farming and animal husbandry led to the settlements. A group of people we call Natufians began building settlements around modern Israel and had houses and infrastructure to the east of the sea of Galilee. They hunted, fished, gathered grain, and buried their dead in a cemetery. 2000 years before the neolithic revolution of farming and herding. We also know that modern hunters and gatherers have a vast knowledge of local environments, including plants and animals. So, it is not too much of a stretch that they would understand how the plants grow and which animals were tame enough to domesticate. Whatever the causes and chain of events, God had a plan for His image bearers and their dominion of the world.

By 10,000 BC there was sheep and goat herding in Turkey. Barley was cultivated by 9000 BC. By 8000 BC, wheat, rye, flax, chickpeas, lentils, and peas were being sown and harvested. Cattle, sheep, goats, and pigs were also domesticated in the Fertile Crescent by this time. Homo Sapiens lived in settled communities and produced culture. They had language and social structures. Religion exists in every human culture that has been studied, so we know man is inherently religious. Every human culture studied has moral codes and concepts of right, wrong, merit, punishment, and justice. Although they do differ. It seems to me there is a lot of guesswork and speculation when trying to ascertain prehistoric religion, but we can be confident of people's ideas when they start writing them down, which was not for another 5000 years.

It is into this period of history and the developed rational and moral faculties of man that God created Adam and Eve around 8000 BC. Adam was miraculously created by God from the dust of the ground, God animated him, and Eve was miraculously created from Adam's side. Adam was designated as the covenantal head of humanity to represent mankind in the covenant of works. This included Adam's descendants and those born after the creation of Adam who were not his biological descendants. God created a garden temple in the then-mostly dry river basin of the Persian Gulf where four rivers came together into a larger river that ran through the garden. Subterranean water sources watered the whole surface of the ground. This man and woman were placed in this sacred space and tasked with expanding the boundaries of the garden and the kingdom of God to the outside common and profane world. To make a way to worship the one true living God and a path to the tree of life. They were created in positive holiness and true knowledge. They had intimate communion with the God who created all things and gave breath to all creatures. These other populations around the globe, estimated to be around 5 million people in 8000 BC, gave birth to the first generation to be imprinted with the image of God on their souls and were included in the covenant of works.

Prior to this time, they did not have the faculties of mind nor social structures to be imprinted with the image of God on their souls. Nor did they have a covenant representative in Adam yet. They stilled owed their existence to the sovereign God who designed, created, and sustained all things, to the One who gave them breath and life, and the God who made sure the earth was not formless and void. To the God who had filled all spheres of creation with the amazing diversity of life, including plants and animals. Who made certain that there was a creature with the capacities for a rational and moral soul capable of reflecting His image. But prior to Adam and Eve, they were not held liable for this appropriation of knowledge because they did not have the image of God, they did not have the moral law of God written on their hearts, and therefore there was no sin.

When Adam sinned, his sin was imputed to all the people and all their descendants who were parties to the covenant of works. God also withheld original righteousness to those after Adam and as a result we are all born into the corruption of sinful natures. People were now held accountable for their rational and moral minds who could perceive God's attributes by what had been created and knew right and wrong in their own moral conscience. God cursed the serpent and Satan, cursed Eve by affecting both her childbearing and relationship with her husband, and temporarily cursed the ground on account of Adam. But Adam and Eve were given the promise of the woman's seed who would crush the serpent's head. They believed God and it was accounted to them for righteousness. They were expelled from the garden temple and the cherubim now guarded the way to the tree of life, rendering physical death sure for everyone.

The elect line of descendants through Seth carried on the hope of the promise of the seed and a Redeemer, and these descendants began to call upon the name of the Lord. Both the line of Seth and of Cain developed culture and civilization and God's common grace was upon all the humans of the Earth. Noah was born in the line of Seth, and he alone was found righteous in God's eyes. God warned him and commanded him to build an ark, because man's sin and violence had reached a crescendo. Instead of Adam and Eve expanding the garden and reflecting God's image to the common world, the world had corrupted the descendants of the original image bearers, the ones who were created upright. The original mission and commission given to the original couple had failed, but God was not going to let his image bearers flail about in darkness, being blinded by the god of this world, Satan. But God is a God of justice, and He flooded the area of southern Mesopotamia with a great flood where Adam and Eve started their mission. This was a local but massive and widespread flood that occurred around 6500 BC. The local peoples who would have been the first to experience the glory and goodness of Yahweh Elohim and to be among the first citizens of the kingdom of God also bore the judgement relating to Adam's sin and the ensuing depravity.

From Noah's three sons, a small culture emerged. They most likely all spoke the language of the Ubaid culture, called by some Proto-Euphratean. Around 6200 BC are the first attested villages in southern Mesopotamia. They seemed to have a well-defined shared culture from about 5500-4000 BC and used irrigation for farming, raised cattle, made textiles, produced pottery, did leatherwork, fished, and drank beer. The following Uruk culture seems to have built upon and elaborated on the Ubaid one that preceded it.

They attempted to build a ziggurat in the city of Eridu (Petrovich, 2020) and God confused their languages and scattered them. Much of the ancient near east was eventually related to them genealogically. Most people think that the claim of the historical accuracy of the table of nations in Genesis 10 and 11 is quite fanciful, but K.A. Kitchen has made some compelling arguments for taking it seriously. (Kitchen, 2003) As have others, like biblical scholar Bruce Waltke. (Waltke, 2001) And I have also argued briefly that the section in Genesis 10 on Shinar (Mesopotamia) makes good sense. Out of these populations of Noah's sons in southern Mesopotamia come the first cities in the whole world. They are known to us as the cities of Sumer. Cities like Eridu, Uruk, Shuruppak, Larak, Sippar, Nippur, Lagash, Kish, and Ur, the city where Abram was from (although much later).

It was almost 5000 years from Noah's flood to the call of Abram out of Ur in Mesopotamia, which I believe happened around 1700 BC. God chose Abram and made a covenant with him that in him all the families of the earth would be blessed. God renamed him Abraham and his descendants of Isaac and Jacob would lead to Jacob's twelve sons, the sons of Israel. The sons of Israel settled in Egypt after one of Israel's sons, Joseph, became vizier in that kingdom. They were later enslaved, and God delivered them from Egypt by the hand of Moses in the Exodus. And God also used this prophet Moses to give them, and us, His word in the first five books of His infallible word. While just the first several chapters of Genesis have occupied most of the present book.

God condescended to communicate to us by the Holy Spirit through the writing of Moses. And He gave us the narrative of creation in Genesis 1 in a form and type of literature that the Israelites were familiar with but was not meant to be interpreted as historical narrative. They were surrounded by polytheistic cultures that had stories of how the gods came to be, how the world was formed, and the place of man in that scheme. This stylized and exalted narrative gave them the correct theology and demystified nature as the material creation of a transcendent God who was separate from it. The heavens and earth were not made up of various deities needing to be placated and propitiated, but was a material creation separated, ordered, and filled by a good God who created man as the crown of creation. Man was not a constant annoyance that was created to do the drudgery of the gods and whose main purpose in life was to feed, clothe, and care for the incarnations of the gods in their temples. Man was made in the image and likeness of the one true living God, and this God delighted in him.

God made a covenant with the nation of Israel that they would be a kingdom of priests, a holy nation, and to be a blessing to the other nations. He had them construct the tabernacle in the wilderness and that was where He met with them. Once again God was dwelling with His people. Although this time they were not positively holy like Adam and Eve, and God emphasized how He could not be approached casually. They needed atonement for their sins and a mediator between God and man. But the mission to reach the people of the world and expand the kingdom of God was not abandoned.

They took possession of the promised land and became a kingdom of their own, originally with God as their King. But the people did what was right in their own eyes and wanted a king like the rest of the nations around them. God first gave them Saul, then a man after His own heart, David, and made a covenant with David that his descendants would always sit on the throne. David's son Solomon took the throne and the extent of Israel's borders, and its wealth, was at its peak. Solomon also built the first temple during his reign and now God met with His people in the holy of holies behind the cherubim embroidered on the veil, by means of sacrifices and the priests in the city of Jerusalem.

But the people were unfaithful to the terms of the covenant and the kingdom was split in two, Israel to the north and Judah to the south. The people by and large broke the covenant and mingled with the surrounding nations and committed idolatry with their gods. They, like Adam and his descendants, did not sanctify themselves to the Lord and married and practiced idolatry with the pagans around them. This led to the curses of the covenant falling on Israel and Judah at different times and in diverse manners, with repentance and restoration sometimes partially happening led by God fearing kings.

God eventually chastised and disciplined the northern kingdom of Israel with the conquest of the Assyrians in 722 BC. Many of the people of Israel were deported to Assyria and some foreigners took their place. After about 140 years, God disciplined the southern kingdom of Judah because of their unfaithfulness to God and His covenant, the temple in Jerusalem was destroyed, and the vast majority of the people still remaining in Judah were deported to Babylon in 586 BC. After 70 years, according to the prophecies of Jeremiah, king Cyrus let the people return to Jerusalem to rebuild the city and the second temple. But even after rebuilding this temple the shekinah glory of the Lord did not fill the temple like He did in Solomon's temple.

The story of the Old Testament ends with what seems to be an unfinished story. But in about 6 BC the eternal Son of God took on human flesh and a human nature in Jesus. This baby, born of a virgin, was the Messiah, God's anointed. The Seed of the woman that would crush the head of the serpent, the Seed of Abraham by whom all the nations of the earth would be blessed, the God who once again would tabernacle among us, the Son of David who would always sit on the throne and whose kingdom would have no end. This God-man would be the covenant head of a new people who would go into all the world, make disciples, and expand the kingdom of God. He would resist temptation, even to the point of death, and live to do His Father's will. He would lay down His life as the spotless lamb of God to atone for sin, and those who trust in Him would be imputed His righteousness. He would conform His covenant people into His image and thereby renew the original image of God in man. This whole universe was created by Him, through Him, and for Him.

He was the last Adam and all those who are found in Him will once again walk with God and freely eat from the tree of life. And all creation will eventually be redeemed after the children of God.

Come Lord Jesus, amen.

So, we have finished our story. The same God who created all things came to Earth to be an atonement for sin in order that the original mission given to Adam would be fulfilled. God the Father sent the Son because of His great love for us and has given this Son all the nations of the earth as His inheritance. And the creation itself will redeemed one day after the children of God.

The Jewish religion was the first true monotheistic religion in the world. It called all other gods false gods. It was the first to reveal the idea of an eternally self-existent God who created all things out of nothing by His will. It was the first to explain the character of this God as wholly different than us. This God was not led by passions and emotions, did not have a female wife or consort goddess, did not sexually reproduce other gods, and did not act arbitrary or selfishly like the other gods of the world. This God is the reason why human beings feel there is more to life than nature. Why we seek to understand and explain the world. Why the universe is ordered, rational, and explainable. As a true grounding and justification for objective morality and meaning. The reason people have dignity and worth. I have argued that science offers no good justification for our curiosity about nature and its workings, emotions that motivate learning and discovery, desires for a coherent and systematic understanding of things, what we ought to do, what is a good life, and meaning and purpose.

This religion was based on history. And it has been verified, despite claims to the contrary. Real people in real space and time with real consequences for the rest of humanity. Christianity has the fulfillment of hundreds of prophecies, many of them regarding the Messiah, the Anointed One, Jesus the Christ. Skeptics have a hard time explaining most of these prophecies. Jesus of Nazareth was a real person who had a real ministry with eyewitnesses galore. His works testified that He was from God and could have been easily discredited if they were not true because His ministry did not take place in private, but in the public eye. He showed us what true humanity should look like. His life, death, and resurrection from the dead is well attested and even acknowledged by skeptics as hard to resolve any other way. His apostles who were eyewitnesses wrote about these matters. The Holy Spirit has revealed to us through human authors how and why this God worked through history.

The image of God stamped upon the human soul is the reason why people universally have a moral conscience and a rational soul. It is the reason why people feel so much moral conflict and cognitive dissonance with the injustice in this world. And the sin of Adam is why the world of humanity looks the way it does. It explains the clouded reason and perception of God and His world by those outside of Christ. Why people practice idolatry and have an innate desire to violate God's moral law and fail to conform to it. Why we feel guilty for not loving our neighbors as ourselves.

But Christianity has had a very substantial and underappreciated influence on the world, and especially the western world. Apart from the universal moral conscience due to the soul and image of God in every person, Christianity is largely responsible for why we in the west seek universal human rights, why we care for the poor and disabled, why we have hospitals and orphanages, why slavery was abolished, why we are charitable with our time and money, why we react to bullies oppressing the weak and vulnerable, why government social programs seem normal and right, and why in-group out-group psychology seems so abhorrent. Why the first will be last and the last first, and why the meek will inherit the Earth. Jesus showed us what the true image of God looks like and even those who do not have eyes to see are indebted to His teachings, influence, and disciples. It is tempting for us in the modern west to think that these moral and social norms are self-evident and for Christians to begin to believe that the United States and the west have a morality that is not much different than ours. With many of them calling for universal human rights, justice, dignity, worth, charity, compassion, empathy, and love. But as historian Tom Holland has ably argued, one big reason for this is the dominance and success of Christian assumptions, beliefs, ethics, and practices in the west. (Holland, 2019) We Christians sometimes feel like we do not stand out from the culture in the same way that the early Christians did in the Roman empire. But one of the reasons (apart from lukewarm and cultural Christianity) for this feeling is precisely because of the success of the Christian worldview and mindset. But even though most of the modern west has unconsciously adopted this moral and social worldview, many of them have failed to acknowledge God or give Him thanks. They still need to be reconciled to God through faith in Jesus Christ.

Although this book was heavy on evidence, we do not need evidence to confirm our faith. Our faith is a gift from God, and it gives us eyes to see. I don't believe that we can argue anyone into the kingdom of God apart from a supernatural work done in their heart. And I have no animosity towards brothers and sisters in Christ who still hold to a young Earth, interpret Genesis 1 as historical narrative, and teach the long-held orthodox theological implications that follow. But I do have a heart for the believer who is constantly assaulted from the world, science, and higher learning. And I want to show them that their faith is rational and firmly rooted in history.

But according to God's word, the spiritual world is just as real as the physical world. And that is where a lot of things are happening to blind the minds of natural man. That is where our Lord and the apostles tell us to put on defensive armor, take up weapons, and be ready to fight. To look through the materialistic explanations of science and see the spiritual enemies that want us to doubt God, His word, and abandon our faith.

2 Corinthians 4:18

18 while we look not at the things which are seen, but at the things which are not seen; for the things which are seen are temporal, but the things which are not seen are eternal.

We all know that the heavens declare the glory of God. We all know that it requires a self-existent Being to call it into existence. We all know that there is an objective moral standard that we regularly fail to live up to. And if we have been born again by the Holy Spirit, know Jesus Christ, have trusted him, and follow Him as Lord, then we also know experientially how He can change lives and make us more human as we are conformed into His image. Let us then give God the glory for all that He has done. Amen.

Bibliography

Augustine, S., & Taylor, J. H. (1982). *Genesi Ad Litteram.* Mahwah, NJ: Paulist Press.

Beale, G. K. (2018). Adam as the First Priest in Eden as the Garden Temple. *The Southern Baptist Journal of Theology 22.2*, 9-24.

Bell, G., & Gonzalez, A. (2009). Evolutionary rescue can prevent extinction following environmental change. *Ecology Letters*.

Berkhof, L. (1996). *Systematic Theology.* Grand Rapids: Wm. B. Eerdmans.

Best, R. M. (1999). *Noah's Ark and the Ziusudra Epic.* Fort Myers: Enlil Press.

Black, J., Cunningham, G., Robson, E., & Zolyomi, G. (2006). *The Literature of Ancient Sumer.* New York: Oxford University Press.

Bodine, J. J. (2009). The Shabaka Stone: An Introduction. *Studia Antiqua*, Volume 7, Number 1, Article 3.

Bunch, T. L. (2021). A Tunguska sized airburst destroyed Tall el-Hammam a Middle Bronze Age city in the Jordan Valley near the Dead Sea. *Nature Sci Rep 11, 18632*, https://doi.org/10.1038/s41598-021-97778-3.

Clay, A. T. (1922). *A Hebrew Deluge Story in Cuneiform.* New Haven: Yale University Press.

Collins, C. J. (2006). *Genesis 1-4: A Linguistic, Literary, And Theological Commentary.* Phillipsburg, NJ: P & R Publishing Company.

Collins, L. G. (2009, September-October). *Reports of the National Center for Science Education | Volume 29 | No. 5 | September-October 2009.* Retrieved from National Center for Science Education: https://ncse.ngo/yes-noahs-flood-may-have-happened-not-over-whole-earth

Collins, S., & Scott, L. C. (2016). *Discovering the City of Sodom.* New York, NY: Simon and Schuster.

Coyne, J. (2009). *Why Evolution is True.* New York, NY: Viking Penguin.

Dalley, S. (2008). *Myths from Mesopotamia.* New York: Oxford University Press.

Dawkins, R. (1986). *The Blind Watchmaker.* New York, NY: W.W. Norton & Company.

Dick, M. B. (1999). *Born in Heaven, Made on Earth: The Making of the Cult Image in the Ancient Near East.* University Park: Penn State University Press.

Dungen, W. v. (2004, October). *The Theology of Memphis*. Retrieved from Sofiatopia: https://maat.sofiatopia.org/memphis.htm

Finkel, I. (2014, January 19). *Noah's Ark: the facts behind the Flood*. Retrieved from The Telegraph: https://www.telegraph.co.uk/culture/books/10574119/Noahs-Ark-the-facts-behind-the-Flood.html

Hamblin, D. J. (1987, May). Has the Garden of Eden been located at last? *Smithsonian Magazine*.

Holland, T. (2019). *Dominion.* New York: Basic Books.

Jacobsen, T. (1987). *The Harps That Once...: Sumerian Poetry in Translation.* New Haven: Yale University Press.

Jonathan Haidt, J. G. (2023, September). *Home*. Retrieved from MoralFoundations.org: https://moralfoundations.org/

Josephus, F. (1999). *The New Complete Works of Josephus.* Grand Rapids, MI: Kregel Academic.

Kidner, D. (1967). *Genesis.* Downers Grove: Inter-Varsity Press.

King, L. W. (1902). *The Seven Tablets of Creation.* London: Luzac.

Kitchen, K. (2003). *On the Relability of the Old Testament.* Grand Rapids, MI: Wm. B. Eerdmans Publishing Co.

Kovacs, M. G. (1989). *The Epic of Gilgamesh.* Stanfod: Stanford University Press.

Lambert, W. G., & Millard, A. R. (1969). *Atra-hasis: The Babylonian Story of the Flood.* Oxford: Clarendon Press.

Lambert, W., & Parker, S. B. (1966). *Enuma eliš: the Babylonian epic of creation.* Oxford: Clarendon Press.

Lichtheim, M. (1973). *Ancient Egyptian Literature Vol. 1.* Berkeley: University of California Press.

MacArthur, J. (1997). *The MacArthur Study Bible.* Nashville, TN: Thomas Nelson Bibles.

Marshall, P. (2015). *Evolution 2.0.* Dallas, TX: Benbella Books, Inc.

Mayr, E. (1982). *The Growth of Biological Thought.* Cambridge, MA: The Belknap Press of Harvard University Press.

McCann, T. (2008). *The Geology of Central Europe.* London, England: Geological Society of London.

Moran, W. (1987). Some Considerations of Form and Interpretation in Atra-Hasis. *Journal of the American Oriental Society*, 245-256.

Munday Jr., J. C. (2013). *Eden's Geography Erodes Flood Geology*. Retrieved from www.godandscience.org: http://www.godandscience.org/youngearth/flood_geology.html

Newport, F. (2014, May 8-11). *In U.S., 42% Believe Creationist View of Human Origins*. Retrieved from Gallup: https://news.gallup.com/poll/170822/believe-creationist-view-human-origins.aspx

Petrovich, D. D. (2020). *Is Genesis History?* Retrieved from YouTube: https://www.youtube.com/watch?v=NurrWE8XX0U

Richter, S. (2018). *Henry Center*. Retrieved from YouTube: https://www.youtube.com/watch?v=WU2dj_S9iM0

Rogers, R. W. (1912). *Cuneiform Parallels to the Old Testament.* London: Oxford University Press.

Rose, J. I. (2010). New Light on Human Prehistory in the Arabo-Persian Gulf Oasis. *Current Anthropology*, 849-883.

Sailhamer, D. J. (1996). *Genesis Unbound: A Provocative New Look at the Creation Account.* Sisters, OR: Multnomah Books.

Sauer, J. A. (1996). The River Runs Dry: Creation Story Preserves Historical Memory. *Biblical Archaeology Review*.

Sproul, Waltke, Silva, & Whitlock. (1995). *The Reformation Study Bible.* Nashville, TN: Thomas Nelson Publishers.

Thompson, R. C. (1928). *The Epic of Gilgamesh.* London: Luzac & Co.

Tigay, J. H. (1982). *The Evolution of the Gilgamesh Epic.* Philadelphia: University of Pennsylvania Press.

Walls, N. H. (2005). *Cult Image and Divine Representation in the Ancient Near East.* Boston: American Schools of Oriental Research.

Waltke, B. K. (2001). *Genesis: A Commentary.* Grand Rapids, MI: Zondervan.

Walton, J. H. (2010). *The Lost World on Genesis One.* Downers Grove, IL: InterVarsity Press.

Westminster. (1646). *The Confession of Faith*. Retrieved from The Orthodox Presbyterian Church: https://opc.org/documents/CFLayout.pdf

Wood, B. G. (1990). Did the Israelites Conquer Jericho? A New Look at the Archaeological Evidence. *Biblical Archaeology Review*, 44-59.